T0207390

Human–Computer Interaction Series

The Human–Computer Interaction Series, launched in 2004, publishes books that advance the science and technology of developing systems which are effective and satisfying for people in a wide variety of contexts. Titles focus on theoretical perspectives (such as formal approaches drawn from a variety of behavioural sciences), practical approaches (such as techniques for effectively integrating user needs in system development), and social issues (such as the determinants of utility, usability and acceptability).

HCI is a multidisciplinary field and focuses on the human aspects in the development of computer technology. As technology becomes increasingly more pervasive the need to take a human-centred approach in the design and development of computer-based systems becomes ever more important.

Titles published within the Human–Computer Interaction Series are included in Thomson Reuters' Book Citation Index, The DBLP Computer Science Bibliography and The HCI Bibliography.

Wolfgang Aigner · Silvia Miksch ·
Heidrun Schumann · Christian Tominski

Visualization of Time-Oriented Data

Second Edition

 Springer

Wolfgang Aigner
Institute of Creative\Media\Technologies
Fachhochschule St. Pölten
Saint Pölten, Austria

Silvia Miksch
Institute of Visual Computing
and Human-Centered Technology
Vienna University of Technology
Vienna, Austria

Heidrun Schumann
Institute for Visual and Analytic Computing
University of Rostock
Rostock, Germany

Christian Tominski
Institute for Visual and Analytic Computing
University of Rostock
Rostock, Germany

ISSN 1571-5035   ISSN 2524-4477 (electronic)
Human–Computer Interaction Series
ISBN 978-1-4471-7529-2   ISBN 978-1-4471-7527-8 (eBook)
https://doi.org/10.1007/978-1-4471-7527-8

This Springer imprint is published by the registered company Springer-Verlag London Ltd., part of Springer Nature.
The registered company address is: The Campus, 4 Crinan Street, London, N1 9XW, United Kingdom

To our families.

Foreword

Time is central to life. We are aware of time slipping away, being used well or poorly, or having a great time. Thinking about time causes us to reflect on the biological evolution over millennia, our cultural heritage, and the biographies of great personalities. It also causes us to think personally about our early life or the business of the past week. But thinking about time is also a call to action since inevitably we must think about the future – the small decisions about daily meetings, our plans for the next year, or our aspirations for the next decades.

Reflections on time for an individual can be facilitated by visual representations such as medical histories, vacation plans for a summer trip, or plans for five years of university study to obtain an advanced degree. These personal reflections are enough justification for research on temporal visualizations, but the history and plans of organizations, communities, and nations are also dramatically facilitated by powerful temporal visual tools that enable exploration and presentation. Even more complex problems emerge when researchers attempt to understand biological evolution, geological change, and cosmic scale events.

For the past 500 years, circular clock faces have been the prime representation for time data. These emphasize the twelve or 24-hour cycles of days, but some clocks include week-day, month or year indicators as well. For longer time periods, timelines are the most widely used by historians as well as geologists and cosmologists.

The rise of computer display screens opened up new opportunities for time displays, challenging but not displacing the elegant circular clock face. Digital time displays are neatly discrete, clear, and compact, but make time intervals harder to understand and compare. Increased use of linear time displays on computers has come with new opportunities for showing multiple time points, intervals, and future events. However, a big benefit of using computer displays is that multiple temporal variables can be shown above or below, or on the same timeline. These kinds of overviews pack far more information in a compact space than was previously possible while affording interactive exploration by zooming and filtering. Users can then see if the variables move in the same or opposite directions, or if one movement consistently precedes the other, suggesting causality.

These rich possibilities have payoffs in many domains including medical histories, financial or economic trends, and scientific analyses of many kinds. However, the design of interfaces to present and manipulate these increasingly complex and large temporal datasets has a dramatic impact on the users' efficacy in making discoveries, confirming hypotheses, and presenting results to others.

This book on Visualization of Time-Oriented Data by Aigner, Miksch, Schumann, and Tominski represents an important contribution for researchers, practitioners, designers, and developers of temporal interfaces as it focuses attention on this topic, drawing together results from many sources, describing inspirational prototypes, and providing thoughtful insights about existing designs. While I was charmed by the historical review, especially the inclusion of Duchamp and Picasso's work, the numerous examples throughout the book showed the range of possibilities that have been tried – successes as well as failures. The analysis of the user tasks and interaction widgets made for valuable reading, provoking many thoughts about the work that remains to be done.

In summary, this second edition extends the coverage and thoughtfulness of the first edition. It continues to be not only about the extensive work that has been done, but it is also a call to action, to build better systems, to help decision makers, and to make a better world.

University of Maryland,   *Ben Shneiderman*
September 2022

Preface

Preface to the Second Edition

The importance of time and time-oriented data remains unbroken. It even increased in recent years. And with this increase, we also see an increased need for effective techniques and tools for understanding temporal phenomena and gaining insight into time-oriented data.

In 2011, the first edition of this book was published, and it was well received by the research community. About a thousand times has the book been cited since then, and more than 10 years later, it is still collecting numerous citations each year. Many colleagues expressed their interest in the book on social media and on research platforms. The success of the first edition motivated us to think about a second edition. At VIS 2019 in Vancouver, we decided to start working on it with the goal to have the second edition ready in 2021, ten years after the first edition. However, as we all know, the world suddenly changed in 2020 and so we did get behind schedule. Today, we are glad that we made it in 2022 and we are excited about the revised and extended second edition.

So, what is new for the second edition? First of all, the book is now published under an open-access license. For the first edition, we received many messages asking us to provide the book's full text, which, unfortunately, we had to deny. Given the great reception in the community, we decided to publish the second edition as an open-access book so that it can be read by anyone interested in the visualization of time and time-oriented data. We gratefully acknowledge the financial support provided by our universities in this regard.

For the second edition, the entire content has been revised and extended. The most obvious change is that we moved the big survey chapter with originally 101 visualization techniques to Appendix A, which now provides more than 150 descriptions and illustrations of classic and contemporary visualization approaches for time and time-oriented data. Also, the chapter on interaction support has been expanded substantially, now including advanced methods for interacting with visual representations of data in Section 5.4, among other new content. Moreover, the book now covers the topics of data quality (see Section 3.4 and Appendix B) as well as segmentation and labeling (see Section 6.3). The completely new Chapter 7 describes how the structured view developed in this book can be used for the guided selection of suitable visualization techniques. Throughout the book, we made further changes to make the content up to date, including clean up, restructuring, and adding state-of-the-art knowledge.

For the second edition, we also revised our *TimeViz Browser*, the digital pendant to the survey of visualization techniques in Appendix A. It includes the same set of techniques as the book, but comes with additional filter and search facilities allowing scientists and practitioners to find exactly the solutions they are interested in. The *TimeViz Browser* is available at `https://browser.timeviz.net`.

We hope you are as excited as we are about the second edition with all its extensions. Happy reading!

St. Pölten University of Applied Sciences,   *Wolfgang Aigner*
TU Wien,   *Silvia Miksch*
University of Rostock,   *Heidrun Schumann*
September 2022   *Christian Tominski*

Preface to the First Edition

Time is an exceptional dimension. We recognize this every day: when we are waiting for a train, time seems to run at a snail's pace, but the hours we spend in a bar with good friends pass by so quickly. There are times when one can wait endlessly for something to happen, and there are times when one is overwhelmed by events occurring in quick succession. Or it can happen that the weather forecast has predicted a nice and sunny summer day, but our barbecue has to be canceled due to a sudden heavy thunderstorm. Our perception of the world around us and our understanding of relations and models that drive our everyday life are profoundly dependent on the notion of time.

As visualization researchers, we are intrigued by the question of how this important dimension can be represented visually in order to help people understand the temporal trends, correlations, and patterns that lie hidden in data. Most data are related to a temporal context; time is often inherent in the space in which the data have been collected or in the model with which the data have been generated. Seen from the data perspective, the importance of time is reflected in established self-contained research fields around temporal databases or temporal data mining. However, there is no such sub-field in visualization, although generating expressive visual representations of time-oriented data is hardly possible without appropriately accounting for the dimension of time.

When we first met, we all had already collected experience in visualizing time and time-oriented data, be it from participating in corresponding research projects or from developing visualization techniques and software tools. And the literature had already included a number of research papers on this topic at that time. Yet despite our experience and the many papers written, we recognized quite early in our collaboration that neither we nor the literature spoke a common (scientific) language. So there was a need for a systematic and structured view of this important aspect of visualization.

We present such a view in this book – for scientists conducting related research as well as for practitioners seeking information on how their time-oriented data can be visualized in order to achieve the bigger goal of understanding the data and gaining valuable insights. We arrived at the systematic view upon which this book is based in the course of many discussions, and we admit that agreeing on it was not an easy process. Naturally, there is still room for arguments to be made and for extensions of the view to be proposed. Nonetheless, we think that we have managed to lay the structural foundation of this area.

The practitioner will hopefully find the many examples that we give throughout the book useful. On top of this, the book offers a substantial survey of visualization techniques for time and time-oriented data. Our goal was to provide a review of existing work structured along the lines of our systematic view for easy visual reference. Each technique in the survey is accompanied by a short description, a visual impression of the technique, and corresponding categorization tags. But visual representations of time and time-oriented data are not an invention of the computer age. In fact, they have ancient roots, which will also be showcased in this book. A discussion of the closely related aspects of user interaction with visual representations and analytical methods for time-oriented data rounds off the book.

We now invite you to join us on a journey through time – or more specifically on a journey into the visual world of time and time-oriented data.

TU Wien &
University of Rostock,
February 2011

Wolfgang Aigner
Silvia Miksch
Heidrun Schumann
Christian Tominski

About the Authors

Wolfgang Aigner is a professor at the Institute of Creative\Media/Technologies at the St. Pölten University of Applied Sciences, Austria where he is responsible for the research area "Data Visualization". He is also an adjunct professor at TU Wien, Austria where he received his habilitation in computer science for his work on "Interactive Visualization and Data Analysis: Visual Analytics with a Focus on Time" in 2013. Dr. Aigner is an expert in information visualization and visual analytics, particularly in the context of time-oriented data, where he has authored and co-authored more than 150 scientific publications. He served as program committee member and chair for various scientific conferences and acts as associate editor for scientific journals. He has received national awards for his research work, was awarded a Top Cited Article 2005–2010 from Pergamon/Elsevier, and received a best paper honorable mention at the IEEE Conference on Visual Analytics Science and Technology (VAST). Since 2002, Wolfgang has been involved in the acquisition and execution of a number of funded basic and applied research projects at the national and international levels.

Silvia Miksch is a university professor and head of the research division "Visual Analytics" (CVAST), Institute of Visual Computing and Human-Centered Technology, TU Wien since 2015. She has been head of the Information and Knowledge Engineering research group, Institute of Software Technology & Interactive Systems, TU Wien from 1998 to 2015. From 2006 to 2010 she was university professor and head of the Department of Information and Knowledge Engineering (ike) at Danube University Krems, Austria. In April 2010 she established the awarded Laura Bassi Centre of Expertise for Visual Analytics Science and Technology (CVAST) funded by the Federal Ministry of Economy, Family and Youth of the Republic of Austria. Silvia has acquired, led, and has been involved in several national and international applied and basic research projects. She served as paper co-chair of several conferences including IEEE VAST 2010, 2011, and 2020 and overall papers chair IEEE VIS 2021 as well as EuroVis 2012, and on the editorial board of several journals including IEEE TVCG and CGF. She acts on various strategic committees, such as the VAST and EuroVis steering committees as well as the VIS executive committee.

In 2020 she was inducted into the IEEE Visualization Academy. Furthermore, she acts as scientific reporter in the board of the Austrian Research Fund (FwF) and is advisory board member of the Vienna Science and Technology Fund (WWTF). She has more than 300 scientific publications and her main research interests are visualization and visual analytics over time and space with particular focus on interaction techniques, network-based, knowledge-assisted, and guidance-enriched methods.

Heidrun Schumann is a professor emeritus at the University of Rostock, Germany, where she was heading the Chair of Computer Graphics at the Institute for Visual & Analytic Computing. She received doctoral degree (Dr.-Ing.) in 1981 and post-doctoral degree (Dr.-Ing. habil.) in 1989. Her research and teaching activities cover a variety of topics related to computer graphics, including information visualization, visual analytics, and rendering. She is interested in visualizing complex data in space and time, combining visualization and terrain rendering, and facilitating visual data analysis with progressive methods. A key focus of Heidrun's work is to intertwine visual, analytic, and interactive methods for making sense of data. Heidrun published more than two hundred articles in top venues and journals. She co-authored the first German textbook on data visualization, a textbook specifically on the visualization of time-oriented data, and a textbook on interactive visual data analysis. In 2014, Heidrun was inducted as a Fellow of the Eurographics Association. In 2020, she was awarded with the Fraunhofer Medal. In the same year, she was inducted to the IEEE Visualization Academy.

Christian Tominski is a professor at the Institute for Visual & Analytic Computing at the University of Rostock, Germany. He received doctoral (Dr.-Ing.) and post-doctoral (Dr.-Ing. habil.) degrees in 2006 and 2015, respectively. In 2021, he was granted the title of professor (apl. Prof.) for human-data interaction. His research is primarily in the area of visualization, with a focus on interacting with visual data representations. He is particularly interested in effective and efficient visual analytics techniques for interactively exploring, analyzing, and editing complex data. Christian has published numerous papers on new visualization and interaction techniques for multivariate data, temporal data, geo-spatial data, and graph data. In his work, he emphasizes conceptual aspects and aims to systematically investigate visual analytics challenges. He co-authored three books, including a book on the visualization of time-oriented data, a book focusing on interaction for visualization, and a more general book about interactive visual data analysis. Christian is the maintainer of several visual analytics systems and visualization tools, including the LandVis system for spatio-temporal health data, the VisAxes tool for time-oriented data, the CGV system for coordinated graph visualization, the Responsive Matrix Cells for exploring and editing multivariate graphs, and the TimeViz Browser for visualization techniques for time-oriented data.

Acknowledgements

Acknowledgements for the Second Edition

We wish wo thank our institutions, the St. Pölten University of Applied Sciences, the TU Wien, and the University of Rostock for their financial support with publishing fees, which allowed us to make this an open-access book. We further thank all colleagues who helped us with figures and permissions for the second edition. We would also like to thank our editor at Springer, Helen Desmond, for her continued support on this project.

This work was funded by the Austrian Ministry for Transport, Innovation and Technology (BMVIT) under the ICT of the Future program via the SEVA project (874018), the Austrian Science Fund (FWF) as part of the projects SoniVis (P33531-N), Vis4Schools (I5622-N), KnoVA (P31419-N31), BSS (P31881-N32), and IMMV (AR 384-G24), the Vienna Science and Technology Fund (WWTF) as part of the project GuidedVA (ICT19-047), the Austrian Research Promotion Agency (FFG) as part of the project DoRIAH (880883) as well as the Land NÖ via the project Dataskop (K3-F-2/015-2019).

This work was further made possible and got inspired through projects funded by the German Research Foundation (UniVA: 380014305, GEMS 2.0: 214484876, VisSect: 310545869, ViES: 289665189, VASSiB: 202784448).

Acknowledgements for the First Edition

Much of the information and insights presented in this book were obtained through the assistance of many of our students and colleagues. We would like to thank all of them for their feedback and discussions. Furthermore, we were kindly supported by our respective universities while developing and writing this book.

We particularly wish to thank all the authors of referenced material who gave feedback and provided images as well as the following publishers for their cooperation and unbureaucratic support in giving permission to reproduce material free of charge: Elsevier, Graphics Press, IEEE Press, International Cartographic Association, Springer, Third Millennium Press, and University of Chicago Press.

Valuable support was provided and insights gained within the VisMaster (Visual Analytics – Mastering the Information Age) project (a coordination action funded by the Future and Emerging Technologies (FET) programme within the Seventh Framework Programme for Research of the European Commission, under FET-Open grant number: 225924) and the various research projects conducted within the research groups of the authors.

Contents

Chapter 1
Introduction

> Computers should also help us warp time, but the challenge here is even greater. Normal experience doesn't allow us to roam freely in the fourth dimension as we do in the first three. So we've always relied on technology to aid our perception of time.
>
> Udell (2004, p. 32)

Space and time are two outstanding dimensions because, in conjunction, they represent the four-dimensional space or simply the world we are living in. Basically, every piece of data we measure is related and often only meaningful within the context of space and time. Consider, for example, the price of a barrel of oil. The data value of $129 alone is not very useful. Only if assessed in the context of where (space) and when (time) is the oil price valid and only then it is possible to meaningfully interpret the cost of $129.

Space and time differ fundamentally in terms of how we can navigate and perceive them. Space can, in principle, be navigated arbitrarily in all three spatial dimensions, and we can go back to where we came from. Humans have senses for perceiving space, in particular, the senses of sight, touch, and hearing. Time is different; it does not allow for active navigation. We are constrained to the unidirectional character of constantly proceeding time. We cannot go back to the past and we have to wait patiently for the future to become present. And above all else, humans do not have senses for perceiving time directly. This fact makes it particularly challenging to visualize time – making the invisible visible.

Time is an important data dimension with distinct characteristics. Time is common across many application domains, for example, medical records, business, science, biographies, history, planning, or project management. In contrast to other quantitative data dimensions, which are usually "flat", time has an inherent semantic structure, which increases time's complexity substantially. The hierarchical structure of granularities in time, for example, minutes, hours, days, weeks, and months, is unlike that of most other quantitative dimensions. Specifically, time comprises different forms of divisions (e.g., 60 minutes correspond to one hour, while 24 hours make up one day), and granularities are combined to form calendar systems (e.g., Gregorian,

© The Author(s) 2023
W. Aigner et al., *Visualization of Time-Oriented Data*, Human–Computer Interaction Series, https://doi.org/10.1007/978-1-4471-7527-8_1

Julian, business, or academic calendars). Moreover, time contains natural cycles and re-occurrences, for example, seasons, but also social (often irregular) cycles, like holidays or school breaks. Therefore, time-oriented data, i.e., data that are inherently linked to time, need to be treated differently than other kinds of data and require appropriate visual, interactive, and computational methods to explore and analyze them.

The human perceptual system is highly sophisticated and specifically suited to spot visual patterns. Visualization strives to exploit these capabilities and to aid in seeing and understanding otherwise abstract and arcane data. Early visual depictions of time series date back as far as the 11th century (see Chapter 2). Today, a variety of visualization methods exist and visualization is applied widely to present, explore, and analyze data. However, many visualization techniques treat time just as a numeric parameter among other quantitative dimensions and neglect time's special character. In order to create visual representations that succeed in assisting people in reasoning about time and time-oriented data, visualization methods have to account for the special characteristics of time. This is also demanded by Shneiderman (1996) in his well-known task by data type taxonomy, where he identifies temporal data as one of seven basic data types most relevant in data analysis scenarios.

Creating good visualizations usually requires good data structures. However, commonly only simple sequences of time-value-pairs $\langle (t_0, v_0), (t_1, v_1), \ldots, (t_n, v_n) \rangle$ are used as the basis for analysis and visualization. Accounting for the special characteristics of time can be beneficial from a data modeling point of view. One can use different calendars that define meaningful systems of granularities for different application domains (e.g., fiscal quarters or academic semesters). Data can be modeled and integrated at different levels of granularity (e.g., months, days, hours, and seconds), enabling, for example, value aggregation along granularities. Besides this, data might be given for time intervals rather than for time points, for example, in project plans, medical treatments, or working shift schedules. Related to this diversity of aspects is the problem that most of the available methods and tools are strongly focused on special domains or application contexts. Silva and Catarci (2000) conclude:

> It is now recognized that the initial approaches, just considering the time as an ordinal dimension in a 2D or 3D visualizations [sic], are inadequate to capture the many characteristics of time-dependent information. More sophisticated and effective proposals have been recently presented. However, none of them aims at providing the user with a complete framework for visually managing time-related information.
>
> Silva and Catarci (2000, p. 9)

The aim of this book is to present and discuss the multitude of aspects that are relevant from the perspective of visualization. We will characterize the dimension of time as well as time-oriented data, and describe tasks that users seek to accomplish using visualization methods. While time and associated data form a part of *what* is being visualized, user tasks are related to the question *why* something is visualized. *How* these characteristics and tasks influence the visualization design will be explained by several examples. These investigations will lead to a systematic categorization of visualization approaches. Because interaction techniques and computational analysis methods play an important role in the exploration of and the reasoning with time-oriented data, these topics will be discussed in dedicated chapters. A large part of this book is devoted to a survey of existing techniques for visualizing time and time-oriented data. This survey presents self-contained descriptions of techniques accompanied by an illustration and corresponding references on a per-page basis.

Before going into detail on visualizing time-oriented data, let us first take a look at the basics and examine general concepts of visualizing data.

1.1 Introduction to Visualization

Visualization is a widely used term. Spence (2007) refers to a dictionary definition of the term: *visualize* – to form a mental model or mental image of something. Visual representations have a long and venerable history in communicating facts and information. But only in 1987, visualization became an independent self-contained research field. In that year, the notion of visualization in scientific computing was introduced by McCormick et al. (1987). They defined the term *visualization* as follows:

> Visualization is a method of computing. It transforms the symbolic into the geometric, enabling researchers to observe their simulations and computations. Visualization offers a method for seeing the unseen. It enriches the process of scientific discovery and fosters profound and unexpected insights.
>
> McCormick et al. (1987, p. 3)

The goal of this new field of research has been to integrate the outstanding capabilities of human visual perception and the enormous processing power of computers to support users in analyzing, understanding, and communicating their data, models, and concepts. In order to achieve this goal, three major criteria have to be satisfied (see Tominski and Schumann, 2020):

- expressiveness,
- effectiveness, and
- efficiency.

Expressiveness refers to the requirement of showing exactly the information contained in the data; nothing more and nothing less must be visualized. *Effectiveness* primarily considers the degree to which users can achieve their analysis goals. An effective visualization addresses the cognitive capabilities of the human visual system, the analysis task at hand, the application background, and other context-related information to obtain intuitively recognizable and interpretable visual representations. Finally, *efficiency* involves a cost-value ratio in order to assess the benefit of using a visualization to accomplish some analysis tasks. While the value of a visual representation is not so easy to determine (see van Wijk, 2006), costs are typically related to the resources required for computation, the display space needed to show the data, and the human effort spent during the data analysis.

Expressiveness, effectiveness, and efficiency are criteria that any visualization should aim to fulfill. To this end, the visualization process, above all else, has to account for two aspects: the data and the task at hand. In other words, we have to answer the two questions: "What has to be presented?" and "Why does it have to be presented?". We will next discuss both questions in more detail.

What? – Specification of the Data

In recent years, different approaches have been developed to characterize data – the central element of visualization. In their overview article, Wong and Bergeron (1997) established the notion of *multidimensional multivariate data* as multivariate data that are given in a multidimensional domain. This definition leads to a distinction between *independent* and *dependent* variables. Independent variables define an n-dimensional domain. In this domain, the values of k dependent variables are measured, simulated, or computed; they define a k-variate dataset. If at least one dimension of the domain is associated with the dimension of time, we call the data *time-oriented data*.

Another useful concept for modeling data along cognitive principles is the *pyramid framework* by Mennis et al. (2000). At the level of data, this framework is based on three perspectives (also see Figure 3.29 on p. 70): *where* (location), *when* (time), and *what* (theme). The perspectives *where* and *when* characterize the data domain, i.e., the independent variables as described above. The perspective *what* describes what has been measured, observed, or computed in the data domain, i.e., the dependent variables as described above. At the level of knowledge, the *what* includes not only simple data values, but also objects and their relationships, where objects and relations may have arbitrary data attributes associated with them.

From the visualization point of view, all three aspects need to be taken into account. The *where* aspect is relevant for representing the spatial frame of reference and associating data values to locations. The *when* aspect is required to show the characteristics of the temporal frame of reference and to associate data values to the time domain. Finally, the *what* aspect takes care of representing individual values or abstractions of a multivariate dataset. As our interest is in time and time-oriented data, this book places special emphasis on the *when* aspect. We will specify the key properties of time and associated data in Chapter 3 and discuss the specific implications for visualization in Chapter 4.

Why? – Specification of the Task

Similar to specifying the data, one also needs to know why the data are visualized and what tasks the user seeks to accomplish with the help of the visualization. On a very abstract level, the following three basic goals can be distinguished (see Ward et al., 2015):

- exploratory analysis,
- confirmatory analysis, and
- presentation of analysis results.

Exploratory analysis can be understood as an undirected search, where no a-priori hypotheses about the data are given. The goal is to get insight into the data, begin extracting relevant information, and come up with hypotheses. In a phase of *confirmatory analysis*, visualization is used to prove or reject hypotheses, which can originate from data exploration or from models associated with the data. In this sense, confirmatory analysis is a form of directed search. When facts about the data have eventually been ascertained, it can be the goal of a *presentation* to communicate and disseminate analysis results.

These three basic visualization goals call for quite different visual representations. This becomes clear when taking a look at two established visualization concepts: filtering and accentuation. The aim of filtering is to visualize only relevant data and to omit less relevant information, and the goal of accentuation is to highlight important information. During an exploratory analysis, both concepts help users to focus on selected parts or aspects of the data. But filtering and accentuation must be applied carefully because it is usually not known upfront which data are relevant or important. Omitting or highlighting information indiscriminately can lead to misinterpretation of the visual representation and to incorrect findings. During a confirmatory analysis, filtering can be applied more easily because we already know which data are relevant and contribute to the hypotheses to be evaluated. Accentuation and de-accentuation are common means to enhance expressiveness and effectiveness, and to fine-tune visual presentations in order to communicate results and insight yielded by an exploratory or confirmatory analysis process.

Although the presentation of results is very important, this book is more about visual analysis and interactive exploration of time-oriented data. Therefore, we will take a closer look at common analysis and exploration tasks. As Bertin (1983) describes, human visual perception has the ability to focus (1) on a particular element of an image, (2) on groups of elements, or (3) on an image as a whole. Based on these capabilities, three fundamental categories of interpretation aims have been introduced by Robertson (1991): *point*, *local*, and *global*. They indicate which values are of interest: (1) values at a given point of the domain, (2) values in a local region, or (3) all values of the whole domain. These basic tasks can be subdivided into more specific, concrete tasks, which are usually given as a list of verbal descriptions. Wehrend and Lewis (1990) define several such low-level tasks: identify or locate data values, distinguish regions with different values or cluster similar data, relate, compare, rank, or associate data, and find correlations and distributions. The task by data type taxonomy by Shneiderman (1996) lists seven high-level tasks that also include the notion of interaction with the data in addition to purely visual tasks:

- Overview: gain an overview of the entire dataset
- Zoom: zoom in on data of interest
- Filter: filter out uninteresting information
- Details-on-demand: select data of interest and get details when needed

- Relate: view relationships among data items
- History: keep a history of actions to support undo and redo
- Extract: allow extraction of data and of query parameters

Yi et al. (2007) further refine the aspect of interaction in information visualization and derive a number of categories of interaction tasks. These categories are organized around the user's intentions to interactively adjust visual representations to the tasks and data at hand. Consequently, a *show me* prefaces six categories:

- show me something else (explore)
- show me a different arrangement (reconfigure)
- show me a different representation (encode)
- show me more or less detail (abstract/elaborate)
- show me something conditionally (filter)
- show me related items (connect)

The *show me* tasks allow for switching between different subsets of the analyzed data (explore), different arrangements of visual primitives (reconfigure), and different visual representations (encode). They also address the navigation of different levels of detail (abstract/elaborate), the definition of data of interest (filter), and the exploration of relationships (connect).

In addition to the *show me* categories, Yi et al. (2007) introduce three further interaction tasks:

- mark something as interesting (select)
- let me go to where I have already been (undo/redo)
- let me adjust the interface (change configuration)

Mark something as interesting (select) subsumes all kinds of selection tasks, including picking out individual data values as well as selecting entire subsets of the data. Supporting users in going back to interesting data or views (undo/redo) is essential during interactive data exploration. Adaptability (change configuration) is relevant when a visualization system is applied by a wide range of users for a variety of tasks and data types.

In summary, the purpose of visualization, that is, the task to be accomplished with visualization, can be defined in different ways. Schulz et al. (2013a) provide a deeper discussion on this topic. The above mentioned visualization and interaction tasks serve as a basic guideline to assist visualization designers in developing representations that effectively support users in conducting visual data exploration and analysis. In Chapter 4 we will come back to this issue and refine tasks with regard to the analysis of time-oriented data. The aspect of interaction will be taken up in Chapter 5.

More than that, the aforementioned tasks are essentially carried out by human users. This makes it necessary to have an understanding of the users who perform these tasks as well as the environment in which they are being conducted. Visualization aims to amplify cognition, but simply producing images is no guarantee that complex visualizations will be understood and are useful for gaining insights.

Therefore, a user-centered approach is essential, i.e., understanding your users with their goals and tasks is a prerequisite for being able to answer the question of how to best visualize the data. Several user-centered design methods do exist for that matter, with the *nested model for visualization design and evaluation* (see Munzner, 2009; Munzner, 2014) being a widely used one. This model is well applicable for designing visualizations of time-oriented data and we encourage our readers to pick up the corresponding details from the available literature.

How? – The Visualization Pipeline

In order to generate effective visual representations, raw data have to be transformed into image data in a data-, user-, and task-specific manner. Conceptually, raw data have to be mapped to geometry and corresponding visual attributes such as color, position, size, or shape, also called *visual variables* (see Bertin, 1983; Mackinlay, 1986). Thanks to the capabilities of our visual system, the perception of visual stimuli is mostly spontaneous. As indicated earlier, Bertin (1983) distinguishes three levels of cognition that can be addressed when encoding information to visual variables. On the first level, elementary information is directly mapped to visual variables. This means that every piece of elementary information is associated with exactly one specific value of a visual variable. The second level involves abstractions of elementary information, rather than individual data values. By mapping the abstractions to visual variables, general characteristics of the data can be communicated. The third level combines the two previous levels and adds representations of further analysis steps and metadata to convey the information contained in a dataset in its entirety.

To facilitate the generation of visual output at all three levels, a flexible mapping procedure is required. This procedure is commonly called the *visualization pipeline*, first introduced by Haber and McNabb (1990). The visualization pipeline consists of three steps (see Figure 1.1a):

1. filtering,
2. mapping, and
3. rendering.

The filtering step prepares the raw input data for processing through the remaining steps of the pipeline. This is done with respect to the given analysis task and includes not only the selection of relevant data, but also operations for data enrichment or data reduction, interpolation, data cleansing, grouping, dimension reduction, and others. The mapping step literally maps the prepared data to suitable graphical marks and visual variables. This is the most crucial step as it largely influences the expressiveness and effectiveness of the resulting visual representation. Finally, based on the output of the mapping step, the rendering step generates actual image data. This general pipeline model is the basis for many visualization systems.

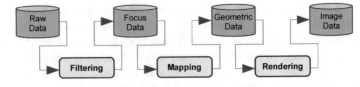

(a) Original variant. ©① *The authors. Adapted from Haber and McNabb (1990).*

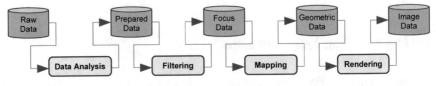

(b) Extended variant. ©① *The authors. Adapted from dos Santos and Brodlie (2004).*

Fig. 1.1: The visualization pipeline.

The basic pipeline model has been refined by dos Santos and Brodlie (2004) in order to better address the requirements of higher dimensional visualization problems. The original filtering has been split up into two separate steps: data analysis and filtering (see Figure 1.1b). The data analysis carries out automatic computations like interpolations, clustering, or pattern recognition. The filtering step then extracts only those pieces of data that are of interest and need to be presented. In the case of large high-dimensional datasets, the filtering step is highly relevant because displaying all information will most likely lead to complex and overloaded visual representations that are hard to interpret. Because interests may vary across users, tasks, and data, the filtering step has to support the interactive refinement of filter conditions. Further input like the specific analysis task or hypothesis as well as application-specific details can be used to steer the data extraction process.

In an effort to formally model the visualization process, Chi (2000) built upon the classic pipeline model and derived the *data state reference model*. This model reflects the step-wise transformation of abstract data into image data through several stages by using operators. While transformation operators transform data from one level of abstraction to another, within stage operators process the data only within the same level of abstraction (see Figure 1.2). This model broadens the capabilities of the visualization process and allows the generation of visual output at all of Bertin's levels. Different operator configurations lead to different views on the data, and thus, to comprehensive insight into the analyzed data. It is obvious that the selection and configuration of appropriate operators to steer the visualization process is a complex problem that depends mainly on the given visualization goal, which in turn is determined by the characteristics of the data and the analysis objectives.

The previous paragraphs may suggest that the image or view eventually generated by a visualization pipeline is an end product. But this is not true. In fact, the user controls the visualization pipeline and interacts with the visualization process in various ways. Views and images are created and adjusted until the user finds them

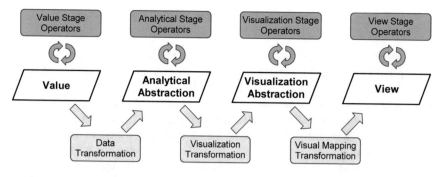

Fig. 1.2: The data state reference model. ©① *The authors. Adapted from Chi (2000).*

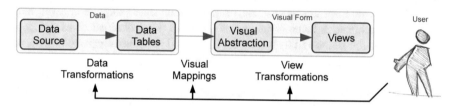

Fig. 1.3: The information visualization reference model. ©① *The authors. Adapted from Card et al. (1999).*

suitable for the task at hand. Therefore, Card et al. (1999) integrate the user in their *information visualization reference model* (see Figure 1.3).

So far, we have mainly touched upon a more abstract perspective of visualization methods and the visualization process itself. For a concrete visualization problem, the *how* aspect not only depends on the *what* and *why* questions discussed earlier, but also on the target device(s) available for a certain environment. Particularly, recent developments in mobile devices such as smartphones, tablets, or smartwatches as well as immersive technologies, for example, head mounted displays for augmented or virtual reality call for adequate visualization methods that make use of the specifics of these display technologies. Apart from the display technologies these devices exhibit, the aspects of available input modalities and computational resources also need to be taken into account. As these questions are to be addressed on a more general level, independent from the context of time-oriented data, we refer the interested reader to Tominski and Schumann (2020) for a more in-depth discussion of these aspects.

Having introduced the very basics of interactive visualization, we now move on to an application example. The goal is to illustrate a concrete visual representation and to demonstrate possible benefits for data exploration and analysis.

1.2 Application Example: Health Record Visualization

A considerable share of physicians' daily work time is devoted to searching and gathering patient-related information to form a basis for adequate medical treatment and decision-making. The amount of information is enormous, often disorganized, and physicians might be overwhelmed by the information provided to them. Electronic health records comprise multiple variables of different data types that are sampled irregularly and independently from each other, as for example quantitative parameters (e.g., blood pressure or body temperature) and qualitative parameters (e.g., events like a heart attack) as well as instantaneous data (e.g., blood sugar measurement at a certain point in time) and interval data (e.g., insulin therapy from January to May). Moreover, the data commonly originate from heterogeneous sources like digital lab systems, hospital information systems, or patient data sheets that are not well integrated. Exploring such heterogeneous time-oriented datasets to get an overview of the current health status and its history for individual patients or a group of patients is a challenging task (Rind et al., 2013b; Rind et al., 2017).

Interactive visualization is an approach to representing a coherent view of such medical data and to catering for easy data exploration. In our example, an interactive discourse of the physician with the visual representation is of major importance because a single static representation typically cannot satisfy task-dependent information needs. In addition to visualizing information intuitively, aiding clinicians in gaining new medical insights about patients' current health status, state changes, trends, or patterns over time is an important aspect.

VisuExplore (↪ p. 339) is an interactive visualization tool for exploring a heterogeneous set of medical parameters over time (see Rind et al. (2011b)). VisuExplore uses multiple views along a common horizontal time axis to convey the different

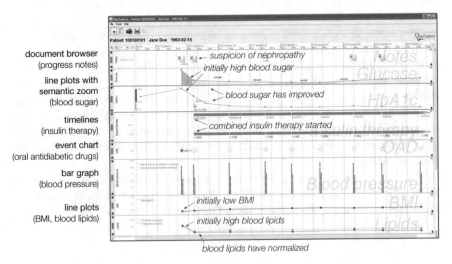

Fig. 1.4: Visualization of medical parameters of a diabetes patient. ⓒⓘ *The authors.*

medical parameters involved. It is based on several visualization methods, including line plots (↪ p. 233), bar graphs (↪ p. 234), event charts, and timelines (↪ p. 258), that are combined and integrated.

Figure 1.4 consists of eight stacked visualization views showing data of a diabetes patient over a period of two years and three months between November 2016 and March 2019. The document browser at the top shows icons for medical documents, like for example diagnostic findings or x-ray images. In the figure, the document browser contains progress notes from the very beginning of treatment, when the physicians suspected renal failure. Below the document browser, a line plot with semantic zoom (see p. 137) shows blood glucose values. Colored areas below the line provide qualitative information about normal (green), elevated (yellow), and high (red) value ranges, which makes this semantic information easy to read. Below that, another line plot shows HbA1c, an indicator of a patient's blood glucose condition. In this case, more vertical space is devoted to the chart, thus allowing more exact readings of the values. Still, semantic information is added as color annotation of the y-axis, using small ticks to indicate when the HbA1c value crosses qualitative range boundaries (e.g., from critically high to elevated as indicated by the horizontal red/yellow line). Below the line plots, there are two timeline charts showing the insulin therapy and oral anti-diabetic drugs. Insulin is categorized into rapid-acting insulin (ALT), intermediate-acting insulin (VZI), and a mixture of these (Misch). Details about the brand name or dosage in the free text are shown as labels that are located below the respective timeline. Oral anti-diabetic drugs are shown via an event chart below. There are also free text details about oral diabetes medication. The sixth view is a bar graph with adjacent bars for systolic and diastolic blood pressure. The bottom two views are line plots related to the body mass index (BMI) and blood lipids with two lines showing triglyceride and cholesterol values.

This arrangement has been chosen because it places views of medical tests directly above views of the related medical interventions. The height of some views has been reduced to fit on a single screen. This is possible because all information that is relevant for the physician's current task can still be recognized in this state.

The shown diabetes case is a 57-year-old patient with initially very high blood sugar values. From the interactive visual representations, several facts about the patient can be inferred as illustrated by the following insights that were gained by a physician using the VisuExplore system. The initially high blood sugar values were examined in detail via tooltips and showed exact values of 428 mg/dl glucose and 14.8% HbA1c. In addition, it can be seen in the bottom panel that blood lipid values are also high (256 mg/dl cholesterol, 276 mg/dl triglyceride). At the same time, the body mass index shown above is rather low (20.1). From the progress notes in the document browser, it can be seen that the physician had a suspicion of nephropathy. But these elevated values are also signs of latent autoimmune diabetes in adults, a special form of type 1 diabetes. After one month, blood sugar has improved (168 mg/dl glucose) and blood lipids have normalized. The patient switched to insulin therapy in a combination of rapid-acting insulin (ALT) and intermediate-acting insulin (VZI). Since April 2017, the insulin dosage has remained stable and concomitant medication is no longer needed. The patient's overall condition has improved through blood sugar management. Furthermore, the physician involved in the case study wondered about the very high HbA1c value of 11.9% in November 2016 and why diabetes treatment had only started four months later.

VisuExplore's interactive features allow physicians to get an overview of multiple medical parameters and focus on interesting parts of the data. Physicians can add views for additional variables and may resize and rearrange them as necessary. Further, it is possible to navigate and zoom across the time dimension by dragging the mouse, using dedicated buttons, or selecting predefined views (e.g., last year). Moreover, the software allows the selection and highlighting of data elements. Other time-based visualization and interaction techniques can extend the system to support special purposes. For example, a document browser shows medical documents (e.g., discharge letters or treatment reports) as document icons (e.g., PDF, Word) that physicians can click to open a document. VisuExplore integrates with the hospital information systems and accesses the medical data stored there.

This example demonstrated that visual representations are capable of providing a coherent view of otherwise heterogeneous and possibly distributed data. The integrative character also supports interactive exploration and task-specific focusing on relevant information.

1.3 Book Outline

With the basics of visualization and an application example, we have set the stage for the next chapters. Before going into detail about the contemporary visualization of time and time-oriented data, some inspiring and thought-provoking historical depictions and images from the arts are given attention in Chapter 2. The characteristics of time and data for modern interactive visualization on computers are the focus of Chapter 3. The actual visualization process, that is, the transformation of abstract data to visual representations, will be discussed in Chapter 4, taking into account the key question words *what*, *why*, and *how* to visualize. In Chapters 5 and 6, we go beyond pure visualization methods and discuss cornerstones of interaction and computational analysis methods to support exploration and visual analysis. Chapter 7 addresses the question of how to select visualization techniques that are appropriate for an application problem at hand. A final summary along with a discussion of open challenges can be found in Chapter 8.

A large part of this book is devoted to a survey of existing visualization techniques for time and time-oriented data in Appendix A. Throughout the book we use the ↪ symbol followed by a page number to refer the reader to a particular technique in the survey. The second Appendix B provides a list of concrete examples for all categories of data quality problems introduced in Chapter 3. Figure 1.5 provides a visual overview of the contents of the book.

Please refer to the companion website of the book for updates and additional resources including links to related material, visualization prototypes, and technique descriptions: https://www.timeviz.net.

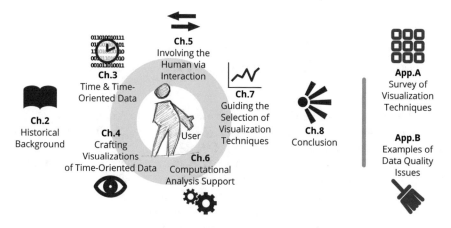

Fig. 1.5: Visual overview of the contents of the book. ©① *The authors.*

References

Bertin, J. (1983). *Semiology of Graphics: Diagrams, Networks, Maps.* Translated by William J. Berg. University of Wisconsin Press.

Card, S., J. Mackinlay, and B. Shneiderman (1999). *Readings in Information Visualization: Using Vision to Think.* Morgan Kaufmann Publishers.

Chi, E. H. (2000). "A Taxonomy of Visualization Techniques Using the Data State Reference Model". In: *Proceedings of the IEEE Symposium Information Visualization (InfoVis).* IEEE Computer Society, pp. 69–76. DOI: 10.1109/INFVIS.2000.885092.

Dos Santos, S. and K. Brodlie (2004). "Gaining Understanding of Multivariate and Multidimensional Data through Visualization". In: *Computers & Graphics* 28, pp. 311–325. DOI: 10.1016/j.cag.2004.03.013.

Haber, R. B. and D. A. McNabb (1990). "Visualization Idioms: A Conceptual Model for Scientific Visualization Systems". In: *Visualization in Scientific Computing.* IEEE Computer Society, pp. 74–93.

Mackinlay, J. (1986). "Automating the Design of Graphical Presentations of Relational Information". In: *ACM Transactions on Graphics* 5.2, pp. 110–141. DOI: 10.1145/22949.22950.

McCormick, B. H., T. A. DeFanti, and M. D. Brown (1987). "Visualization in Scientific Computing". In: *ACM SIGGRAPH Computer Graphics* 21.6, pp. 3–3. DOI: 10.1145/41997.41998.

Mennis, J. L., D. J. Peuquet, and L. Qian (2000). "A Conceptual Framework for Incorporating Cognitive Principles into Geographical Database Representation". In: *International Journal of Geographical Information Science* 14.6, pp. 501–520. DOI: 10.1080/136588100415710.

Munzner, T. (2009). "A Nested Process Model for Visualization Design and Validation". In: *IEEE Transactions on Visualization and Computer Graphics* 15.6, pp. 921–928. DOI: 10.1109/TVCG.2009.111.

Munzner, T. (2014). *Visualization Analysis and Design*. A K Peters/CRC Press. DOI: 10.1201/b17511.

Rind, A., W. Aigner, S. Miksch, S. Wiltner, M. Pohl, T. Turic, and F. Drexler (2011b). "Visual Exploration of Time-Oriented Patient Data for Chronic Diseases: Design Study and Evaluation". In: *Information Quality in e-Health*. Springer, pp. 301–320. DOI: 10.1007/978-3-642-25364-5_22.

Rind, A., P. Federico, T. Gschwandtner, W. Aigner, J. Doppler, and M. Wagner (2017). "Visual Analytics of Electronic Health Records with a Focus on Time". In: *New Perspectives in Medical Records: Meeting the Needs of Patients and Practitioners*. Springer, pp. 65–77. DOI: 10.1007/978-3-319-28661-7_5.

Rind, A., T. D. Wang, W. Aigner, S. Miksch, K. Wongsuphasawat, C. Plaisant, and B. Shneiderman (2013b). "Interactive Information Visualization to Explore and Query Electronic Health Records". In: *Foundations and Trends in Human-Computer Interaction* 5.3, pp. 207–298. DOI: 10.1561/1100000039.

Robertson, P. K. (1991). "A Methodology for Choosing Data Representations". In: *IEEE Computer Graphics and Applications* 11.3, pp. 56–67. DOI: 10.1109/38.79454.

Schulz, H.-J., T. Nocke, M. Heitzler, and H. Schumann (2013a). "A Design Space of Visualization Tasks". In: *IEEE Transactions on Visualization and Computer Graphics* 19.12, pp. 2366–2375. DOI: 10.1109/TVCG.2013.120.

Shneiderman, B. (1996). "The Eyes Have It: A Task by Data Type Taxonomy for Information Visualizations". In: *Proceedings of the IEEE Symposium on Visual Languages*. IEEE Computer Society, pp. 336–343. DOI: 10.1109/VL.1996.545307.

Silva, S. F. and T. Catarci (2000). "Visualization of Linear Time-Oriented Data: A Survey". In: *Proceedings of the International Conference on Web Information Systems Engineering (WISE)*. IEEE Computer Society, pp. 310–319. DOI: 10.1109/WISE.2000.882407.

Spence, R. (2007). *Information Visualization: Design for Interaction*. 2nd edition. Prentice-Hall.

Tominski, C. and H. Schumann (2020). *Interactive Visual Data Analysis*. AK Peters Visualization Series. CRC Press. DOI: 10.1201/9781315152707.

Udell, J. (2004). "Space, Time, and Data". In: *InfoWorld* 26, p. 32. URL: https://www.infoworld.com/article/2665760/space--time--and-data.html.

Van Wijk, J. J. (2006). "Views on Visualization". In: *IEEE Transactions on Visualization and Computer Graphics* 12.4, pp. 421–433. DOI: 10.1109/TVCG.2006.80.

Ward, M. O., G. Grinstein, and D. Keim (2015). *Interactive Data Visualization: Foundations, Techniques, and Applications*. 2nd edition. A K Peters/CRC Press. DOI: 10.1201/b18379.

Wehrend, S. and C. Lewis (1990). "A Problem-Oriented Classification of Visualization Techniques". In: *Proceedings of the IEEE Visualization Conference (Vis)*. IEEE Computer Society, pp. 139–143. DOI: 10.1109/VISUAL.1990.146375.

Wong, P. C. and R. D. Bergeron (1997). "30 Years of Multidimensional Multivariate Visualization". In: *Scientific Visualization: Overviews, Methodologies, and Techniques*. Edited by Nielson, G. M., Hagen, H., and Müller, H. IEEE Computer Society, pp. 40–62.

Yi, J. S., Y. ah Kang, J. T. Stasko, and J. A. Jacko (2007). "Toward a Deeper Understanding of the Role of Interaction in Information Visualization". In: *IEEE Transactions on Visualization and Computer Graphics* 13.6, pp. 1224–1231. DOI: 10.1109/TVCG.2007.70515.

Chapter 2
Historical Background

> There is a magic in graphs. The profile of a curve reveals in a flash a whole situation – the life history of an epidemic, a panic, or an era of prosperity. The curve informs the mind, awakens the imagination, convinces.

Henry D. Hubbard in Brinton (1939, Preface)

Long before computers even appeared, visualization was used to represent time-oriented data. Probably the oldest time-series representation to be found in literature is the illustration of planetary orbits created in the 10th or possibly 11th century (see Figure 2.1). The illustration is part of a text from a monastery school and shows inclinations of the planetary orbits as a function of time.

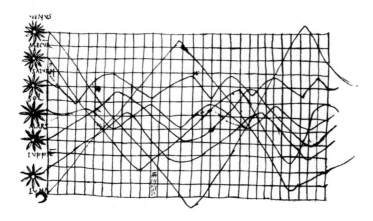

Fig. 2.1: Time-series plot depicting planetary orbits (10th/11th century). The illustration is part of a text from a monastery school and shows the inclinations of the planetary orbits over time. © *1936 University of Chicago Press. Reprinted, with permission, from Funkhouser (1936, p. 261).*

© The Author(s) 2023
W. Aigner et al., *Visualization of Time-Oriented Data*, Human–Computer Interaction Series, https://doi.org/10.1007/978-1-4471-7527-8_2

In human history, keeping track of the passage of time, the seasons of the year, and planning ahead for them has been one of the most important tasks of even the earliest human civilizations. Being essential for central elements of life, for example, for agriculture or religious acts, a variety of tools such as bone engravings as well as visual representations of calendars can be found throughout history. Figure 2.2 shows an example of a perpetual calendar from 1594 designed by Ortensio Toro that shows the Gregorian calendar for a 400-year cycle. It allowed date calculations far into the future.

An example of a particularly interesting artifact by Native American people is the Time Ball shown in Figure 2.3. Unlike calendars that are valid for larger parts of people, the time ball is a mnemonic device, i.e., a tangible, personal record of developments and life events over the course of its owner's life. It acted as a memory aid usually kept by women where simple knots recorded individual days and meaningful occasions, such as births, deaths, marriages, or days of bounty, hardship, and even conflicts were highlighted using special markers. These included glass beads, shells, cloth fragments, or human hair.

To broaden the view beyond computer-aided visualization and provide background information on the history of visualization methods, we present historical and application-specific representations. They mostly consist of historical techniques of the pre-computer age, such as the works of William Playfair, Étienne-Jules Marey, or Charles Joseph Minard.

Furthermore, we will take the reader on a journey through the arts. Throughout history, artists have been concerned with the question of how to incorporate the dynamics of time and motion in their artworks. We present a few outstanding art movements and art forms that are characterized by a strong focus on representing temporal concepts. We believe that art can be a valuable source of inspiration; concepts or methods developed by artists might even be applicable to information visualization, possibly improving existing techniques or creating entirely new ones.

2.1 Classic Ways of Graphing Time

Representing business data graphically is a broad application field with a long tradition. William Playfair (1759–1823) can be seen as the protagonist and founding father of modern statistical graphs. He published the first known time-series depicting economic data in his *Commercial and Political Atlas* of 1786 (Playfair and Corry, 1786). His works contain basically all of the widely-known standard representation techniques (see Figure 2.4, 2.5, 2.6, and 2.7) such as the pie chart, the silhouette graph (↪ p. 281) , the bar graph (↪ p. 234), and the line plot (↪ p. 233).

Playfair's work widely popularized new graphic forms and many other economists and statisticians built upon this to develop them further. One example from 1874 can be seen in Figure 2.8. It shows a fiscal chart of the United States by Francis Amasa Walker (1840–1897) that uses a symmetric layout to contrast state revenues

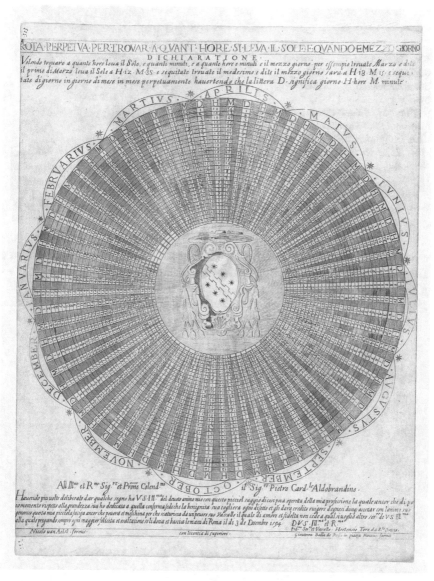

Fig. 2.2: Perpetual calendar (1594). Gregorian calendar designed for a 400-year cycle. Ⓢ *Ortensio Toro (Italian, active 16th century). Retrieved from Cooper Hewitt, Smithsonian Design Museum.*

Fig. 2.3 Time Ball
(mnemonic device) of the Native American Tribe Yakama
by Vivian Harrison (1945).
ⓒ *Courtesy of the National
Museum of the American In-
dian, Smithsonian Institution
(26/9725), from NMAI Photo
Services.*

with state expenditures over a period from 1789–1870 in absolute (center) as well as
relative terms (stacked bars left and right).

In Figure 2.7 multiple heterogeneous time-oriented variables are integrated within
a single view: the weekly wages of a good mechanic as a line plot, the price of a
quarter of wheat as a bar graph, as well as historical context utilizing timelines (↪
p. 258). Playfair himself credits the usage of timelines to Joseph Priestley (1733–
1804) who created a graphical representation of the life spans of famous historical

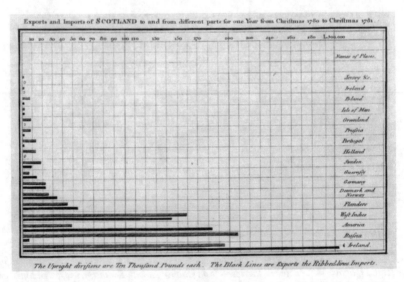

Fig. 2.4: Bar graph from the *Commercial and Political Atlas* by Playfair and Corry (1786) representing exports and imports of Scotland during one year. Ⓢ *1786 Playfair and Corry. Retrieved
from Wikimedia Commons.*

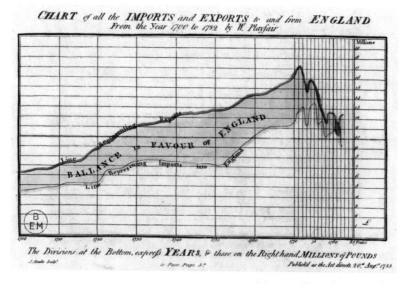

Fig. 2.5: Line plot from the *Commercial and Political Atlas* by Playfair and Corry (1786) repre-senting imports and exports of England from 1700 to 1782. The yellow line on the bottom shows imports into England and the red line at the top exports from England. Color shading is added between the lines to indicate positive (light blue) and negative (red; around 1781) overall balances. Ⓢ *1786 Playfair and Corry. Retrieved from Wikimedia Commons.*

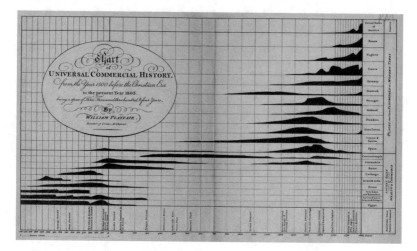

Fig. 2.6: Silhouette graph used by William Playfair (1805) to represent the rise and fall of nations over a period of more than 3000 years. A horizontal time scale is shown at the bottom that uses a compressed scale for the years before Christ on the left. Important events are indicated textually above the time scale. Countries are grouped vertically into Ancient Seats of Wealth & Commerce (bottom), Places that have Flourished in Modern Times (center), and America (top). Ⓢ *1805 Playfair. Retrieved from Wikimedia Commons.*

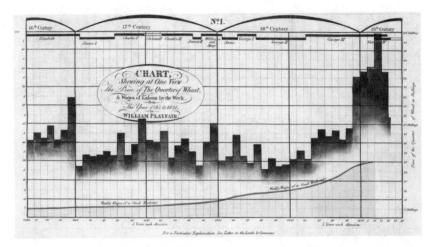

Fig. 2.7: Information rich chart of William Playfair (1821) that depicts the weekly wages of a good mechanic (line plot at the bottom), the price of a quarter of wheat (bar graph in the center), as well as historical context (timeline at the top) over a time period of more than 250 years. Ⓢ *1821 Playfair. Retrieved from Wikimedia Commons.*

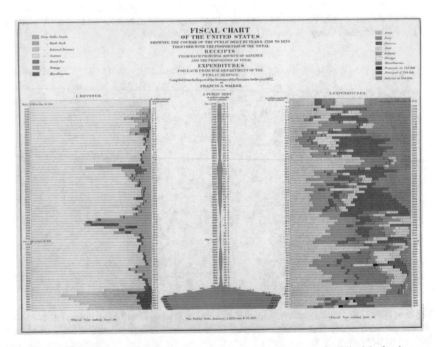

Fig. 2.8: Fiscal chart of the United States showing the the development of public debt for the years 1789 to 1870 together with the proportions of receipts and expenditures. ⓒⓘⓢⓞ *1874 Walker. Retrieved from David Rumsey Map Collection, David Rumsey Map Center, Stanford Libraries.*

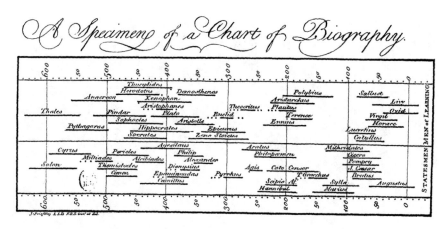

Fig. 2.9: *Chart of biography* by Joseph Priestley (1765) that portrays the life spans of famous historical persons using timelines. ⓢ *1765 Priestley. Retrieved from Wikimedia Commons.*

persons divided into two groups of Statesmen and Men of Learning (see Figure 2.9). The usage of a horizontal line to represent an interval of time might seem obvious to us nowadays, but in Priestley's days this was certainly not the case. This is reflected in the fact that he devoted four pages of text to describe and justify his technique to his readers. A remarkable detail of Priestley's graphical method is that he acknowledged the importance of representing temporal uncertainties and provided a solution to deal with them using dots. Even different levels of uncertainty were taken into account, ranging from dots below lines to lines and dotted lines.

Even earlier than both Priestley and Playfair, Jacques Barbeu-Dubourg (1709–1779) created the earliest known modern timeline. His *carte chronographique* (Barbeu-Dubourg, 1753) consisted of multiple sheets of paper that were glued together and add up to a total length of 16.5 meters (see Figure 2.10). A rare version of the chart is available at Princeton University Library where the paper is mounted on two rollers in a foldable case that can be scrolled via two handles (see Ferguson (1991) for a detailed description).

Another prominent example of a graphical representation of historical information via annotated timelines is *Deacon's synchronological chart of universal history* which was originally published in 1890 and was drawn by Edmund Hull (see Figure 2.11). Various reprints and books extending the original historic facts to the present and adaptations for specialized areas like for example inventions and explorations can be found in the literature (e.g., Third Millennium Press, 2001). A slightly different layout approach for depicting historical data is Willard's Chronographer of American History (see Figure 2.12). In contrast to the example before, Emma Willard uses a botanical tree metaphor to structure historical periods combined with a round time scale on the outside.

Apart from calendars, maps have also been an essential tool in human history for thousands of years. Combining time-oriented data with cartographic maps allows for

Fig. 2.10: *Carte chronographique* by Jacques Barbeu-Dubourg (1753) that shows the known history from the beginning of time up to 1760. Multiple sheets of paper were glued together and mounted in the *chronology machine* which allows to manually scroll back and forth in time using two handles. © *Courtesy of Rare Book Division, Department of Rare Books and Special Collections, Princeton University Library, from Princeton University Library Catalog.*

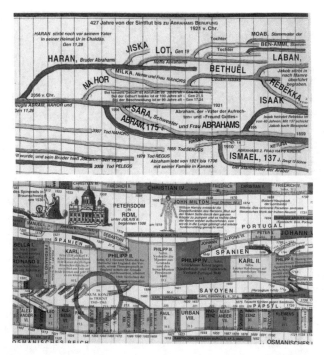

Fig. 2.11: Parts of Deacon's synchronological chart of universal history. © *2001 Third Millennium Press Ltd. Reprinted, with permission, from Third Millennium Press (2001).*

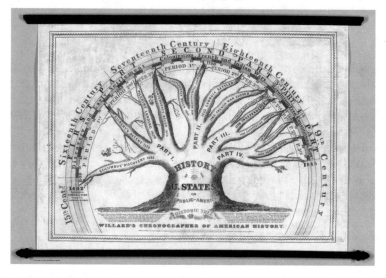

Fig. 2.12: Chronographer of American History by Emma Willard (1845). Wall map for representing important events in American history. © *1845 Willard. Retrieved from David Rumsey Map Collection, David Rumsey Map Center, Stanford Libraries.*

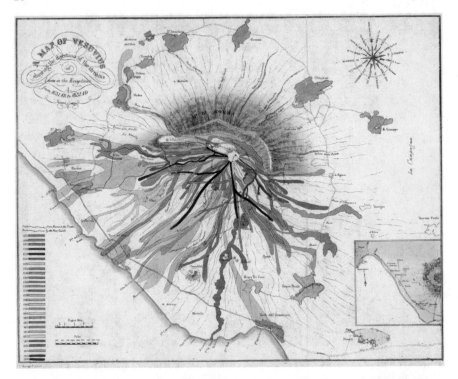

Fig. 2.13: Map of Vesuvius by John Auldjo (1833). It shows the direction of the streams of lava in the eruptions from 1631 to 1831. ⓒ①⑤◎ *1833 Auldjo. Retrieved from David Rumsey Map Collection, David Rumsey Map Center, Stanford Libraries.*

depicting both, spatial relationships as well as temporal developments. A remarkable approach for depicting time-orented data on maps is the map of Vesuvius created by John Auldjo (1805–1886) in 1833 (see Figure 2.13). In his map, he uses different color hues to represent time, i.e., the years of eruptions over a period of 200 years, resulting in an ordinal time scale.

Charles Joseph Minard (1781–1870) created a masterpiece of the visualization of historical information in 1869. His graphical representation of *Napoleon's Russian campaign* of 1812 is extraordinarily rich in information, conveying no less than six different variables in two dimensions (see Figure 2.14). Tufte (1983) comments on this representation as follows:

It may well be the best statistical graphic ever drawn.

Tufte (1983, p. 40)

The basis of the representation is a 2-dimensional map on which a band symbolizing Napoleon's army is drawn. The width of the band is proportional to the army's size; the direction of movement (advance or retreat) is encoded by color.

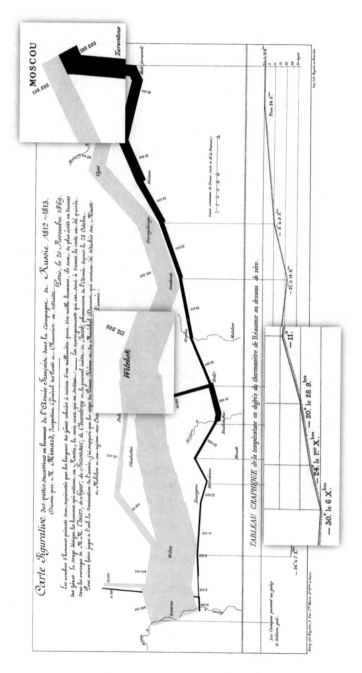

Fig. 2.14: Napoleon's Russian campaign of 1812 by Charles Joseph Minard (1869). A band visually traces the army's location during the campaign, whereby the width of the band indicates the size of the army and the color encodes advance and retreat. Labels and a parallel temperature chart provide additional information. ⓒⓘ *The authors. Adapted from Minard (1869) via Wikimedia Commons.*

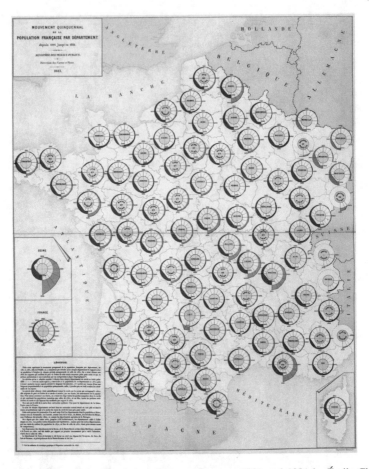

Fig. 2.15: Movement of the population of France between 1801 and 1881 by Émile Cheysson (1883). ©①⊜◎ *1883 Cheysson. Retrieved from David Rumsey Map Collection, David Rumsey Map Center, Stanford Libraries.*

Furthermore, various important dates are plotted and a parallel line graph shows the temperature over the course of time.

An early example of combining statistical graphics that use a cyclic time axis with maps is shown in Figure 2.15. The representation of Émile Cheysson created in 1883 shows the movement of the population for each department of France between 1801 and 1881. To make different absolute population values better comparable, the data shown is indexed at the time midpoint 1841 and shown relative to that. Different color hues are used to fill the circular silhouette graph (↪ p. 281) depending on whether the population is below (red) or above (gray) the value of the indexing point. This map is part of a series of graphs created for the French Ministry of Public Works and was inspired by the earlier work of Charles Joseph Minard.

Also in the 19th century, the prominent historic figure Florence Nightingale used a statistical graph to show numbers and causes of deaths over time during the Crimean War. When Nightingale was sent to run a hospital near the Crimean battlefields to care for British casualties of war, she made a devastating discovery: many more men were dying from infectious diseases they had caught in the filthy hospitals of the military than from wounds. By introducing new standards of hygiene and diet, and most importantly, by ensuring proper water treatment, deaths due to infectious diseases fell by 99% within a year. Florence Nightingale tediously recorded mortality data for two years and created a novel diagram to communicate her findings. Figure 2.16 shows two of these *rose charts*. This representation is also called *polar area graph* and consists of circularly arranged wedges that convey quantitative data. Unlike pie charts, all the segments of rose charts have the same angle. Bringing the data in this form clearly revealed the horrible fact that many more soldiers were dying because of preventable diseases they had caught in the hospitals than from wounds sustained in battle. Not only this fact was communicated, but also how this situation could be improved by the right measures; these can be seen from the left rose chart in Figure 2.16. Through this diagram, which was more a call to action than merely a presentation of data, she persuaded the government and the Queen to introduce wide-reaching reforms, thus bringing about a revolution in nursing, health care, and hygiene in hospitals worldwide.

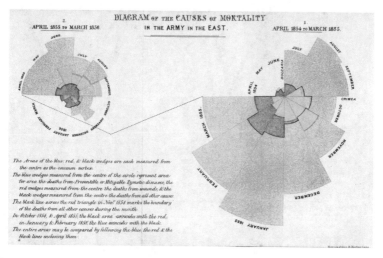

Fig. 2.16: Rose charts showing number of casualties and causes of death in the Crimean War by Florence Nightingale (1858). Red shows deaths from wounds, black represents deaths from accidents and other causes, and blue shows deaths from preventable infectious diseases soldiers caught in hospitals. The chart on the right shows the first year of the war and the chart on the left shows the second year after measures of increased hygiene, diet, and water treatment had been introduced. Ⓢ *1858 Nightingale. Retrieved from Wikimedia Commons.*

A quite different approach to representing historical information is the illustration of the *Cuban missile crisis* during the Cold War by Bertin (1983). The diagram shows decisions, possible decisions, and the outcomes thereof over time (see Figure 2.17). This representation is similar to the *decision chart* ($\hookrightarrow$ p. 237). Chapple and Garofalo (1977) provided an illustration of *Rock'n'Roll history* shown in Figure 2.18 that depicts protagonists and developments in the area as curved lines that are stacked according to the artists' percentage of annual record sales. The *ThemeRiver™* technique ($\hookrightarrow$ p. 293) can be seen as a further more formal development of this idea.

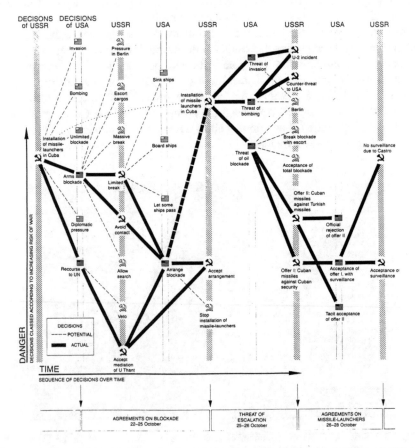

Fig. 2.17: Cuban missile crisis (threat level and decisions over time). The diagram shows decisions, possible decisions, and the outcomes thereof over time. © *1983 The University of Wisconsin Press. Reprinted, with permission, from Bertin (1983, p. 264).*

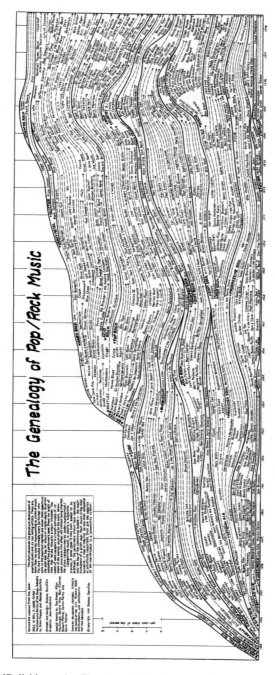

Fig. 2.18: Rock'n'Roll history by Chapple and Garofalo (1977) that depicts protagonists and developments in the area as curved lines that are stacked according to the artists' percentage of annual record sales. © *Courtesy of Reebee Garofalo.*

With the advance of industrialization in the late 19th and early 20th century, optimizing resources and preparing time schedules became essential requirements for improving productivity. One of the main protagonists of the study and optimization of work processes was Frederick Winslow Taylor (1856–1915). His associate Henry Laurence Gantt (1861–1919) studied the order of steps in work processes and developed a family of timeline-based charts as an intuitive visual representation to illustrate and record time-oriented processes (see Figures 2.19 and 2.20). Widely known as *Gantt charts* (↪ p. 253), these representations are such powerful analytical instruments that they are used nearly unchanged in modern project management.

Other interesting representations of work-related data can be seen in Figures 2.21 and 2.22. A record of hours worked per day by an employee is shown in Figure 2.21. It is interesting to note that both axes are used for representing different granularities

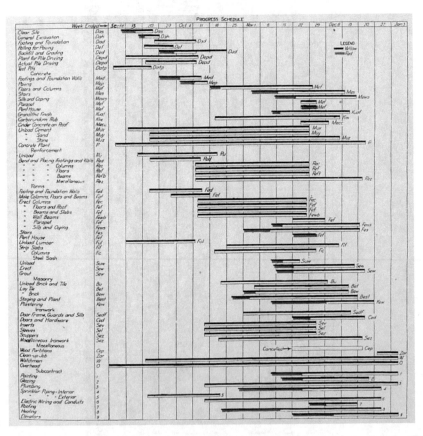

Fig. 2.19: Progress schedule based on the graphical method of Henry L. Gantt (see Brinton, 1939, p. 259). Different work packages are depicted as horizontal lines. Black lines indicate the planned timings; the actual quantity of work done is shown below in red. © *1917 Engineering News-Record. Retrieved from Internet Archive.*

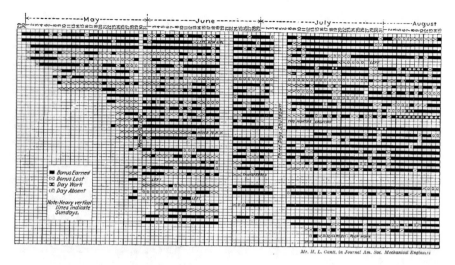

Fig. 2.20: Record of work carried out in one room of a Worsted Mill by Henry L. Gantt (see Brinton, 1914, p. 52). Each row represents one worker and gives information about whether a bonus was earned and if the worker was present. ⑤ *1914 Gantt. Retrieved from Internet Archive.*

Fig. 2.21 Exact hours and days worked in 1929 by an employee at the Oregon ports (see Brinton, 1939, p. 250). Days are mapped on the horizontal axis and hours per day worked are represented as bars on the vertical axis. The representation shows extreme irregularities in working hours. ⑤ *1934 Foisie. Retrieved from Internet Archive.*

of time, i.e., days on the horizontal axis and hours per day on the vertical axis. Figure 2.22 employs a radial layout of the time and allows a reading on multiple levels: the outer ring shows days without work and the inner rings show hours worked during the day, whereas the green areas indicate night hours.

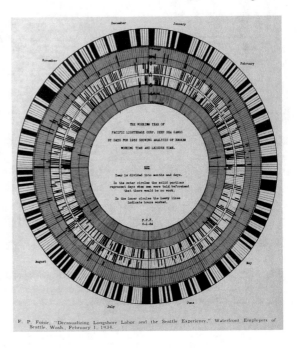

Fig. 2.22 An analysis of working time and leisure time in 1932 (see Brinton, 1939, p. 251). Uses a radial layout of time and allows a reading on multiple levels: the outer ring shows days without work and the inner rings show hours worked during the day, whereas the green areas indicate night hours. Ⓢ *1934 Foisie. Retrieved from Internet Archive.*

F. P. Foisie, "Decasualizing Longshore Labor and the Seattle Experience," Waterfront Employers of Seattle, Wash., February 1, 1934.

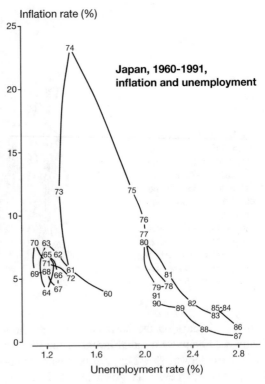

Fig. 2.23 Phillips curve. Unemployment rate (horizontal axis) is plotted against inflation rate (vertical axis). Each point in the plot corresponds to one year and is labeled accordingly. The markers of subsequent years are linked to create a visual trace of time. ⓒⓘ *The authors. Adapted, with permission of Graphics Press, from Tufte (1997, p. 60)*

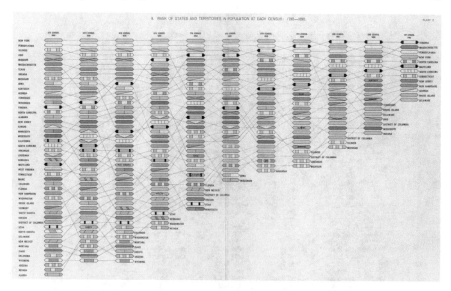

Fig. 2.24: Rank of states and territories in population at each census from 1790 to 1890 by Henry Gannett (1898). ⓒ①⑤◎ *1898 Gannett. Retrieved from David Rumsey Map Collection, David Rumsey Map Center, Stanford Libraries.*

A quite unique representation of economic data is the so-called *Phillips curve* – a 2D plot based on an economic theory that shows unemployment vs. inflation in a Cartesian coordinate system. In this representation, time is neither mapped to the horizontal nor the vertical axis, but is rather shown textually as labeled data points on the curve. This way, the dimension of time is slightly de-emphasized in favor of showing the relationship of two time-dependent variables (see Figure 2.23). Each year's combination of the two variables unemployment rate and inflation rate leads to a data point in 2D space that is marked by the digits of the corresponding year. The markers of subsequent years are connected by a line resulting in a path over the course of time.

For representing positional changes within a set of elements, *rank charts* were already introduced in early statistical publications, for example, by Henry Gannett (1846–1914) or Willard Brinton (1880–1957) (see Figures 2.24 and 2.25). Elements are ordered according to their ranking and displayed next to each other in columns for different points in time. The positional change of individual elements is emphasized by connecting lines. This way, the degree of rank change is represented by the angles of the connecting lines, thus making big changes in rank stand out visually by the use of very steep lines. Note that the two examples differ in the direction of their time axes. While the chart of Henry Gannett (Figure 2.24) uses a time axis from right to left, the example of Willard Brinton (Figures 2.25) employs the more frequently used order from left to right.

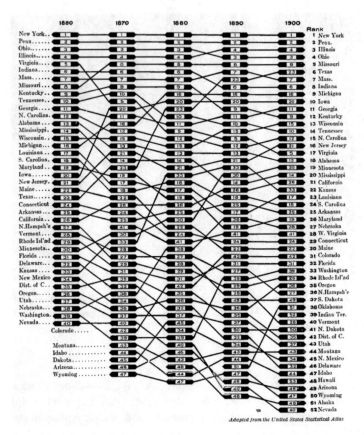

Fig. 2.25: Rank of states and territories in population at different census years from 1860 to 1900 by Willard Brinton (1914, p. 65). © *1914 Brinton. Retrieved from Internet Archive.*

A remarkable representation of time-oriented information was created by Étienne-Jules Marey (1830–1904) in the 1880s (see Figure 2.26). It shows the train schedule for the track Paris to Lyon graphically. Basically, a 2D diagram is used which places the individual train stops according to their distance in a list on the vertical axis, while time is represented on the horizontal axis. Thus, horizontal lines are used to identify the individual stops and a vertical raster is used for timing information. The individual trains are represented by diagonal lines running from top-left to bottom-right (Paris–Lyon) and bottom-left to top-right (Lyon–Paris), respectively. The slope of the line gives information about the speed of the train – the steeper the line, the faster the respective train is traveling. Moreover, horizontal sections of the trains' lines indicate if the train stops at the respective station at all and how long the train stops. On top of that, the density of the lines provides information about the frequency of trains over time. This leads to a clear and powerful representation showing complex information at a glance while allowing for in-depth analysis of

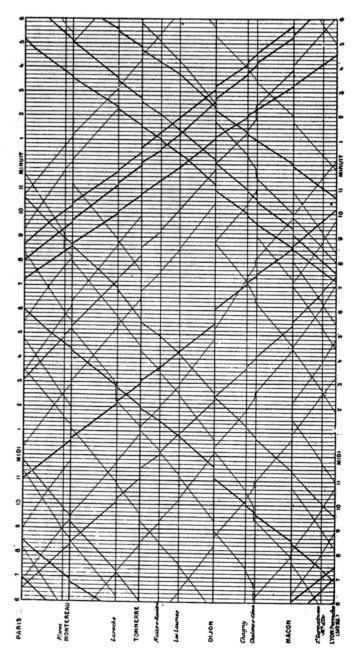

Fig. 2.26: Train schedule by Étienne-Jules Marey (1875, p. 260). Individual train stops are placed according to their distance in a list on the vertical axis, while time is represented on the horizontal axis (figure above is rotated by 90°). The individual trains are represented by diagonal lines running from top-left to bottom-right (Paris–Lyon) and bottom-left to top-right (Lyon–Paris) respectively.
Ⓢ *1875 Marey. Retrieved from Internet Archive.*

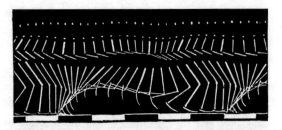

Fig. 2.27 A person walking. Studies of movement by Étienne-Jules Marey (1894, p. 61). Ⓢ *1894 Marey. Retrieved from Internet Archive.*

Fig. 2.28 Chronophotography. A photo of flying pelican taken by Étienne-Jules Marey around 1882. Ⓢ *1882 Marey. Retrieved from Wikimedia Commons.*

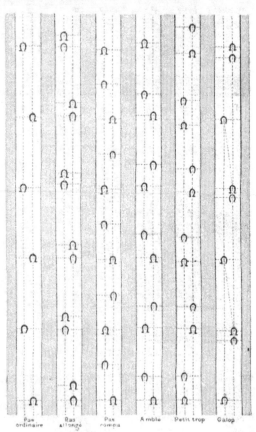

Fig. 2.29 Horse gaits. Studies of movement by Étienne-Jules Marey (1875, p. 147). Ⓢ *1875 Marey. Retrieved from Wikimedia Commons.*

the data. Similar representations have also been used for the Japanese Shinkansen train line and the Javanese Soerabaja-Djokjakarta train line where the track's terrain profile is additionally shown. The basic idea of this representation even stood the test of time and interactive versions are still used today in modern software systems of railway companies to support train scheduling or in ViDX (↪ p. 364) to visualize automated assembly lines.

Étienne-Jules Marey not only created the fabulous train schedule, but was also very interested in exploring all kinds of movement. Born in 1830 in France, he was a trained physician and physiologist. His interest in internal and external movements in humans and animals, such as blood circulation, human walking, horse gaits, or dragonfly flight, led to the decomposition of these movements via novel photography and representation methods (see Figures 2.27, 2.28, and 2.29). This photography method, which is called *chronophotography*, paved the way for the birth of modern film-making at the end of the nineteenth century.

Today, Marey is still a valuable source of inspiration. Reason enough to speak highly of him and his work:

> Tirelessly, this brilliant visionary stopped the passage of time, accelerated it, slowed it down to "see the invisible," and recreated life through images and machines.
>   La maison du cinema and Cinematheque Francaise (2000)

In medicine, large amounts of information are generated which mostly have to be processed by humans. Graphical representations which help to make this myriad of information comprehensible play a crucial role in the workflow of healthcare personnel. These representations range from the *fever curves* of the nineteenth century (see Figure 2.30) and EEG time-series plots (see Figure 2.31) to information-rich patient status overviews (see Figure 2.32). Especially the graphical summary of patient status by Powsner and Tufte (1994) makes use of concepts such as *small multiples* (↪ p. 359), *focus+context* (see p. 137), or the integration of textual and graphical information. It manages to display information on a single page that would otherwise fill up entire file folders and would require serious effort to summarize.

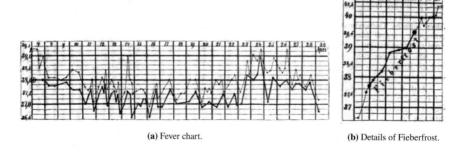

(a) Fever chart.   (b) Details of Fieberfrost.

Fig. 2.30: Fever charts created by Carl August Wunderlich (1870, p. 161, 167). Ⓢ *1870 Wunderlich. Retrieved from Internet Archive.*

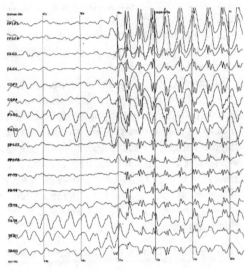

Fig. 2.31 EEG time-series plot. ©①① *2005 Der Lange. Retrieved from Wikimedia Commons.*

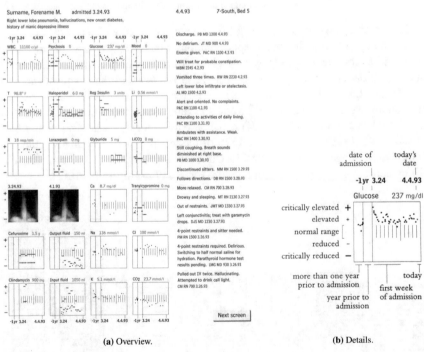

(a) Overview.

(b) Details.

Fig. 2.32: Graphical summary of patient status by Powsner and Tufte (1994). Concise summary of patient information. Uses *small multiples*, *focus+context*, and integrates textual as well as graphical information. © *1997 Graphics Press. Reprinted, with permission, from Tufte (1997, pp. 110–111).*

Weather in 1980

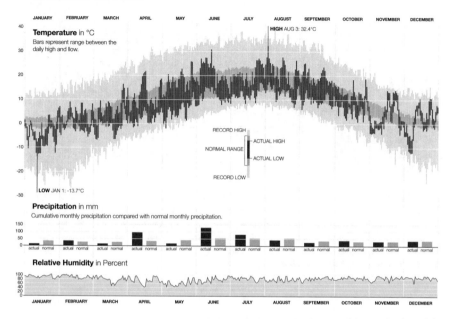

Fig. 2.33: Weather statistics for 1980. Aggregated values are displayed along with detailed information on temperature, humidity, and precipitation. Similar illustrations have been printed annually by the New York Times for more than 30 years. ⓒ① *The authors. Generated with Protovis.*

Weather and climate are further well-known application areas dealing with time-oriented data. Here, developments over time are of greater interest than single snapshots. Figure 2.33 shows the adaptation of an extremely information-rich illustration provided by the New York Times for more than 30 years to show New York City's weather developments for a whole year. Monthly and yearly aggregates are displayed along with more detailed information on temperature, humidity, and precipitation. All in all, more than 2500 numbers are shown in this representation in a very compact and readable form. An even earlier example of a visual representation of the weather data of New York City is shown in Figure 2.34. Here, temperatures, wind velocity, relative humidity, wind direction, and the weather conditions of a single month (December, 1912) are displayed.

Considering the long history of visualizing time-oriented data, two main metaphors for representing time can be identified: *arrow/line* and *river*. First, a vast majority of visualization techniques uses lines or arrows to depict time (see Davis, 2012). Commonly, a left-to-right direction is applied where later points in time are shown toward the right. Second, the metaphor of a river was frequently used already in historic depictions (see Rendgen, 2019). This metaphor is also used in contemporary visualization techniques, less often though, for example in ThemeRiver (↪ p. 293) and stream graphs (↪ p. 286).

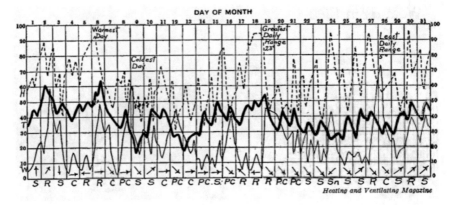

Fig. 2.34: Record of the Weather in New York City for December, 1912 (see Brinton, 1914, p. 93). The bold line indicates temperature in degrees Fahrenheit. The light solid line shows wind velocity in miles per hour. The dotted line depicts relative humidity in percentage from readings taken at 8 a.m. and 8 p.m. Arrows portray the prevailing direction of the wind. Initials at the base of the chart show the weather conditions as follows: S, clear; PC, partly cloudy; C, cloudy; R, rain; Sn, snow. Ⓢ *1914 Brinton. Retrieved from Internet Archive.*

2.2 Time in Visual Storytelling & Arts

Two disciplines that are seldomly connected to time-oriented information are *visual explanations* and *visual storytelling*. Although ubiquitously used in various forms in daily life, they are rarely considered for visualizing abstract information. Visual explanations are often used in manuals for home electronics, furniture assembly, car repair, and many more (see Figures 2.36 and 2.37). Often, they are used to illustrate

Fig. 2.35: In comics, time and space are one and the same. © *1993, 1994 HarperCollins Publishers. Reprinted, with permission, from McCloud (1994, p. 100).*

1. Rip off the
wrist band along
the perforation.

2. Remove
protective foil
and fix around
your wrist.

3. Cheer for
your favorite!

Fig. 2.36: Visual explanation to illustrate a stepwise process as used in Tomitsch et al. (2007). ©① *The authors.*

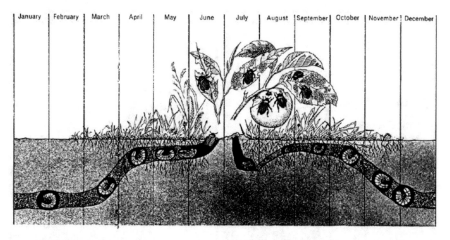

Fig. 2.37: Life Cycle of the Japanese Beetle (Newman, 1965, p. 104–105). © *1990 Graphics Press. Reprinted, with permission, from Tufte (1990, p. 43).*

stepwise processes visually to an international audience to support the often poorly translated textual instructions. The stepwise nature conveys a temporal aspect and might also be applied to represent abstract information. Even older than everything we presented previously is the craft of *storytelling*, especially visual storytelling, starting from caveman paintings and Egyptian hieroglyphs to picture books and comic strips (see Figure 2.35). Time is the central thread that ties everything together in visual storytelling. Many interesting techniques and paradigms exist that might be applicable to visualization in general (see for example Gershon and Page, 2001) as well as to the representation of time-oriented information in particular.

Comics The art of *comics* is often dubbed as *visual storytelling over time* or *sequential art* (a term used by Will Eisner) because temporal flows are represented in

(a) *Classical comic layout representing an ordered sequence of scenes in juxtaposed panels.* © *Courtesy of Greg Dean, from RealLife Comics.*

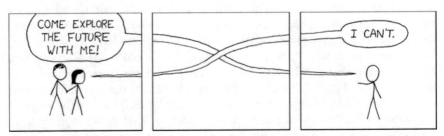

(b) *Exploration of the duality of space and time in comic panels.* ⊜①⑤ *"Future" by Randall Munroe. Retrieved from xkcd.com.*

Fig. 2.38: Comics where temporal flows are represented in juxtaposed canvases on a page.

Fig. 2.39: A single comic panel contains more than a frozen moment in time. © *1993, 1994 HarperCollins Publishers. Reprinted, with permission, from McCloud (1994, p. 95).*

juxtaposed canvases on a page (see Figure 2.38). These descriptions already suggest that comics incorporate many concepts of time, while still retaining a static, 2-dimensional form. McCloud (1994) analyzed many of the methods and paradigms of comics, concluding that powerful means of representing time, dynamics, and movement are applied which differ from those applied in painting or photography. Comics allow for the seamless representation of many temporal concepts that may be also applicable to visualization. Basically, the course of time is represented in comics via juxtaposition of panels. But the individual panels portray more than single frozen moments in time and are more than photos placed side by side. Rather, single panels contain whole scenes whose temporal extent may span from milliseconds to arbitrary lengths (see Figure 2.39). Not only the content of a panel sheds light on the length of its duration but also the shape of the panel itself can affect our perception of time. Even more freedom in a temporal sense is given by the transition from one panel to the next or by the space between panels, respectively (see Figure 2.40). Here, time might be compressed, expanded, and rewound; deja vu's might be incorporated and much more. This also implies that comics are not just simply linearly told stories. Comics are very versatile and much more powerful in incorporating time in comparison to paintings, photographs, and even film. Besides the purely temporal aspect, motion is another important topic in comics. Several visual techniques, such as motion lines or action lines with additional effects like multiple images, streaking effects, or blurring are applied (see Figure 2.41). In part, these techniques are borrowed from photography. Research work on generating these comic-like effects from motion pictures has been conducted, for example, in Markovic and Gelautz (2006).

Music & dance   Music notes are a notation almost everybody is aware of, but it is one which is rarely seen in conjunction with time-oriented information (see Figure 2.42). Nevertheless, music notes are clearly a visual representation of temporal information

Fig. 2.40: Transitions between panels might span intervals of arbitrary length. © *1993, 1994 HarperCollins Publishers. Reprinted, with permission, from McCloud (1994, p. 100).*

Fig. 2.41: Techniques to represent movement in comics (motion lines, streaking, multiple images, background streaking). © *1993, 1994 Harper-Collins Publishers. Reprinted, with permission, from McCloud (1994, p. 114).*

AMAZING GRACE

John Newton, 1779

Fig. 2.42: Music notation of "Amazing Grace". A rich set of symbols, lines, and text visualizes beat, rhythm, pitch, note length, pausing, instrument tuning, and parallelism. ⑤ *2007 HenryLi. Retrieved from Wikimedia Commons.*

– even more than that. A rich set of different symbols, lines, and text constitute a very powerful visual language. Beat, rhythm, pitch, note length, pausing, instrument tuning, and parallelism are the most important visualized parameters. In fact, it is hard to imagine any other way of representing musical compositions than via music notes. Related to that, special notations are used for recording dance performances statically on paper (see Figure 2.43).

Movies One art form that is only touched upon briefly here, but which might also offer interesting ideas for visualization, is *film*. We will present movies that exemplify

Fig. 2.43: Dance notation. Used for recording dance performances statically on paper. © *1990 Graphics Press. Reprinted, with permission, from Tufte (1990, p. 117).*

how moviemakers are able to transport highly non-linear stories in the temporally linear medium of film. These examples pertain to the plot of a film, and not to filming or cutting techniques.

Run Lola Run[1] is a movie that presents several possible successions of events sequentially throughout the film (compare *branching time* in Section 3.1.1). The individual episodes begin at the same point in time and show different possible strands of events.

The movie *Pulp Fiction*[2] comprises an even more complicated and challenging plot. It is a collection of different episodes that are semantically as well as temporally linked. Moreover, the movie ends by continuing the very first scene in the movie, thus closing the loop.

A further example of the use of interesting temporal constellations in film is the movie *Memento*.[3] The main character of the movie is a man who suffers from short-term memory loss, and who uses notes and tattoos to hunt for his wife's killer. What makes the storytelling so challenging is the fact that time flows backward from scene to scene (i.e., the end is shown at the beginning and the story progresses to the beginning from there).

Music videos are also often used as an innovation playground where directors can experiment with unconventional temporal flows such as the *reverse narrative* as used in Coldplay's *The Scientist*.[4]

Paintings A very interesting approach to overcoming the limitations of time can be found in *Renaissance* paintings. Here, sequences of different temporal episodes are shown in a single composition. Figure 2.44 for example shows a painting by

---

[1] Run Lola Run (Lola rennt), written and directed by Tom Twyker, 1998.

[2] Pulp Fiction, written by Quentin Tarantino et al., directed by Quentin Tarantino, 1994.

[3] Memento, written by J. and C. Nolan, directed by Christopher Nolan, 2000.

[4] The Scientist, recorded by Coldplay, music video directed by Jamie Thraves, 2001.

Fig. 2.44: Masolino da Panicale, Curing the Crippled and the Resurrection of Tabitha (Brancacci Chapel, S. Maria del Carmine, Florence, Italy), 1420s. Different stages or episodes of a single person are shown within a unifying scenery. ⓢ *1424 Masolino da Panicale. Retrieved from Wikimedia Commons.*

Masolino da Panicale that presents two scenes in the life of St. Peter within a single scenery. While this method of showing different stages or episodes within a unifying scenery was well understood by the people at that time (the Middle Ages), it might not be as easily understood by a modern viewer. In his article, Jones (2020) provides an overview of how paintings depict time and mentions that:

> Paint is usually thought to be a static medium, capable of depicting only frozen instants of time. Yet with a little inventiveness, it's possible for paint to represent the passage of time too.
>
> Jones (2020)

The beginning of the 20th century was characterized by new findings and break-throughs in the natural sciences, especially in mathematics and physics, such as Einstein's theory of relativity. But not only the world of science was shaken by these developments; artists also addressed these topics in their own way. Foremost among these were the protagonists of the art movement of *Cubism*, who focused on incorporating time in their artworks. They coined the term *Four-dimensional Art*. In his book, Miller (2001) gives an overview of the history of this movement.

As already mentioned, the concept of the n-dimensional space in mathematics and physics inspired artists to think about 4D space. Figure 2.45 shows Marcel Duchamp's painting *Nude Descending a Staircase* which incorporates the dimension of time in a very interesting way by overlaying different stages of a person's movement. Another example is Pablo Picasso's *Portrait of Ambroise Vollard* (see Figure 2.46), where many different observations are composed and partly overlaid to form a single picture. The artists wanted to put emphasis on the *process* of looking and recording over time (in contrast to taking a photo). These new ways of bringing the fourth dimension into the static domain of pictures are still a challenge to viewers today.

Fig. 2.45 Marcel Duchamp, Nude Descending a Staircase (No. 2), 1912. The dimension time is incorporated by overlaying different stages of a person's movement. © *2010 VBK, Vienna.*

Fig. 2.46 Pablo Picasso, Portrait of Ambroise Vollard, 1910. Many different observations are composed and partly overlaid to form a single picture. © *2010 Succession Picasso/VBK, Vienna.*

2.3 Summary

We have provided a brief review of relevant historical and application-specific visualization techniques and representations of time in the visual arts. Our aim was to provide historical context for developments in this area and to present some ideas from related fields that might act as a further source of inspiration for designing visualizations. Furthermore, this chapter has demonstrated the enormous breadth of the topic which we are only able to cover in part.

Readers interested in more information about historical representations of time-oriented data and historical representations in general are referred to the wonderful books of Tufte (1983), Tufte (1990), Tufte (1997), Tufte (2006), Wainer (2005), Rosenberg and Grafton (2010), Davis (2017), Rendgen (2019), and Dick (2020). Michael Friendly's great work on the history of data visualization can be studied in numerous articles such as (Friendly, 2008) as well as online in his Data Visualization Gallery[5] and the Milestones Project.[6] Additionally, interesting historic facts related to time representations are discussed on the Chronographics Weblog[7] of Stephen Boyd Davis.

Now, after setting the stage and considering various concepts and ideas from related disciplines, we will narrow our focus and present a systematic view of the visualization of time-oriented data. In this sense, we will first discuss important aspects that make the handling of time and time-oriented data possible. Following that, the visualization problem itself will be systematically explained and discussed.

References

Barbeu-Dubourg, J. (1753). *Carte chronographique*. Paper roll in scroll case, 40 cm x 16.5 m. Princeton University Library.

Bertin, J. (1983). *Semiology of Graphics: Diagrams, Networks, Maps*. Translated by William J. Berg. University of Wisconsin Press.

Brinton, W. C. (1914). *Graphic Methods for Presenting Facts*. New York, NY, USA: The Engineering Magazine Company.

Brinton, W. C. (1939). *Graphic Presentation*. New York, NY, USA: Brinton Associates.

Chapple, S. and R. Garofalo (1977). *Rock 'N' Roll is Here to Pay: The History and Politics of the Music Industry*. Chicago, IL, USA: Burnham Inc Pub.

Davis, S. B. (2012). "History on the Line: Time as Dimension". In: *Design Issues* 28.4, pp. 4–17. DOI: 10.1162/DESI_a_00171.

Davis, S. B. (2017). "Early Visualizations of Historical Time". In: *Information Design*. Routledge. DOI: 10.4324/9781315585680-8.

---

[5] https://www.datavis.ca/gallery
[6] https://www.datavis.ca/milestones
[7] https://chronographics.blogspot.com

Dick, M. (2020). *The Infographic: A History of Data Graphics in News and Communications*. MIT Press.

Ferguson, S. (1991). "The 1753 Carte chronographique of Jacques Barbeu-Dubourg". In: *Princeton University Library Chronicle* 52.2, pp. 190–230. DOI: 10.2307/26404421.

Friendly, M. (2008). "A Brief History of Data Visualization". In: *Handbook of Data Visualization*. Edited by Chen, C.-h., Härdle, W., and Unwin, A. Springer, pp. 15–56.

Funkhouser, H. G. (1936). "A Note on a Tenth Century Graph". In: *Osiris* 1.1, pp. 260–262. DOI: 10.1086/368425.

Gannett, H. (1898). *Statistical Atlas of the United States, Based upon Results of the Eleventh Census (1890)*. United States Census Office.

Gershon, N. and W. Page (2001). "What Storytelling Can Do for Information Visualization". In: *Communications of the ACM* 44.8, pp. 31–37. DOI: 10.1145/381641.381653.

Jones, C. P. (2020). *How Paintings Depict Time*. URL: https://medium.com/thinksheet/how-paintings-depict-time-33850ff344f4.

La maison du cinema and Cinematheque Francaise (2000). *Étienne-Jules Marey: Movement in Light*. URL: https://web.archive.org/web/20060209080334/http://www.expo-marey.com/ANGLAIS/home.html.

Marey, É.-J. (1875). "La Méthode Graphique dans les Sciences Expérimentales (Suite)". In: *Physiologie Expérimentale: travaux du Laboratoire de M. Marey*. Vol. 1. Paris, France: Masson, G., pp. 255–278.

Marey, É.-J. (1894). *Le mouvement*. Paris, France: Masson, G.

Markovic, D. and M. Gelautz (2006). "Comics-Like Motion Depiction from Stereo". In: *Proceedings of the International Conference in Central Europe on Computer Graphics, Visualization and Computer Vision (WSCG)*. University of West Bohemia, pp. 155–160.

McCloud, S. (1994). *Understanding Comics*. New York, NY, USA: HarperPerennial.

Miller, A. I. (2001). *Einstein, Picasso: Space, Time, and Beauty That Causes Havoc*. Basic Books.

Newman, L. H. (1965). *Man and Insects*. London, UK: Aldus Books.

Playfair, W. (1805). *An Inquiry into the Permanent Causes of the Decline and Fall of Powerful and Wealthy Nations*. London, UK: Greenland and Norris. URL: https://archive.org/details/inquiryintoperma00play.

Playfair, W. (1821). *A Letter on our Agricultural Distresses, their Causes and Remedies*. London, UK: William Sams.

Playfair, W. and J. Corry (1786). *The Commercial and Political Atlas: Representing, by Means of Stained Copper-Plate Charts, the Progress of the Commerce, Revenues, Expenditure and Debts of England during the Whole of the Eighteenth Century*. London, UK: printed for J. Debrett; G. G. et al.

Powsner, S. M. and E. R. Tufte (1994). "Graphical Summary of Patient Status". In: *The Lancet* 344.8919, pp. 386–389. DOI: 10.1016/S0140-6736(94)91406-0.

Priestley, J. (1765). *A Chart of Biography*. London, UK: Johnson, J.

Rendgen, S. (2019). *History of Information Graphics*. Edited by Wiedemann, J. Taschen.

Rosenberg, D. and A. Grafton (2010). *Cartographies of Time: A History of the Timeline*. Princeton Architectural Press.

Third Millennium Press (2001). *Zeittafel der Weltgeschichte. Den letzen 6000 Jahren auf der Spur*. Cologne, Germany: Könemann Verlagsgesellschaft mbH.

Tomitsch, M., W. Aigner, and T. Grechenig (2007). "A Concept to Support Seamless Spectator Participation in Sports Events Based on Wearable Motion Sensors". In: *Proceedings of the 2nd International Conference on Pervasive Computing and Applications (ICPCA)*. IEEE Computer Society, pp. 209–214. DOI: `10.1109/icpca.2007.4365441`.

Tufte, E. R. (1983). *The Visual Display of Quantitative Information*. Graphics Press. URL: `https://www.edwardtufte.com/tufte/books_vdqi`.

Tufte, E. R. (1990). *Envisioning Information*. Graphics Press. URL: `https://www.edwardtufte.com/tufte/books_ei`.

Tufte, E. R. (1997). *Visual Explanations*. Graphics Press. URL: `https://www.edwardtufte.com/tufte/books_visex`.

Tufte, E. R. (2006). *Beautiful Evidence*. Graphics Press. URL: `https://www.edwardtufte.com/tufte/books_be`.

Wainer, H. (2005). *Graphic Discovery: A Trout in the Milk and Other Visual Adventures*. Princeton University Press.

Walker, F. A. (1874). *Statistical Atlas of the United States Based on the Results of the Ninth Census 1870 with Contributions from Many Eminent Men of Science and Several Departments of the Government*. United States Census Office.

Wunderlich, C. A. (1870). *Das Verhalten der Eigenwärme in Krankheiten*. 2nd edition. Leipzig, Germany: Otto Wigand.

Chapter 3
Time & Time-Oriented Data

> What, then, is time?
> If no one asks me, I know what it is.
> If I wish to explain it to him who asks, I do not know.

Saint Augustine (AD 354-430, The Confessions)

The fundamental phenomenon of time has always been of interest to mankind. Many different theories for characterizing the physical dimension of time have been developed and discussed over literally thousands of years in philosophy, mathematics, physics, astronomy, biology, and many other disciplines. As reported by Whitrow et al. (2003), a 1981 literature survey by J. T. Fraser found that the total number of entries judged to be potentially relevant to the systematic study of time reached about 65,000. This illustrates the breadth of the topic and the restless endeavor of man to uncover its secrets. What can be extracted as the bottom line across many theories is that time is *unidirectional* (arrow of time) and that time gives *order* to events.

The most influential theories for the natural sciences are probably Newton's concepts of absolute vs. relative time and Einstein's four-dimensional spacetime. Newton assumed an absolute, true, mathematical time that exists in itself and is not dependent on anything else. Together with space, it resembles a container for all processes in nature. This image of an absolute and independent dimension prevailed until the beginning of the 20th century. Then, Einstein's relativity theory made clear that time in physics depends on the observer. Thus, Einstein introduced the notion of *spacetime*, where space and time are inherently connected and cannot be separated. That is, each event in the universe takes place in four-dimensional space at a location that is defined by three spatial coordinates at a certain time as the fourth coordinate (see Lenz, 2005). Both Newton's notion of absolute time and Einstein's spacetime are concepts that describe time as a fundamental characteristic of the universe. In contrast to that, the way humans deal with time in terms of deriving it essentially from astronomical movements of celestial bodies or phenomena in nature is what Newton called relative time.

The first signs of the systematic use of tools for dealing with time have been found in the form of bone engravings that resembled simple calendars based on the

© The Author(s) 2023
W. Aigner et al., *Visualization of Time-Oriented Data*, Human–Computer
Interaction Series, https://doi.org/10.1007/978-1-4471-7527-8_3

cycle of the moon. In this regard, the most fundamental natural rhythm perceived by humans is the day. Consequently, it is the basis of most calendars and was used to structure the simple life of our ancestors who lived in close contact with nature (see Lenz, 2005). More complex calendars evolved when man moved away from the life of hunter-gatherers and settled into communities to live from agriculture. Until very late in human history, time was kept only very roughly. Industrialization and urban civilization brought about the need for more precise, regular, and synchronized overall timekeeping.

Today, the most commonly used calendric system is the Gregorian calendar. It was introduced by Pope Gregory XII in 1582, primarily to correct the drift of the previously used Julian calendar, which was slightly too long in relation to the astronomical year and the seasons.[1] Apart from this calendric system, many other systems are in use around the world, such as the Islamic, the Chinese, or the Jewish calendars, or calendars for special purposes, like academic (semester, trimester, etc.) or financial calendars (quarter, fiscal year, etc.).

In this book, we will not look at the physical dimension of time itself and its philosophical background, how time is related to natural phenomena, or how clocks have been developed and used. We focus on how the physical dimension of time and associated data can be modeled in a way that facilitates interactive visualization using computer systems. As a next step, we are now going to examine the design aspects for modeling time.

3.1 Modeling Time

First of all, it is important to make a clear distinction between the physical dimension of time and a model of time in information systems. When modeling time in information systems, the goal is not to perfectly imitate the physical dimension time, but to provide a model that is best suited to reflect the phenomena under consideration and support the analysis tasks at hand. Moreover, as Frank (1998) states, there is nothing like a single correct model or taxonomy of time – there are many ways to model time in information systems and time is modeled differently for different applications depending on the particular problem. Extensive research has been conducted in order to formulate the notion of time in many areas of computer science, including artificial intelligence, data mining, simulation, modeling, databases, and more. A theoretical overview which includes many references to fundamental publications is provided by Hajnicz (1996). However, as she points out, the terminology is not consistent across the different fields, and hence, does not integrate well with visualization. Moreover, as Goralwalla et al. (1998) note, most research focuses on the development of specialized models with different features for particular domains. But apart from the many time models created for specific purposes and applications, attempts have been made to capture the major design aspects underlying all specific

---

[1] Interestingly, much more precise calendars were known hundreds of years earlier in other cultures, such as those developed by the Mayas and the Chinese.

instances, as for example by Frank (1998), Goralwalla et al. (1998), Peuquet (1994), Peuquet (2002), Furia et al. (2010), and Furia et al. (2012).

In the context of our book, we want to present the overall design aspects of modeling time, and not a particular model. To do this, we will describe a number of major design aspects and their features which are particularly important when visualizing time. Application-specific models can be derived from these as particular configurations.

3.1.1 Design Aspects

To define the design aspects relevant to time, we adapted the works of Frank (1998) and Goralwalla et al. (1998), where principal orthogonal aspects are presented to characterize different types of time. Next, the aspects of scale, scope, arrangement, and viewpoint will be described in detail.

Scale: ordinal vs. discrete vs. continuous Let us first consider the scale along which elements of time are given. In an *ordinal* time domain, only relative order relations are present (e.g., before, after). For example, statements like "Valentina went to sleep before Arvid arrived" and "Valentina woke up after a few minutes of sleep" can be modeled using an ordinal scale. Note that only relative statements are given and we cannot discern from the given example whether Valentina woke up before or after Arvid arrived (see Figure 3.1). This might be sufficient if only qualitative temporal relationships are of interest or no quantitative information is available.

In *discrete* time domains, it is possible to consider temporal distances. Time values can be mapped to a set of integers which enables quantitative modeling of time values (e.g., quantifiable temporal distances). Discrete time domains are based on a smallest possible unit (e.g., seconds or milliseconds as in UNIX time) and they are the most commonly used time models in information systems (see Figure 3.2). *Continuous* time models are characterized by a possible mapping to real numbers, i.e., between any two points in time, another point in time exists (also known as dense time, see Figure 3.3).

Examples of visualization techniques capable of representing the three types of scale are the *point and figure chart* (see Figure 3.4) for an ordinal scale, *tile maps* (see Figure 3.5 and ↪ p. 269) for a discrete scale, and the *circular silhouette graph* (see Figure 3.6 and ↪ p. 281) for a continuous time scale.

Scope: point-based vs. interval-based Secondly, we consider the scope of the basic elements that constitute the structure of the time domain. *Point-based* time domains can be seen in analogy to discrete Euclidean points in space, i.e., having a temporal extent equal to zero. Thus, no information is given about the region between two points in time. In contrast to that, *interval-based* time domains relate to subsections of time having a temporal extent greater than zero. This aspect is also closely related to the notion of granularity, which will be discussed in Section 3.1.2. For example,

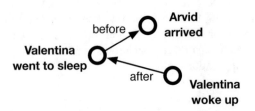

Fig. 3.1 Ordinal scale. Only relative order relations are present. At this level, it is not possible to discern whether Valentina woke up before or after Arvid arrived. ⓒ① *The authors.*

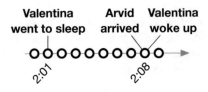

Fig. 3.2 Discrete scale. Smallest possible unit is minutes. Although Arvid arrived and Valentina woke up within the same minute, it is not possible to model the exact order of events. ⓒ① *The authors.*

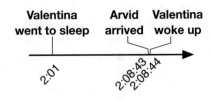

Fig. 3.3 Continuous scale. Between any two points in time, another point in time exists. Here, it is possible to model that Arvid arrived shortly before Valentina woke up. ⓒ① *The authors.*

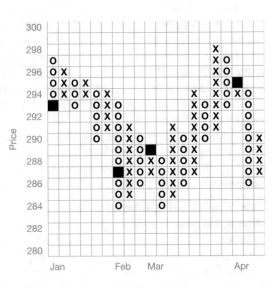

Fig. 3.4 Point and figure chart. Visualization technique tracking price and price direction changes. Uses an *ordinal time scale*. ∘...positive price change of a certain amount, ×...negative price change of a certain amount, ■...begin/end of a trading period. ⓒ① *The authors. Adapted from Harris (1999).*

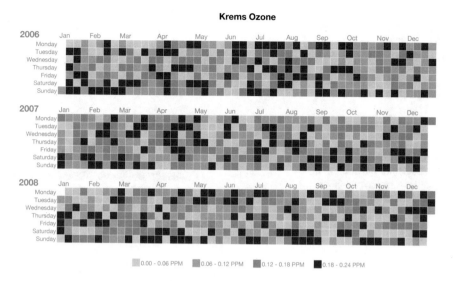

Fig. 3.5: Tile maps showing average daily ozone measurements (scale: *discrete*, scope: *interval-based*) for three years. ⓒⓘ *The authors. Adapted from Mintz et al. (1997).*

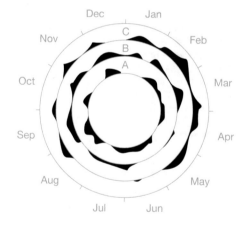

Fig. 3.6 Circular silhouette graph. Enables the representation of time along a *continuous scale* with a *cyclic arrangement*. The representation emphasizes the visual impression by filling the area below the plotted line in order to create a distinct silhouette. This eases comparison when placed side by side. ⓒⓘ *The authors. Adapted from Harris (1999).*

the time value October 23, 2012 might relate to the single instant October 23, 2012 00:00:00 in a point-based domain, whereas the same value might refer to the interval [October 23, 2012 00:00:00, August 23, 2012 23:59:59] in an interval-based domain (see Figures 3.7 and 3.8).

Examples of visualization techniques capable of representing the two types of scope are the *TimeWheel* (see Figure 3.9 and ↪ p. 298) for a point-based domain and *tile maps* (see Figure 3.5 and ↪ p. 269) for an interval-based time domain.

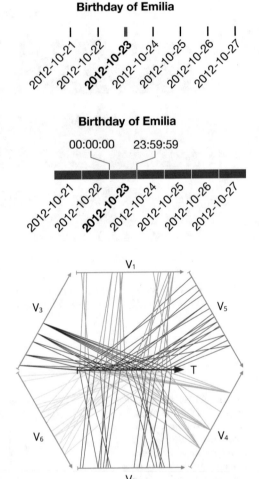

Fig. 3.7 Time value "October 23, 2012" for the birthday of Emilia in a point-based domain. No information is given in between two time points. ©① *The authors.*

Fig. 3.8 Time value "October 23, 2012" for the birthday of Emilia in an interval-based domain. Each element covers a subsection of the time domain greater than zero. ©① *The authors.*

Fig. 3.9 TimeWheel. Axes of time-dependent variables are arranged around a central horizontal time axis. Lines connect the time points on the time axis with the corresponding data values on the variable axes. Colors indicate different variables. ©① *The authors. Adapted from Tominski et al. (2004).*

Arrangement: linear vs. cyclic As the third design aspect, we look at the arrangement of the time domain. Corresponding to our natural perception of time, we mostly consider time as proceeding *linearly* from the past to the future, i.e., each time value has a unique predecessor and successor (see Figure 3.10). However, periodicity is very common in all kinds of data, for example, seasonal variations, monthly averages, and many more. In a *cyclic* organization of time, the domain is composed of a set of recurring time values (e.g., the seasons of the year, see Figure 3.11). Hence, any time value *A* is preceded and succeeded at the same time by any other time value *B* (e.g., winter comes before summer, but winter also succeeds summer). In order to enable meaningful temporal relationships in cyclic time, Frank (1998) suggests the use of the relations *immediately before* and *immediately after*. Strictly cyclic data, where the linear progression of time from past to future is neglected, is very rare

Fig. 3.10 Linear time. Time proceeds linearly from past to future. ⓒⓘ *The authors.*

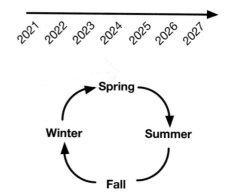

Fig. 3.11 Cyclic time. Set of recurring time values such as the seasons of the year. ⓒⓘ *The authors.*

(e.g., records for the day of the week not considering month or year). The combination of periodic and linear progression denoted by the term *serial periodic data* (e.g., monthly temperature averages over a couple of years) is much more common. Periodic time-oriented data in this sense includes both strictly cyclic data and serial periodic data.

Examples of visualization techniques capable of representing the two types of arrangement are the *TimeWheel* (see Figure 3.9 and ↪ p. 298) for linear time and the *circular silhouette graph* (see Figure 3.6 and ↪ p. 281) for cyclic time.

Viewpoint: ordered vs. branching vs. multiple perspectives The fourth subdivision is concerned with the views of time that are modeled. *Ordered* time domains consider things that happen one after the other. On a more detailed level, we might also distinguish between totally ordered and partially ordered domains. In a totally ordered domain, only one thing can happen at a time. In contrast to this, simultaneous or overlapping events are allowed in partially ordered domains, i.e., multiple time primitives at a single point or overlapping in time. A more complex form of time domain organization is the so-called *branching* time (see Figure 3.12). Here, multiple strands of time branch out and allow the description and comparison of alternative scenarios (e.g., in project planning). This type of time supports decision-making processes where only one of the alternatives will actually happen. Note that branching is not only useful for future scenarios but can also be applied for investigating the past, e.g., for modeling possible causes of a given decision. In contrast to branching time where only one path through time will actually happen, *multiple perspectives* facilitate simultaneous (even contrary) views of time, which are necessary, for instance, to structure eyewitness reports. A further example of multiple perspectives is stochastic multi-run simulations. For a single experiment, there might be completely different output data progressions depending on the respective initialization.

Temporal databases usually take a multi-perspective viewpoint as well. They consider the two perspectives of *valid time* and *transaction time* (see Figure 3.13). The valid time perspective of a fact is the time when the fact is true in the modeled reality (e.g., "Vincent was born on August 8, 2006"). In contrast to that, the transaction

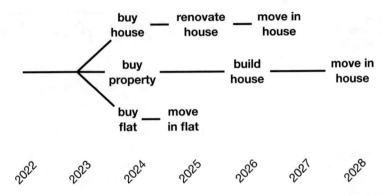

Fig. 3.12: Branching time. Alternative scenarios for moving to a different place. ©① *The authors.*

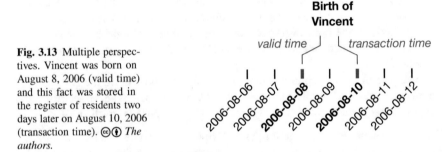

Fig. 3.13 Multiple perspectives. Vincent was born on August 8, 2006 (valid time) and this fact was stored in the register of residents two days later on August 10, 2006 (transaction time). ©① *The authors.*

time perspective of a fact denotes when it was stored in the database (e.g., the birth of Vincent is stored in the register of residents after filling out a form two days after his birth). In practice, it is often necessary to condense multiple perspectives into a single consistent view of time (see for example Wolter et al., 2009).

Both branching time and multiple perspectives introduce the need to deal with probability (or uncertainty), to convey, for example, which path through time will most likely be taken, or which evidence is believable. The *decision chart* (see Figure 3.14 and ↪ p. 237) is an example of a visualization technique capable of representing branching time.

3.1.2 Granularities & Time Primitives

The previous section introduced design aspects to adequately model the time domain's scale, scope, and arrangement as well as possible viewpoints onto the time domain. Besides these general aspects, the hierarchical organization of time as well as the definition of concrete time elements used to relate data to time need to be specified. In the following, we will discuss this facet in more detail.

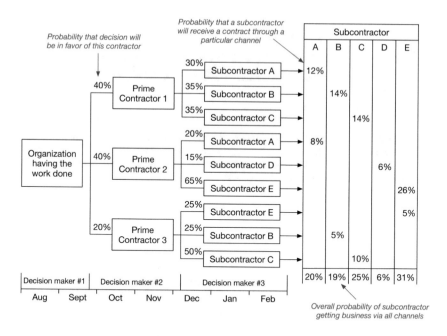

Fig. 3.14: Decision chart. Example of a visualization technique capable of representing branching time. Future decisions and potential alternative outcomes along with their probabilities can be depicted over time. ⓒⓘ *The authors. Adapted from Harris (1999).*

Granularity and calendars: none vs. single vs. multiple To tame the complexity of time, it is practical to consider different levels of granularity. Basically, granularities can be thought of as human-made abstractions of time (e.g., minutes, hours, days, weeks, months). More generally, granularities describe mappings from time values to larger or smaller conceptual units (see Figure 3.15 for an example of time granularities and their relationships). A comprehensive overview and formalization of time granularity concepts is given by Bettini et al. (2000).

Most information systems that deal with time-oriented data are based on a discrete time model that uses a fixed smallest granularity also known as *bottom granularity* (e.g., Java's `java.time` package uses nanoseconds as the smallest granularity). Consequently, the underlying time domain corresponds to a sequence of non-decomposable, consecutive time intervals of identical duration, so-called *chronons* (see Jensen et al., 1998). A point in time can then be specified simply as the number of chronons relative to a reference point (e.g., milliseconds since January 1, 1970 00:00:00 GMT as for Unix time).

Chronons may be grouped into larger segments, termed *granules*. That said, a granularity is basically a non-overlapping mapping of granules to subsets of the time domain (see Dyreson et al., 2000). Granularities are related in the sense that the granules in one granularity may be further aggregated to form larger granules

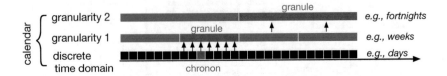

Fig. 3.15: Example of a discrete time domain with multiple granularities. The smallest possible unit (chronon) is one *day*. Based on this, the granularity *weeks* contains granules that are defined as being a set of seven consecutive days. Moreover, the granularity *fortnights* consists of granules that are a set of two consecutive weeks. ©① *The authors.*

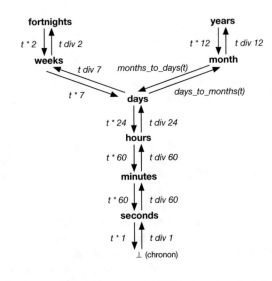

Fig. 3.16 Annotated granularity lattice of the Gregorian calendar that contains regular and irregular mappings (leap seconds are not considered in the granularity lattice). ©① *The authors.*

belonging to a coarser granularity. For example, 60 consecutive seconds are mapped to one minute.

A system of multiple granularities in lattice structures is referred to as a *calendar* (see Figure 3.16 for the granularity lattice of the Gregorian calendar). More precisely, it is a mapping between human-meaningful time values and an underlying time domain. Thus, a calendar consists of a set of granularities including mappings between pairs of granularities that can be represented as a graph (see Dyreson et al., 2000). Calendars most often include cyclic elements, allowing human-meaningful time values to be expressed succinctly. For example, dates in the common Gregorian calendar may be expressed in the form <day, month, year> where each of the fields day, month, and year circle as time passes (see Jensen et al., 1998). To help users in grasping the complexities of a calendar, a visual notation based on icons and glyphs has been developed by Dudek and Blaise (2013) for comparing different calendars to each other.

Moreover, mappings between granularities might be regular or irregular. A regular mapping exists for example between seconds and minutes where one minute always maps to 60 seconds.[2] In contrast to that, the mapping of days to months is irregular because a month might be composed of 28, 29, 30, or 31 days depending on the context (particular year and month).

To work effortlessly with granularities and calendars, an appropriate infrastructure of data models and operators is required. This includes not only the definition of granularities and calendars, but also methods for converting from one granularity to another or for combining calendars. Particularly, conversion operations can be quite complex due to the irregularities in granularities, for example when converting from days to months. Many programming languages and their corresponding standard libraries implement the described functionalities for the Gregorian calendar following the ISO 8601 standard (e.g., `java.time`). More sophisticated implementations with support for alternative calendars (e.g., `java.time.chrono`) and multiple (user-defined) granularities are becoming increasingly important in a globalized world (see Dyreson et al., 2000; Lee et al., 1998).

Finally, it is worth mentioning that granularities influence equality relationships. Take for example two events A and B that happened on December 31, 2020 and January 2, 2021 (see Figure 3.17). At the granularity of days, the two events are on different days. Yet, at the granularity of weeks, both events are within the same granule. At the still coarser granularity of years, A and B are again different. Note that this is contradictory to the naive assumption that when an equality relationship holds true on a fine granularity it also holds true on a coarser one.

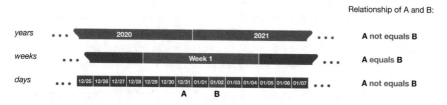

Fig. 3.17: Granularities influence equality relationships. The times of A and B are not equal on the granularity of days, but are equal on the granularity of weeks, and then again are not equal on the coarser granularity of years. ⓒ① *The authors.*

The concepts of chronon, granule, granularity, and calendar help us organize the time domain. If a visualization makes use of granularities or calendar systems, it is categorized as supporting *multiple* granularities. Besides this complex variant, a visualization's time model might support only a *single* granularity (e.g., every time value is given in terms of milliseconds) or *none* at all (e.g., abstract ticks). An example of a visualization technique that uses time granularities is the *cycle plot* (see Figure 3.18 and ↪ p. 268).

---

[2] We are not considering the exception of leap seconds here.

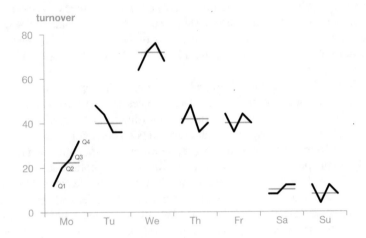

Fig. 3.18: Cycle plot. Visualization technique that utilizes two time granularities to represent cycles and trends. The example shows trends of measurements of weekdays over quarters. For example, on Mondays, the values show an increasing trend over the year while on Tuesdays the trend is decreasing. Furthermore, the general shape of a week's cycle is visible. ©① *The authors. Adapted from Cleveland (1993).*

Time primitives: instant vs. interval vs. span Next, we present a set of basic elements used to relate data to time, so-called time primitives: instant, interval, and span. These time primitives can be seen as an intermediary layer between data elements and the time domain. Basically, time primitives can be divided into anchored (absolute) and unanchored (relative) primitives. Instant and interval are primitives that belong to the first group, i.e., they are located at a fixed position in the time domain. In contrast to that, a span is a relative primitive, i.e., it has no absolute position in time.

An *instant*[3] is a single point in time, e.g., May 23, 1977. Depending on the scope, i.e., whether a point-based or interval-based time model is used (see previous section), an instant might also have a duration (see Figure 3.19 and Figure 3.20). Time primitives can be defined at all levels of granularity representing chronons, granules, or sets of both. Examples of instants are the date of birth "May 23, 1977" and the beginning of a presentation on "January 10, 2023 at 2 p.m." whereas the first instant (date of birth) is given at a granularity of *days* and the second (beginning of presentation) at a granularity of *hours*.

An *interval* is a portion of the time domain that can be represented by two instants, one denoting the beginning of the interval and the other its end. Intervals being defined in this way usually correspond to closed intervals that include the beginning and the end instant (e.g., [August 7, 2022; August 10, 2022] as in Figure 3.21). Alternatively, intervals can be specified via a beginning instant plus a duration (positive span), or via a duration (positive span) plus an end instant.

---

[3] Oftentimes also referred to as *time point*.

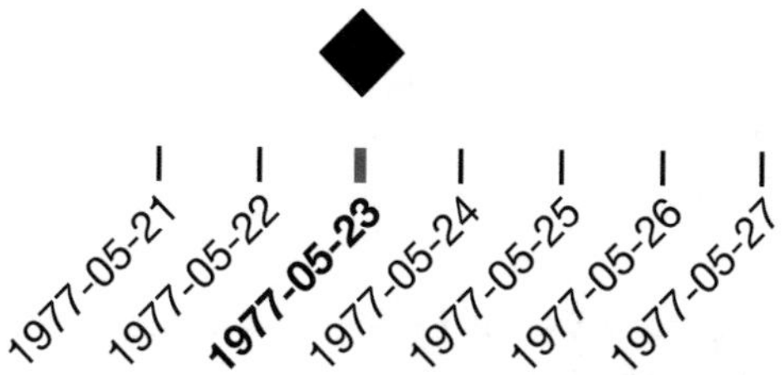

Fig. 3.19 Instant in a point-based time model, where instants have no duration. ⓒⓘ *The authors.*

Fig. 3.20 Instant in an interval-based time model, where instants have a duration that depends on their granularity. ⓒⓘ *The authors.*

[August 7, 2022; August 10, 2022]

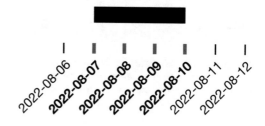

Fig. 3.21 Interval [August 7, 2022; August 10, 2022] in a point-based time model. ⓒⓘ *The authors.*

The *span* is the only unanchored primitive. A time span is defined as a directed, unanchored primitive that represents a directed amount of time in terms of a number of granules in a given granularity. Examples of spans are the length of a vacation of "10 days" and the duration of a lecture of "150 minutes". Figure 3.22 illustrates this graphically by showing an example span of "four days" which is a count of four granules of the granularity of *days*. A span is either positive, denoting the forward motion of time, or negative, denoting the backward motion of time (see Jensen et al., 1998). In the case of irregular granularities, the exact length of a span is not known precisely. Consider for example the granularity of *months*, where a span of "two months" might be 59, 60, 61, or 62 days depending on the particular time context. This implies that the exact length of spans within irregular granularities can only be determined exactly if the spans are related absolutely to the time domain (anchored). Otherwise, as a last resort, mean values might be used for calculations (e.g., mean month and mean year).

Fig. 3.22 Span. Example of
the span "four days" which
is formed by four granules of
the granularity *days*. ⓒⓘ *The*
authors.

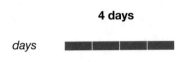

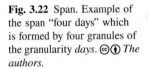

In terms of visualizing time primitives, most of the previously given visualization
examples are suited for time instants. The *Gantt chart* (↪ p. 253) is an example
of a visualization technique that is designed particularly to show time intervals (see
Figure 3.23).

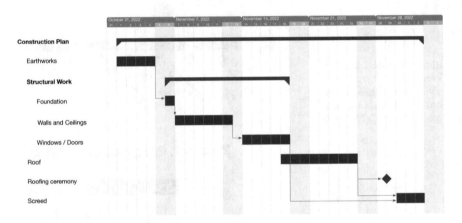

Fig. 3.23: Gantt chart. Example of a visualization technique capable of representing intervals. The
tasks of a project plan are displayed as a list in the left part of the diagram. For each task, a horizontal
bar (timeline) displays the extent of the task in time. ⓒⓘ *The authors.*

Relations between time primitives Between individual time primitives, relations
might exist. Temporal relations are important concepts, especially when reasoning
about time (see Peuquet, 1994). Depending on the involved types of primitives,
different relations make sense.

Between two instants *x* and *y*, three relationships are possible (see Figure 3.24):

- *x before y*
- *x after y*
- *x equals y*

Similarly, for time spans, which are amounts of time, there are three possible
relations. Given two time spans *s* and *t*, one of the following relations can hold: *s*
shorter than t, *s longer than t*, or *s as long as t*.

For relations between time intervals *A* and *B*, things get more complex. Allen
(1983) defined a set of thirteen basic relations that are very common in time modeling
(see Figure 3.25):

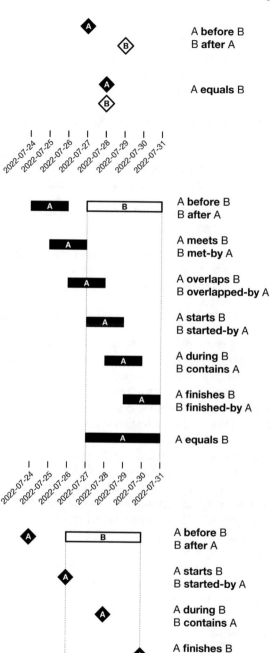

Fig. 3.24 Instant relations. Instants can be related in three different ways. ⓒⓘ *The authors.*

Fig. 3.25 Interval relations. Intervals can be related in thirteen different ways. ⓒⓘ *The authors.*

Fig. 3.26 Instant+interval relations. Instants and intervals can be related in eight different ways. ⓒⓘ *The authors.*

- *A before B* (or *B after A*): Interval *A* ends before interval *B* starts.
- *A meets B* (or *B met-by A*): Interval *A* ends right when interval *B* starts.
- *A overlaps B* (or *B overlapped-by A*): Intervals *A* and *B* overlap whereas interval *A* ends during interval *B*.
- *A starts B* (or *B started-by A*): Intervals *A* and *B* start at the same time but interval *A* ends earlier.
- *A during B* (or *B contains A*): Interval *A* starts later than interval *B* and ends before interval *B* ends.
- *A finishes B* (or *B finished-by A*): Interval *A* and *B* end at the same time but interval *A* starts later.
- *A equals B*: Intervals *A* and *B* start and end at the same time.

When looking at relations between an instant *x* and an interval *A*, eight options exist (see Figure 3.26):

- *x before A* (or *A after x*): Instant *x* is before the start of interval *A*.
- *x starts A* (or *A started-by x*): Instant *x* and the start of interval *A* are the same.
- *x during A* (or *A contains x*): Instant *x* is after the start and before the end of interval *A*.
- *x finishes A* (or *A finished-by x*): Instant *x* and the end of interval *A* are the same.

Determinacy: determinate vs. indeterminate In addition to the set of possible relations, further design aspects are relevant in the context of time-oriented data. Uncertainty is one such aspect. If there is no complete or exact information about time specifications or if time primitives are converted from one granularity to another, uncertainties are introduced and have to be dealt with. Therefore, the *determinacy* of the given time specification needs to be considered.

A determinate specification is present when there is complete knowledge of all temporal aspects. Prerequisites for determinate specification are either a continuous time domain or only a single granularity within a discrete time domain. Information that is temporally indeterminate can be characterized as *don't know when* information, or more precisely, *don't know exactly when* information (see Jensen et al., 1998). Examples of this are inexact knowledge (e.g., "time when the earth was formed"), future planning data (e.g., "it will take 2-3 weeks"), or imprecise event times (e.g., "one or two days ago").

Notice that temporal indeterminacy as well as the relativity of references to time are mainly qualifications of statements rather than of the events they denote. Indeterminacy might be introduced by explicit specification (e.g., earliest beginning and latest beginning of an interval) or is implicitly present in the case of multiple granularities. Consider for example the statement "Activity A started on July 25, 2022 and ended on July 31, 2022" – this statement can be modeled by the beginning instant "July 25, 2022" and the end instant "July 31, 2022" both at the granularity of *days*. If we look at this interval from a granularity of *hours*, the interval might begin and end at any point in time between 0 a.m. and 12 p.m. of the specified day (see Figure 3.27).

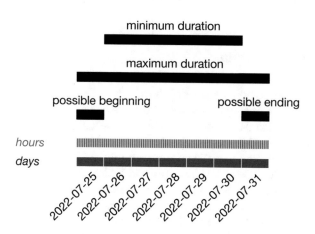

Fig. 3.27: Indeterminacy. Implicit indeterminacy when representing the interval [July 25, 2022; July 31, 2022] that is given at a granularity of *days* on the finer granularity of *hours*. ⓒⓘ *The authors.*

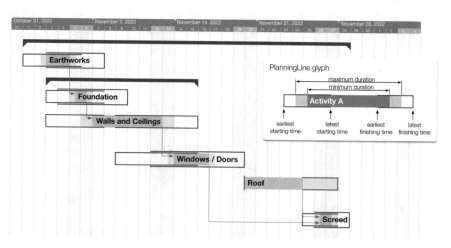

Fig. 3.28: PlanningLines allow the depiction of temporal indeterminacy via a glyph consisting of two encapsulated bars representing minimum and maximum duration. The bars are bounded by two caps that represent the start and end intervals. ⓒⓘ *The authors. Adapted from Aigner et al.* (2005).

Examples of time models that consider temporal indeterminacy are HMAP[4] by Combi and Pozzi (2001) and the time model underlying the time annotations used in the medical treatment plan specification language Asbru by Shahar et al. (1998). A visualization technique capable of depicting temporal indeterminacy is for example *PlanningLines* (see Figure 3.28 and ↪ p. 260).

3.2 Characterizing Data

After discussing the question of modeling the time domain itself, we now move on to the question of characterizing time-oriented data. When we speak of time-oriented data, we basically mean data that are somehow connected to time. More precisely, we consider data values that are associated with time primitives.

The available modeling approaches are manifold and range from considering continuous to discrete data models (see Tory and Möller, 2004). In the former case, time is seen as an observational space and data values are given relative to it (e.g., a time series in form of time-value pairs (t, v)). For the latter, data are modeled as objects or entities which have attributes that are related to time (e.g., calendar events with attributes *beginning* and *end*). Moreover, certain analytic situations even demand domain transformations, such as a transformation from the time domain into the frequency domain (Fourier transformation).

A useful concept for modeling time-oriented data along cognitive principles is the *pyramid framework* (see Figure 3.29) by Mennis et al. (2000), which has already been mentioned briefly in Section 1.1. The model is based on the three

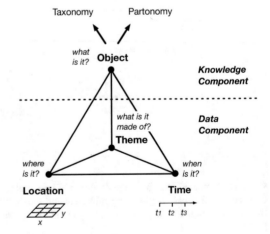

Fig. 3.29 Pyramid framework. Data are conceptualized along the three perspectives of location, time, and theme. Derived interpretations form objects on the cognitively higher level of knowledge. ©① *The authors. Adapted from Mennis et al. (2000).*

---

[4] The word HMAP is not an abbreviation, but it is the transliteration of the ancient Greek poetical word *day*.

perspectives location (*where* is it?), time (*when* is it?), and theme (*what* is it made of?) at the level of data. Derived interpretations of these data aspects form objects (*what* is it?) on the cognitively higher level of knowledge, along with their taxonomy (classification; super-/subordinate relationships) and partonomy (interrelationships; part-whole relationships).

Depending on the phenomena under consideration and the purpose of the analysis, different points of view can be taken. An example of this would be considering distinct conceptual entities that are related to time (objects) vs. the observation of a continuous phenomenon, like temperature over time (values). There cannot be a single model that is ideal for all kinds of applications. However, certain fundamental design alternatives can be identified to characterize time-oriented data. In the context of this book, we focus on the data component, i.e., the lower part of the pyramid framework as depicted in Figure 3.29.

Scale: quantitative vs. qualitative In terms of data scale, we distinguish between quantitative and qualitative variables. Quantitative variables are based on a metric (discrete or continuous) range that allows numeric comparisons. In contrast, the scale of qualitative variables includes an unordered (nominal) or ordered (ordinal) set of data values. It is of fundamental importance to consider the characteristics of the data scale to design appropriate visual representations.

Frame of reference: abstract vs. spatial It further makes sense to distinguish abstract and spatial data. By abstract data we mean a data model that does not include the *where* aspect with regard to the pyramid framework, i.e., abstract data are not connected per se to some spatial location. In contrast to this, spatial data contain an inherent spatial layout, i.e., the underlying data model includes the *where* aspect. The distinction between abstract and spatial data reflects the way in which time-oriented data should be visualized. For spatial data, the inherent spatial information can be exploited to find a suitable mapping of data to screen. The *when* aspect has to be incorporated into that mapping, where it is not always easy to achieve an emphasis on the time domain. For abstract data, no a-priori spatial mapping is given. Thus, first and foremost an expressive spatial layout has to be found. This spatial layout should be defined such that the time domain is exposed.

Kind of data: events vs. states This criterion refers to the question of whether events or states are dealt with. Events can be seen as markers of state changes, like for example the departure of a plane. States can be characterized as phases of continuity between events (e.g., plane is in the air). As one can see, states and events are two sides of the same coin. However, it should be clearly communicated whether states or events, or even a combination of both, are visualized.

Number of variables: single vs. multiple This criterion concerns the number of time-dependent variables. In principle, it makes a difference if we have to represent data where each time primitive is associated with only one single data value (i.e., univariate data) or if multiple data values (i.e., multivariate data) must be represented. Compared to univariate data, for which many methods have been developed, the range of methods applicable for multivariate data is substantially smaller.

3.3 Relating Data & Time

Aspects regarding time dependency of data have been extensively examined in the field of temporal databases (see Liu and Özsu, 2018). Here, we adapt the notions and definitions developed in that area. According to Steiner (1998), any dataset is related to two temporal domains:

- internal time $\mathfrak{T}_i$ and
- external time $\mathfrak{T}_e$.

Internal time is considered to be the temporal dimension inherent in the data model. Internal time describes when the information contained in the data is valid. Conversely, *external time* is considered to be extrinsic to the data model. The external time is necessary to describe how a dataset evolves over time. Depending on the number of time primitives in internal and external time, time-related datasets can be classified as shown in Figure 3.30.

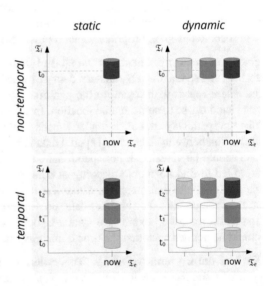

Fig. 3.30 Temporal characteristics of data. A dataset is related to the two temporal domains internal time $\mathfrak{T}_i$ and external time $\mathfrak{T}_e$. ⊛① *The authors. Adapted from Steiner (1998).*

Static non-temporal data If both internal and external time are each comprised of only a single time primitive, the data are completely independent of time. A fact sheet containing data about the products offered by a company is an example of static non-temporal data. This kind of data is not addressed in this book.

Static temporal data If the internal time contains more than one time primitive, while the external time contains only one, then the data can be considered dependent on time. Since the values stored in the data depend on the internal time, static temporal data can be understood as a historical view of how the real world or some model developed over time. Common time series are a prominent example of static temporal data. Most of today's visualization approaches that explicitly consider time as a special data dimension address static temporal data, for instance, the TimeSearcher (see Hochheiser and Shneiderman (2004) and ↪ p. 290).

Dynamic non-temporal data If the internal time contains only one, but the external time is composed of multiple time primitives, then the data depend on the external time. To put it simply, the data themselves change over time, i.e., they are dynamic. Dynamic data that change at high rate are often referred to as *streaming data*. Since the internal time is not considered, only the current state of the data is preserved; no historical view is maintained. There are fewer visualization techniques available that explicitly focus on dynamic non-temporal data. These techniques are mostly applied in monitoring scenarios, for instance, to visualize process data (see Matković et al. (2002) and ↪ p. 331). However, since internal time and external time can usually be mapped from one to the other, some of the known visualization techniques for static temporal data can be applied for dynamic non-temporal data as well.

Dynamic temporal data If both internal and external time are comprised of multiple time primitives, then the data are considered to be bi-temporally dependent. In other words, the data contain variables depending on (internal) time, and the actual state of the data changes over (external) time. Usually, in this case, internal and external time are strongly coupled and can be mapped from one to the other. Examples of such data could be health data or climate data that contain measures depending on time (e.g., daily number of cases of influenza or daily average temperature), and that are updated every 24 hours with new data records of the passed day. An explicit distinction between internal and external time is usually not made by current visualization approaches, because considering both temporal dimensions for visualization is challenging. Therefore, dynamic temporal data are beyond the scope of this book.

3.4 Considering Data Quality

When talking about data, it is also important to consider aspects of data quality (see Rahm and Do, 2000). The variability and dynamic changes inherent in time-oriented data make them particularly prone to various types of errors and failures. Data suffering from data quality issues, often called 'dirty data' (see Kim et al., 2003), can lead to all sorts of problems such as wrong results, misleading statistics, or inapplicability of visualization and analysis methods. It is often the case that severe data quality problems are only discovered as soon as one tries to visualize the data. Steele and Iliinsky (2010) point out the famous 80:20 rule, according to which oftentimes as much as 80% of the effort needs to go into dealing with data quality issues, whereas only 20% actually goes into the core visualization. In the following, we briefly consider typical data quality problems with time-oriented data and outline procedures to tackle them appropriately.

Taxonomy of dirty time-oriented data A first step for getting to grips with dirty time-oriented data is to understand the potential problems. Gschwandtner et al. (2012) provide a systematic overview of various quality problems with time-oriented data. An adapted version of their taxonomy is depicted in Figure 3.31. With this structured view, developers and users of visual analysis methods for time-oriented data are able to systematically check and mitigate possible data quality issues.

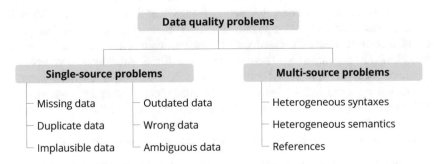

Fig. 3.31: Classification of data quality problems. ©① *The authors. Adapted from Gschwandtner et al. (2012).*

On the top level, the taxonomy distinguishes data quality problems related to a single source and problems arising from multiple sources of data. For *single-source problems*, one can differentiate the following possible problem types: missing data, duplicate data, implausible data, outdated data, wrong data, and ambiguous data. Missing data can be individual values or entire data tuples. Sometimes missing data are marked by special values (e.g., *null* or −999). If this is not the case, missing data can easily go unnoticed, such as a missing February 29 in leap years. Duplicates can cause inconsistencies when, for example, for the same date, two different values are present in the data. Implausible data are data that are outside of the expected value range or indicative of unexpected behavior. For example, many subsequent repetitions of one and the same value in sensor data might hint at a broken sensor. Outdated data are literally not up-to-date and might not reflect the current situation. Wrong data are plainly incorrect, for example, when a time interval's start is after its end. For ambiguous data, there are several valid, but potentially conflicting interpretations of a given datum, and it is unclear which interpretation to use. For example, the date value '06-03-05' may be March 5, 2006 or March 6, 2005. Without additional information, we cannot tell.

In contrast to single-source problems, *multi-source problems* occur when multiple data sources have to be integrated, and the different sources use inconsistent formats or have overlapping and contradicting data. For multi-source problems, the taxonomy distinguishes the problem types of heterogeneous syntaxes, heterogeneous semantics, as well as references. Heterogeneous syntaxes are a problem caused by the use of different data formats. For example, data tables might have different structures where one table contains date and time in two separate columns, while another table stores date and time together in a single column. Heterogeneous semantics stem from inconsistent interpretations of time values. While in one data table, the duration of intervals is given as whole hours only, another table might store durations as the number of minutes. Finally, reference may cause trouble when their referential integrity is violated, for example, when the time instance being referred to does not exist.

In Appendix B, we provide more details on the individual problem types as well as more concrete examples. With this information, we have a kind of checklist that can be used for scrutinizing time-oriented data before engaging in any visual data analysis activities. Steps that are helpful for cleaning the data are described next.

Data cleansing Data cleansing (also called data cleaning, data scrubbing, or data wrangling) is the process of detecting and correcting dirty data, which is typically a prerequisite for interactive visualization. Müller and Freytag (2003) describe data cleansing as a four-step process:

1. Data auditing
2. Workflow specification
3. Workflow execution
4. Post-processing/control

The first step of data auditing is concerned with detecting different types of anomalies contained in the data. The taxonomy of dirty time-oriented data can be harnessed to carry out this step systematically. For the actual data cleansing, a workflow of data correction operations is specified based on the identified quality issues. To actually rectify the anomalies contained in the data, the workflow is executed. Finally, the corrected data need to be verified one more time to verify their correctness after carrying out the specified operations.

Another important task of the data cleansing is the transformation of a given data source into a table structure that is suited for subsequent processing steps, such as splitting/merging of columns (e.g., for time and date), removing additional rows (e.g., summary rows and comments), or the aggregation of temporal tuples into consistent uniform intervals. To aid these transformations, a number of software products are available, such as Tableau Prep (Tableau Software, 2021), Trifacta Wrangler (Trifacta, 2021), or OpenRefine (Huynh, 2021).

However, the majority of dirty data problem types require intervention by a domain expert to be cleansed. Thus, it is advisable to combine automated data transformation steps operating on the whole data with interactive visual interfaces and semi-automatic data correction steps during which domain expertise is employed to solve specific problems in particular parts of the data. In fact, data cleansing is not only a prerequisite for interactive visualization of time-oriented data, but vice versa, interactive visualization can also be employed as a tool to facilitate data cleansing. Examples for such approaches are described by Bernard et al. (2012), Gschwandtner et al. (2014), and Arbesser et al. (2017).

3.5 Summary

In this chapter, we structured and specified the characteristics of time and time-oriented data. We considered four perspectives: the dimension of time, the characteristics of data, the relation of time and data, and the quality of time-oriented data. Figures 3.32 and 3.33 summarize these perspectives and their corresponding aspects.

The first perspective mainly addressed time and the complexity of modeling time. We clarified the concepts of scale, scope, arrangement, and viewpoints of time and then discussed granularity and calendars, time primitives, temporal relations, and temporal determinacy. Building upon this understanding of time and its models, the second perspective focused on relevant aspects of the data variables. Specifically, we discussed the data scale, the frame of reference, the kind of data, and the number of variables. The third perspective showed us how time and data are related. We presented basic options of how data variables can be linked to internal and external time. Finally, we looked at time-oriented data from a quality perspective. Here, we considered a taxonomy of single-source and multi-source data quality problems and briefly outlined the process of data cleaning.

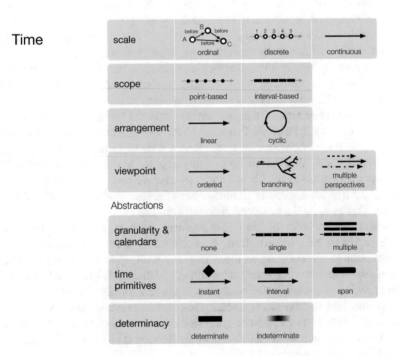

Fig. 3.32: Design aspects of the dimension of time. ©① *The authors.*

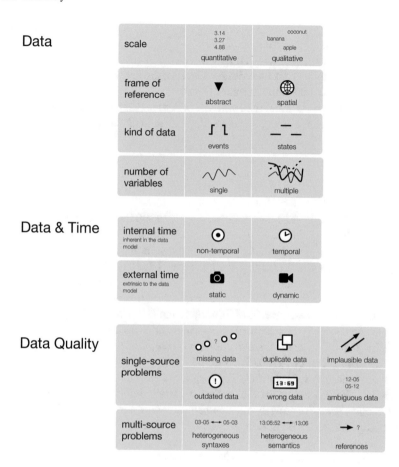

Fig. 3.33: Design aspects of time-oriented data. ⓒⓘ *The authors.*

The key take-home message of this chapter is that all of these perspectives need to be considered when visualizing and analyzing data that are related to time. We took the rather hard road through the data jungle, which required the reader to digest a number of models, characterizations, and quality concerns, because we are convinced that developing visualization methods specifically for time-oriented data requires a clear understanding of the specifics of such data. A data modeling concept and reference implementation to support these special characteristics is *TimeBench* by Rind et al. (2013a). It provides foundational data structures and algorithms for time-oriented data in visual analytics.

Given this book's focus on time aspects, we did not discuss other issues regarding data structures and the relationships between different data variables that are not strictly related to time. We are aware that the relationships between data variables

are of importance as well. These aspects have been widely discussed in database
and data modeling theories. Also, many useful modeling alternatives and reference
models have been developed and can be adopted, such as continuous models using
scalars, vectors, or tensors, etc. (see Wright, 2007) or discrete models using structures
like trees, graphs, etc. (see Shneiderman, 1996).

While this chapter was concerned with the data to be visualized, the next chapter,
we will discuss how time and time-oriented data can actually be represented visually.

References

Aigner, W., S. Miksch, B. Thurnher, and S. Biffl (2005). "PlanningLines: Novel
   Glyphs for Representing Temporal Uncertainties and Their Evaluation". In: *Pro-
   ceedings of the International Conference Information Visualisation (IV)*. IEEE
   Computer Society, pp. 457–463. DOI: 10.1109/IV.2005.97.
Allen, J. F. (1983). "Maintaining Knowledge about Temporal Intervals". In: *Com-
   munications of the ACM* 26.11, pp. 832–843. DOI: 10.1145/182.358434.
Arbesser, C., F. Spechtenhauser, T. Mühlbacher, and H. Piringer (2017). "Visplause:
   Visual Data Quality Assessment of Many Time Series Using Plausibility Checks".
   In: *IEEE Transactions on Visualization and Computer Graphics* 23.1, pp. 641–
   650. DOI: 10.1109/tvcg.2016.2598592.
Bernard, J., T. Ruppert, O. Goroll, T. May, and J. Kohlhammer (2012). "Visual-
   Interactive Preprocessing of Time Series Data". In: *Proceedings of the An-
   nual Conference of the Swedish Computer Graphics Association (SIGRAD)*. 81.
   Linköping University Electronic Press, pp. 39–48. URL: https://ep.liu.se/
   ecp/081/006/ecp12081006.pdf.
Bettini, C., S. Jajodia, and X. S. Wang (2000). *Time Granularities in Databases, Data
   Mining, and Temporal Reasoning*. 1st edition. Springer. DOI: 10.1007/978-3-
   662-04228-1.
Cleveland, W. S. (1993). *Visualizing Data*. Hobart Press.
Combi, C. and G. Pozzi (2001). "HMAP – A Temporal Data Model Managing
   Intervals with Different Granularities and Indeterminacy from Natural Lan-
   guage Sentences". In: *The VLDB Journal* 9.4, pp. 294–311. DOI: 10.1007/
   s007780100033.
Dudek, I. and J.-Y. Blaise (2013). "Visualising Time with Multiple Granularities:
   A Generic Framework". In: *Proceedings of the Annual Conference of Computer
   Applications and Quantitative Methods in Archaeology*. Amsterdam University
   Press, pp. 470–481. DOI: 10.1515/9789048519590-050.
Dyreson, C. E., W. S. Evans, H. Lin, and R. T. Snodgrass (2000). "Efficiently
   Supporting Temporal Granularities". In: *IEEE Transactions on Knowledge and
   Data Engineering* 12.4, pp. 568–587. DOI: 10.1109/69.868908.
Frank, A. U. (1998). "Different Types of "Times" in GIS". In: *Spatial and Temporal
   Reasoning in Geographic Information Systems*. Edited by Egenhofer, M. J. and
   Golledge, R. G. Oxford University Press, pp. 40–62.

Furia, C. A., D. Mandrioli, A. Morzenti, and M. Rossi (2010). "Modeling Time in Computing: A Taxonomy and a Comparative Survey". In: *ACM Computing Surveys* 42.2, 6:1–6:59. DOI: 10.1145/1667062.1667063.

Furia, C. A., D. Mandrioli, A. Morzenti, and M. Rossi (2012). *Modeling Time in Computing*. Springer. DOI: 10.1007/978-3-642-32332-4.

Goralwalla, I. A., M. T. Özsu, and D. Szafron (1998). "An Object-Oriented Framework for Temporal Data Models". In: *Temporal Databases: Research and Practice*. Edited by Etzion, O., Jajodia, S., and Sripada, S. Springer, pp. 1–35. DOI: 10.1007/bfb0053696.

Gschwandtner, T., W. Aigner, S. Miksch, J. Gärtner, S. Kriglstein, M. Pohl, and N. Suchy (2014). "TimeCleanser: A Visual Analytics Approach for Data Cleansing of Time-oriented Data". In: *Proceedings of the International Conference on Knowledge Technologies and Data-driven Business (i-KNOW)*. ACM Press, pp. 1–8. DOI: 10.1145/2637748.2638423.

Gschwandtner, T., J. Gärtner, W. Aigner, and S. Miksch (2012). "A Taxonomy of Dirty Time-Oriented Data". In: *Multidisciplinary Research and Practice for Information Systems*. Edited by Quirchmayr, G., Basl, J., You, I., Xu, L., and Weippl, E. Springer, pp. 58–72. DOI: 10.1007/978-3-642-32498-7_5.

Hajnicz, E. (1996). *Time Structures: Formal Description and Algorithmic Representation*. Vol. 1047. Lecture Notes in Computer Science. Springer. DOI: 10.1007/3-540-60941-5.

Harris, R. L. (1999). *Information Graphics: A Comprehensive Illustrated Reference*. Oxford University Press. URL: https://global.oup.com/academic/product/information-graphics-9780195135329.

Hochheiser, H. and B. Shneiderman (2004). "Dynamic Query Tools for Time Series Data Sets: Timebox Widgets for Interactive Exploration". In: *Information Visualization* 3.1, pp. 1–18. DOI: 10.1057/palgrave.ivs.9500061.

Huynh, D. (2021). *OpenRefine*. URL: https://openrefine.org/ (visited on 02/26/2021).

Jensen, C. S., C. E. Dyreson, M. H. Böhlen, J. Clifford, R. Elmasri, S. K. Gadia, F. Grandi, P. J. Hayes, S. Jajodia, W. Käfer, N. Kline, N. A. Lorentzos, Y. G. Mitsopoulos, A. Montanari, D. A. Nonen, E. Peressi, B. Pernici, J. F. Roddick, N. L. Sarda, M. R. Scalas, A. Segev, R. T. Snodgrass, M. D. Soo, A. U. Tansel, P. Tiberio, and G. Wiederhold (1998). "The Consensus Glossary of Temporal Database Concepts – February 1998 Version". In: *Temporal Databases: Research and Practice*. Edited by Etzion, O., Jajodia, S., and Sripada, S. Springer, pp. 367–405. DOI: 10.1007/bfb0053710.

Kim, W., B.-J. Choi, E.-K. Hong, S.-K. Kim, and D. Lee (2003). "A Taxonomy of Dirty Data". In: *Data Mining and Knowledge Discovery* 7.1, pp. 81–99. DOI: 10.1023/a:1021564703268.

Lee, J. Y., R. Elmasri, and J. Won (1998). "An Integrated Temporal Data Model Incorporating Time Series Concept". In: *Data and Knowledge Engineering* 24.3, pp. 257–276. DOI: 10.1016/S0169-023X(97)00034-7.

Lenz, H. (2005). *Universalgeschichte der Zeit*. Wiesbaden, Germany: Marixverlag, p. 575.

Liu, L. and M. T. Özsu, eds. (2018). *Encyclopedia of Database Systems*. 2nd edition. Springer. DOI: 10.1007/978-1-4614-8265-9.

Matković, K., H. Hauser, R. Sainitzer, and E. Gröller (2002). "Process Visualization with Levels of Detail". In: *Proceedings of the IEEE Symposium Information Visualization (InfoVis)*. IEEE Computer Society, pp. 67–70. DOI: 10.1109/INFVIS.2002.1173149.

Mennis, J. L., D. J. Peuquet, and L. Qian (2000). "A Conceptual Framework for Incorporating Cognitive Principles into Geographical Database Representation". In: *International Journal of Geographical Information Science* 14.6, pp. 501–520. DOI: 10.1080/136588100415710.

Mintz, D., T. Fitz-Simons, and M. Wayland (1997). "Tracking Air Quality Trends with SAS/GRAPH". In: *Proceedings of the 22nd Annual SAS User Group International Conference (SUGI)*. SAS, pp. 807–812. URL: https://support.sas.com/resources/papers/proceedings/proceedings/sugi22/INFOVIS/PAPER173.PDF.

Müller, H. and J.-C. Freytag (2003). *Problems, Methods, and Challenges in Comprehensive Data Cleansing*. Tech. rep. HUB-IB-164. Humboldt University Berlin.

Peuquet, D. J. (1994). "It's about Time: A Conceptual Framework for the Representation of Temporal Dynamics in Geographic Information Systems". In: *Annals of the Association of American Geographers* 84.3, pp. 441–461. DOI: 10.1111/j.1467-8306.1994.tb01869.x.

Peuquet, D. J. (2002). *Representations of Space and Time*. The Guilford Press.

Rahm, E. and H. H. Do (2000). "Data Cleaning: Problems and Current Approaches". In: *IEEE Data Engineering Bulletin* 23.4, pp. 3–13. URL: http://sites.computer.org/debull/A00DEC-CD.pdf.

Rind, A., T. Lammarsch, W. Aigner, B. Alsallakh, and S. Miksch (2013a). "TimeBench: A Data Model and Software Library for Visual Analytics of Time-Oriented Data". In: *IEEE Transactions on Visualization and Computer Graphics* 19.12, pp. 2247–2256. DOI: 10.1109/TVCG.2013.206.

Shahar, Y., S. Miksch, and P. Johnson (1998). "The Asgaard Project: A Task-Specific Framework for the Application and Critiquing of Time-Oriented Clinical Guidelines". In: *Artificial Intelligence in Medicine* 14.1-2, pp. 29–51. DOI: 10.1016/s0933-3657(98)00015-3.

Shneiderman, B. (1996). "The Eyes Have It: A Task by Data Type Taxonomy for Information Visualizations". In: *Proceedings of the IEEE Symposium on Visual Languages*. IEEE Computer Society, pp. 336–343. DOI: 10.1109/VL.1996.545307.

Steele, J. and N. Iliinsky (2010). *Beautiful Visualization: Looking at Data through the Eyes of Experts*. O'Reilly Media, Inc.

Steiner, A. (1998). "A Generalisation Approach to Temporal Data Models and their Implementations". PhD thesis. Swiss Federal Institute of Technology.

Tableau Software (2021). *Tableau Prep*. URL: https://www.tableau.com/products/prep (visited on 02/26/2021).

Tominski, C., J. Abello, and H. Schumann (2004). "Axes-Based Visualizations with Radial Layouts". In: *Proceedings of the ACM Symposium on Applied Computing (SAC)*. ACM Press, pp. 1242–1247. DOI: 10.1145/967900.968153.

Tory, M. and T. Möller (2004). "Rethinking Visualization: A High-Level Taxonomy". In: *Proceedings of the IEEE Symposium Information Visualization (InfoVis)*. IEEE Computer Society, pp. 151–158. DOI: 10.1109/INFVIS.2004.59.

Trifacta (2021). *Trifacta Wrangler*. URL: https://www.trifacta.com/ (visited on 02/26/2021).

Whitrow, G. J., J. T. Fraser, and M. P. Soulsby (2003). *What is Time? The Classic Account of the Nature of Time*. Oxford University Press.

Wolter, M., I. Assenmacher, B. Hentschel, M. Schirski, and T. Kuhlen (2009). "A Time Model for Time-Varying Visualization". In: *Computer Graphics Forum* 28.6, pp. 1561–1571. DOI: 10.1111/j.1467-8659.2008.01314.x.

Wright, H. (2007). *Introduction to Scientific Visualization*. Springer. DOI: 10.1007/978-1-84628-755-8.

Chapter 4
Crafting Visualizations of Time-Oriented Data

> The graphical method has considerable superiority for the exposition
> of statistical facts over the tabular. A heavy bank of figures is
> grievously wearisome to the eye, and the popular mind is as capable
> of drawing any useful lessons from it as of extracting sunbeams from
> cucumbers.

Farquhar and Farquhar (1891, p. 55)

Many different types of data are related to time. Meteorological data, financial data, census data, medical data, simulation data, news articles, photo collections, or project plans, to name only a few examples, all contain temporal information. In theory, because all these data are time-oriented, they should be representable with one and the same visualization approach. In practice, however, the data exhibit specific characteristics and hence each of the above examples requires a dedicated visualization. For instance, stock exchange data can be visualized with flocking boids (see Vande Moere (2004) and ↪ p. 333), census data can be represented with Bubbles (see Gapminder Foundation (2021) and ↪ p. 330), and simulation data can be visualized efficiently using MOSAN (see Unger and Schumann (2009) and ↪ p. 316). News articles (or keywords therein) can be analyzed with ThemeRiver (see Havre et al. (2002) and ↪ p. 293) and project plans can be made comprehensible with PlanningLines (see Aigner et al. (2005) and ↪ p. 260). Finally, meteorological data are visualized for us in the daily weather show. Apparently, this list of visualizations is not exhaustive. The aforementioned approaches are just examples from a substantial body of techniques that recognize the special role of the dimension of time in visualization contexts. A more complete list is provided in the rich survey of visualization techniques in Appendix A.

Besides these dedicated techniques, time-oriented data can also be visualized using generic approaches. Since time is mostly seen as a quantitative dimension or at least can be mapped to a quantitative domain (natural or real numbers), general visualization frameworks such as Tableau (see Loth, 2019) or Power BI (see Knight et al., 2018) as well as standard diagrams and charts, as surveyed by Harris (1999), are applicable for visualizing time-oriented data. For simple data and basic analysis

© The Author(s) 2023
W. Aigner et al., *Visualization of Time-Oriented Data*, Human–Computer
Interaction Series, https://doi.org/10.1007/978-1-4471-7527-8_4

tasks, these approaches outperform specialized techniques, because they are easy to learn and understand (e.g., common line plot). However, in many cases, time is treated just as one quantitative variable among many others, not more, not less. Therefore, generic approaches usually do not support establishing a direct visual connection between multiple variables and the time axis, they do not communicate the specific aspects of time (e.g., the different levels of temporal granularity), and they are limited in terms of direct interactive exploration and browsing of time-oriented data, which are essential for a successful visual analysis.

The bottom line is that time must be specifically considered to support the visual analysis (also see Wills, 2012). Different types of time-oriented data need to be visualized with dedicated methods. As the previous examples suggest, a variety of concepts for analyzing time-oriented data are known in the literature (see for example the work by Silva and Catarci, 2000; Müller and Schumann, 2003; Daassi et al., 2005; Aigner et al., 2008; Bach et al., 2017; Fang et al., 2020). This variety makes it difficult for researchers to assess the current state of the art, and for practitioners to choose visualization approaches most appropriate to their data and tasks.

What is required is a systematic and comprehensive view on the diverse options of visualizing time-oriented data (see Aigner et al., 2007). In this chapter, we will develop such a view. The different design options derived from the systematic view will be discussed and illustrated by a number of visualization examples.

4.1 Characterization of the Visualization Problem

In the first place, we need a structure to organize our systematic view. But instead of using formal or theoretical constructs, we decided to present a structure that is geared to three practical questions that are sufficiently specific for researchers and at the same time easy to understand for practitioners:

1. *What* is presented? – *Time & data*
2. *Why* is it presented? – *User tasks*
3. *How* is it presented? – *Visual representation*

Because any visualization originates from some data, the first question addresses the structure of time and the data that have been collected over time. The motivation for generating a visualization is reflected by the second question. It relates to the aim of the visualization and examines the tasks carried out by the users. How the data are represented is covered by the third question. The following sections will provide more detailed explanations and refinements for each of these questions.

4.1.1 What? – Time & Data

It goes without saying that the temporal dimension itself is a crucial aspect that any visualization approach for representing time and time-oriented data has to consider. It is virtually impossible to design expressive visual representations without knowledge about the characteristics of the given data and time domain. The characteristics of time and data as well as corresponding design aspects have already been explained in detail in Sections 3.1 and 3.2. Here, we will only briefly summarize these aspects.

Characteristics of time The following list briefly reiterates the key criteria of the dimension of time that are relevant for visualization:

- *Scale – ordinal vs. discrete vs. continuous*: In an ordinal time model, only relative order relations are present (e.g., before, during, after). In discrete and continuous domains, temporal distances can also be considered. In discrete models, time values can be mapped to a set of integers based on a smallest possible unit (e.g., seconds). In continuous models, time values can be mapped to the set of real numbers, and hence, between any two points in time, another point can be inserted.
- *Scope – point-based vs. interval-based*: Point-based time domains have basic elements with a temporal extent equal to zero. Thus, no information is given about the region between two points in time. Interval-based time domains relate to subsections of time having a temporal extent greater than zero.
- *Arrangement – linear vs. cyclic*: Linear time corresponds to an ordered model of time, i.e., time proceeds from the past to the future. Cyclic time domains are composed of a finite set of recurring time elements (e.g., the seasons of the year).
- *Viewpoint – ordered vs. branching vs. multiple perspectives*: Ordered time domains consider things that happen one after the other. In branching time domains, multiple strands of time branch out and allow for description and comparison of alternative scenarios, but only one path through time will actually happen (e.g., in planning applications). Multiple perspectives facilitate simultaneous (even contrary) views of time (as for instance required to structure eyewitness reports).

In addition to these criteria, which describe the dimension of time, aspects regarding the presence or absence of different levels of granularity, the time primitives used to relate data to time, and the determinacy of time elements are relevant (see Section 3.1 in the previous chapter).

Characteristics of time-oriented data Like the time domain, the data have a major impact on the design of visualization approaches. Let us briefly reiterate the key criteria for data that are related to time:

- *Scale – quantitative vs. qualitative*: Quantitative data are based on a metric scale (discrete or continuous). Qualitative data describe either unordered (nominal) or ordered (ordinal) sets of data elements.

- *Frame of reference – abstract vs. spatial*: Abstract data (e.g., a bank account) have been collected in a non-spatial context and are not per se connected to some spatial layout. Spatial data (e.g., census data) contain an inherent spatial layout, e.g., geographical positions.
- *Kind of data – events vs. states*: Events, on the one hand, can be seen as markers of state changes, whereas states, on the other hand, characterize the phases of continuity between events.
- *Number of variables – single vs. multiple*: Univariate data contain only a single data value per temporal primitive, whereas in the case of multivariate data each temporal primitive holds multiple data values.

We see that time-oriented data can differ significantly in their structure and basic properties. The visualization design must take these properties into account in order to provide appropriate visual representations. The defined primary categories capture the key aspects to be considered when answering the *what* question of our systematic view. We will demonstrate this in more detail in Section 4.2.1.

Yet, having characterized what has to be visualized is just a first step. The subsequent step is to focus on the *why* question.

4.1.2 Why? – User Tasks

It is commonly accepted that software development has to start with an analysis of the problem domain users work in (see Hackos and Redish, 1998; Courage and Baxter, 2005). To specify the problem domain, so-called task models are widely used in the related field of human-computer interaction (see Constantine, 2003). A prominent example of such task models is the ConcurTaskTree (CTT) by Paternò et al. (1997). It describes a hierarchical decomposition of a goal into tasks and subtasks. Four specific types of tasks are supported in the CTT notation: abstract tasks, interaction tasks, user tasks, and application tasks. Abstract tasks can be further decomposed into subtasks (including abstract subtasks). Leaf nodes are always interaction tasks, user tasks, or application tasks. They have to be carried out either by the user, by the application system, or by the interaction between the user and the system. The CTT notation is enriched with a set of temporal operators that define temporal relationships among tasks and subtasks (e.g., independent concurrency, concurrency with information exchange, disabling, and enabling).

The development of solutions for visual data analysis, and thus the design of visualizations for time-oriented data also starts with the analysis of the application, the given data, and the tasks to be accomplished. Munzner's nested model reflects this strategy (see Munzner, 2009). The model consists of four nested levels, which describe the path from problem specification to implementation. The first two levels address the visualization problem. The first level refers to the characterization of the application domain, while the second level refers to the abstraction of data and tasks. We already examined the specification of data in the previous section. Now we take a closer look at the description of the tasks. To do this, we will refer to the task abstraction by Tominski and Schumann (2020), which characterizes tasks by four key aspects: goals, analytic questions, targets, and means.

Goals describe the overarching intent with which the analysis tasks are performed. Possible goals are to explore, describe, explain, confirm, or present the data. By exploring the data, we want to make observations, such as identifying trends or outliers. The goal of description is to characterize the discovered observations, while explanation means to identify all contributing data and detect the main reasons for the observations, which allows us to establish hypotheses. Confirmation is about verifying the hypotheses, and with presentation, we communicate confirmed results.

Analytical questions specify what is actually to be investigated in a particular step of the analysis. According to Andrienko and Andrienko (2006), we can distinguish between two fundamental categories: elementary and synoptic questions. Elementary questions refer to one or more data elements, which are examined individually. Elementary questions can be for example the following: Identify: What is the value? Locate: Where is the value? Compare: Is it less or more? Synoptic questions refer to groups of data in order to characterize sets of data elements. Identify, locate, and compare also apply to sets of data values. Additionally, we can ask more specific synoptic questions as follows: Group: Do data values belong together? Correlate: Are there any dependencies? Trends: Do groups of values develop systematically? Outliers: Are some data values special with respect to the rest?

Targets tells us where in the data a task should be performed. The notion of targets allows us to narrow down which specific data we need to look at in order to complete the task. Targets can be specific time-dependent variables or particular time primitives of interest.

Means describe how a task is performed. We distinguish between visual, interactive, and computational means. Visual means subsume all types of visual inspection, while interactive means refer to interactive information retrieval. In both cases, the tasks are performed by human users. In contrast, computational means stand for calculations, which are performed by the machine.

While goals, targets, and means are more or less generic, the particular analytic questions to be answered depend on the characteristics of the data to be investigated. An accepted formulation of analytic questions addressing time-oriented data has been introduced by MacEachren (1995). He describes the following types of questions:

- *Existence of data element*
  Question: Does a data element exist at a specific time?
  Starting point: time point or time interval
  Search for: data element at that time
  Example: "Was a measurement made in June, 1960?"
- *Temporal location*
  Question: When does a data element exist in time?
  Starting point: data element

Search for: time point or time interval

Example: "When did the Olympic Games in Vancouver start?"

- *Time interval*

  Question: How long is the time span from beginning to end of the data element?

  Starting Point: data element

  Search for: duration, i.e., length of time of a data element from its beginning to its end

  Example: "How long was the processing time for dataset A?"

- *Temporal pattern*

  Question: How often does a data element occur?

  Starting point: interval in time

  Search for: frequency of data elements within a certain portion of time and based on this the detection of a pattern

  Example: "How often was Jane sick last year?"

- *Rate of change*

  Question: How fast is a data element changing or how much difference is there from data element to data element over time?

  Starting point: data element

  Search for: magnitude of change over time

  Example: "How did the price of gasoline vary in the last year?"

- *Sequence*

  Question: In what order do data elements occur?

  Starting point: data elements

  Search for: temporal order of different data elements

  Example: "Did the explosion happen before or after the car accident?"

- *Synchronization*

  Question: Do data elements exist together?

  Starting point: data elements

  Search for: occurrence at the same point or interval in time

  Example: "Is Jill's birthday on Easter Monday this year?"

This list of tasks covers two basic scenarios. First, given one or more time primitives, the user seeks to discern the data values associated with them. Second, given one or more data values, the user is searching for time primitives that exhibit these values. Both cases reflect the well-established distinction between *identification* tasks (i.e., identify data values) and *location* tasks (i.e., locate when and where data values occur in time and space).

From a practical perspective, the verbal descriptions of analytical questions by MacEachren (1995) are very helpful because they are easy to understand. They can serve as a guideline when designing visual representations of time and time-oriented data. However, in order to automate the design process, a more abstract description would be desirable. For this purpose, we introduced three levels of analytical questions based on Andrienko and Andrienko (2006). The first level deals with the fundamental categories. It is about whether individual data values (elementary questions) or data subsets (synoptic questions) are to be answered. The second level distinguishes whether we aim to determine the values of data (lookup) or to compare them (comparison). Finally, the third level considers whether we want to identify or locate data values. In the next section, we will apply this categorization in order to examine the influence of user tasks on the visualization design. But before, we want to complete the description of visualization aspects by focusing on the *how* perspective.

4.1.3 How? – Visual Representation

The answers to the questions of what the data input is and why the data are analyzed very much determine the answer to the last remaining question: How can time-oriented data be represented visually? More precisely, the question is how time and associated data are to be represented. Appendix A shows that a large variety of visual approaches provide very different answers to this question. To abstract from the subtle details of this variety, we concentrate on two fundamental criteria: the mapping of time and the dimensionality of the presentation space.

Mapping of Time

Like any data variable that is to be visualized, the dimension of time has to pass the mapping step of the visualization pipeline. Usually, abstract data are made visually comprehensible by mapping them to some geometry (e.g., two-dimensional shapes) as marks and corresponding visual attributes (e.g., color) as channels in the presentation space. On top of this, human perception has an intrinsic understanding of time that emphasizes the progression of time, and visualization can make use of this fact by mapping the dimension of time to the dynamics of a visual representation.

So practically, there are two options for mapping time: the mapping of time to space and the mapping of time to time. When speaking of a mapping from time to space, we mean that time and data are represented in a single coherent visual representation using a spatial substrate. This representation does not automatically change over time, which is why we call such visualizations of time-oriented data *static*. In contrast to that, *dynamic* representations utilize the physical dimension of time as a temporal substrate to convey the time dependency of the data, that is, time is mapped to time. This results in visualizations that change over time automatically (e.g., slide shows or animations). Note that the presence or absence of interaction facilities to navigate in time has no influence on whether a visualization approach is categorized as static or dynamic.

Static representations For static representations, the time axis is embedded into the visual representation. The visual encoding of the time axis needs to be designed in such a way that the temporal relation to other data variables can be easily recognized. There are various ways of mapping time to visual variables (see Bertin (1983) and Figure 4.1). Most visualization approaches that implement a time-to-space mapping use one display dimension to represent the time axis. Classic examples are charts where time is often mapped to the horizontal x-axis and time-dependent variables are mapped to the vertical y-axis (see Figure 4.2). More complex mappings are possible when two or more display dimensions are used for representing time. For example, Perin et al. (2018) performed a study to assess the graphical perception of mapping time and speed to 2D trajectories. They compared nine different combinations of line width, brightness, as well as tick mark mappings to encode the two variables time and speed. For encoding speed, using brightness and for encoding time, segment length between ticks are recommended. When both speed and time should be encoded, the authors advise to use segment length whenever possible. Besides, mappings that generate two-dimensional spirals or three-dimensional helices are examples that emphasize the cyclic character of time. The different granularities of time are often illustrated by a hierarchical subdivision of the time axis.

The actual data can then be visualized in manifold ways. It is practical to use a data mapping that is orthogonal to the mapping of time. For example, point plots (↪ p. 232), line plots (↪ p. 233), or bar graphs (↪ p. 234) map data values to position or size relative to the time axis. Time dependency is immediately perceived and can be recognized easily, which facilitates the interpretation of the temporal character of the data. In fact, for quantitative variables (discrete or continuous time and data), using position or length is more effective than using color or other visual variables such as texture, shape, or orientation (see Mackinlay, 1986). For categorical variables, color-coding is a good alternative. Each point or interval on the time axis can be visualized using a unique hue from a color scale. However, care must be taken when using color for the visualization of ordinal data (see Silva et al., 2007). It is absolutely mandatory that the applied color scale be capable of communicating order[1]. Only then are users able to interpret the visualization and to relate data items to their temporal context easily. It is also possible to enhance the visual representation of the time-dependent data by using a composite visual encoding (see Jabbari et al., 2018) where the same data variable is mapped to two or more visual variables (e.g., length plus color).

Because time is often considered to be absolute, position or length encodings are predominant, and only rarely is time mapped to other visual variables. When time is interpreted relatively rather than absolutely, for instance, when considering the age of a data item or the duration between two occurrences of a data item, then visual variables such as transparency, color, and others gain importance. An example of encoding duration to color is given in Figure 4.3.

Instead of encoding data to basic graphical primitives such as points, lines, or bars that are aligned with the time axis, one can also create fully fledged visual representations and align multiple thumbnails of them along the time axis – a concept that Tufte (1983) refers to as *small multiples* (↪ p. 359). The advantage is that a single thumbnail may contain much more visual information than basic graphical primitives. But this comes at the price that the number of time primitives (i.e., the number of thumbnails) that can be shown on screen simultaneously is limited. This

---

[1] Borland and Taylor (2007) warn that this is not the case for the most commonly used rainbow color scale. The ColorBrewer tool by Harrower and Brewer (2003) is a good source of useful color scales.

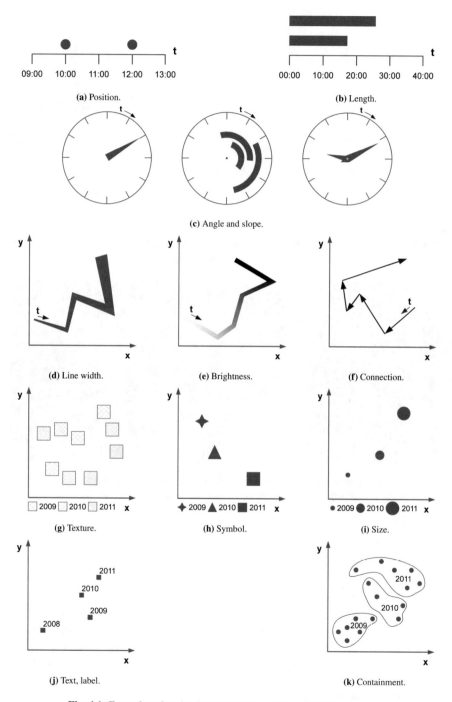

Fig. 4.1: Examples of static visual mappings of time. ⓒⓘ *The authors.*

Fig. 4.2: Mapping time to position. The horizontal axis of the chart encodes the positions of points in time, whereas the vertical axis encodes data values. ⓒ① *The authors.*

Fig. 4.3: Mapping time to color. Color encodes the time it takes to travel from a location on our planet to the nearest major city. ⓒ① *Weiss et al. (2018b), also see Weiss et al. (2018a).*

reflects the general need to find a good trade-off between the complexity of the visual encoding of time and that of the data. Appendix A makes apparent that a variety of suitable solutions exist, each with an individually determined trade-off depending on the addressed data and tasks.

Dynamic representations In cases where much screen space is required to convey characteristics and relationships of data items (e.g., geographical maps, multivariate data visualization, and visualization of graph structures), it is difficult to embed the time axis into the display space as well. As an alternative, physical time can be utilized to encode time. To this end, several visualizations (also called *frames*) are rendered successively for the time steps in the data. In theory, a one-to-one mapping between time steps and frames can be implemented, which means that the dynamic visualization represents time authentically. In practice, however, this is only rarely possible. More often, dynamic visualizations perform interpolation to compute intermediate results in cases where only a few time steps are present, or perform aggregation or sampling to compress the length of an animation in cases where large numbers of time steps have to be visualized (see Wolter et al., 2009).

Self-evidently, dynamic approaches have to take human perception into account when representing a series of successively generated visualization frames. Depending on the number of images shown per second, dynamic visualizations are either perceived as *animations* or as slide shows. Animations usually show between 15 and 25 frames per second, while slide shows usually show a new frame every 2 to 4 seconds. On the one hand, data that contain only a few snapshots of the underlying phenomenon should preferably be represented as slide shows to avoid creating false impressions of dynamics. On the other hand, large numbers of observations of highly dynamic processes are best represented using animations, because they communicate quite well the underlying dynamics in the data. Animations provide us with a qualitative overview on the data. However, quantitative statements are hardly possible. For example, we cannot easily capture the concrete data values for a specific point in time. For this reason, it is important that the user can control the animation flow. Figure 4.4 gives an example of a typical VCR-like widget for controlling the mapping of time to time in an animation.

Fig. 4.4 A typical animation widget to control the mapping of time to time. ⓒ① *The authors.*

When to use static or dynamic representations The distinction between the mapping of time to space and that of time to time is crucial because different visualization tasks and goals are supported by these mappings. Dynamic representations are well suited to convey the general development and the major trends in the analyzed data. However, there are also critical assessments of animations used for the purpose of visualization (see Tversky et al., 2002; Simons and Rensink, 2005; Thompson et al., 2020). Especially when larger multivariate time series have to be visualized, animation-based approaches reach their limits. In such cases, users are often unable to follow all of the changes in the visual representation, or the animations simply take too long and users face an indigestible flood of information. This problem becomes aggravated when using animations in multiple views. On the other hand, if animations are designed well and if they can be steered interactively by the user (e.g., slow motion or fast forward), mapping the dimension of time to the physical time can be beneficial (see Robertson et al., 2008). This is not only the case from the point of view of perception, but it is also because using physical time for visual mapping implies that the spatial dimensions of the presentation space can be used exclusively to visualize the time-dependent data.

This is not the case, however, for static representations. In contrast to animations, static representations require screen real estate to represent the time axis itself and the data in an integrated fashion. On the one hand, the fact that static representations show all of the information on one screen is advantageous because one can fully concentrate on the dependency of time and data. Especially visual comparison of different parts of the time axis can be accomplished easily using static representations. On the other hand, the integration of time and data in one single view tends to lead to overcrowded representations that are hard to interpret. In the face of larger time-oriented datasets, interaction and analytical methods (see Chapters 5 and 6) are mandatory to avoid visual clutter.

Finally, it is worth mentioning that any (non-temporal) data visualization can be extended to a visualization for time-oriented data simply by repetition. Repetition in time leads to dynamic representations, where each frame shows a snapshot of the data. Repetition in space leads to static multiple-view representations (or Tufte's *small multiples*, ↪ p. 359), where each view shows an individual part of the time axis. While static representations always have to deal with the issue of finding a good layout for the views, dynamic representations encode time linearly in a straightforward manner. Perhaps this is the reason why many visualization solutions resort to simple animations, even though these might not be the best option for the data and tasks at hand.

Dimensionality of the Presentation Space

The presentation space of a visualization can be either two-dimensional or three-dimensional, or 2D or 3D for short. Two-dimensional visualizations address the spatial dimensions of computer displays, that is, the x-axis and the y-axis. All graphical elements are described with respect to x- and y-coordinates. Dots, lines, circles, or arcs are examples of 2D geometry. The semantics of the data usually determine the layout of the geometry on screen. 3D visualizations use a third dimension, the z-axis, for describing geometry. This allows the visualization of more complex and volumetric structures. As human perception is naturally tuned to the three-dimensional world around us, 3D representations potentially communicate such structures better than 2D approaches. This is especially true if the output device used supports immersive analytics and 3D representations such as stereoscopic displays or augmented-reality headsets. However, on a 2D computer display, the z-axis does not physically exist, so projection is required before rendering 3D visualizations. The projection is commonly realized by standard computer graphics methods that do not require additional effort. Hence, it is usually not transparent to the user.

In Figure 4.1, various 2D presentations of time-oriented data were shown. Visualization approaches using such a 2D presentation typically map the time axis to a visual axis on the display (provided that the approach is not dynamic). In many cases, the time axis is aligned with either coordinate axis of the display. However, this is not necessarily always the case. In particular, time axes representing a cyclic time domain are usually depicted by a radial visualization (see Draper et al., 2009).

Fig. 4.5 Mapping to 2D.
Data are visualized as height-
varying bars along a spiral
time axis. ©① *The authors.*

Radial time axes (e.g., the spiral in Figure 4.5) use polar coordinates, which actually can be mapped to Cartesian coordinates and vice versa. It is also possible to apply affine transformations to the time axis.

Because one dimension of the display space is usually occupied for the representation of the dimension of time, the possibilities of encoding the data depending on time are restricted. One data variable can be encoded to the remaining spatial dimension of the presentation space, as for instance in a bar graph, where the x-axis encodes time and the y-axis, more precisely the height of bars, encodes a time-dependent variable. In order to visualize multiple variables, further graphical attributes like shape, texture, or color can be used.

Multidimensional data, that is, data with more independent variables than just the dimension of time, are hard to visualize in 2D without introducing overlap and visual clutter. In this case, it is therefore often useful to use a 3D representation. Particularly, data with a spatial frame of reference can benefit from the additional dimension. This allows us to apply the so-called *space-time cube* concept (see Kraak (2003) and ↪ p. 377), according to which the z-axis encodes time and the x- and y-axes represent two independent variables (e.g., longitude and latitude). Further variables, dependent or independent, are then encoded to color, size, shape, or other visual attributes (see Figure 4.6 and ↪ p. 389).

The question of whether or not it makes sense to exploit three dimensions for visualization has been discussed at length by the research community (see Card et al., 1999; Dübel et al., 2014; Munzner, 2014). One camp of researchers argues that two dimensions are sufficient for effective visual data analysis and that the third dimension introduces unnecessary difficulties (e.g., information hidden on back faces, information lost due to occlusion, or information distorted through perspective projection) that are much less or not at all relevant for 2D representations. But having just two dimensions for the visual mapping might not be enough for large and complex datasets.

Fig. 4.6: Mapping to 3D. Three-dimensional helices represent time axes for individual regions of a map and associated data are encoded by color. ⓒⓘ *The authors.*

This is where the other camp of researchers makes their arguments. They see the third dimension as an additional possibility to naturally encode further information. Undoubtedly, certain types of data (e.g., geo-spatial data) might even require the third dimension for expressive data visualization, because there exists a one-to-one mapping between the data dimensions and the dimensions of the presentation space. Moreover, human perception is by nature adapted to the three-dimensional character of our physical world.

We do not argue for either position in general. The question of whether to use 2D or 3D is rather a question of which data have to be visualized and what are the analytic goals to be achieved. The application background and user preferences also influence the decision for 2D or 3D. But definitely, when developing 3D visualizations, the previously mentioned disadvantages of a three-dimensional presentation space have to be addressed, for example, by providing ways to cope with occlusion as suggested by Elmqvist and Tsigas (2007) or Röhlig et al. (2017). Moreover, intuitive interaction techniques are mandatory and additional visual cues are usually highly beneficial. The field of *immersive analytics* studies the advantages and potential issues of interactive immersive 3D visual representations of data in detail (see Marriott et al., 2018; Kraus et al., 2021).

In this section, we discussed different options for the visual mapping of time and questions related to the dimensionality of the representation space. While our focus was on representing time and time-oriented data, further aspects can also play a role and are worth mentioning. For example, Section 3.4 of Chapter 3 listed uncertainty as an aspect of data quality that might also be relevant to communicate visually, for which dedicated solutions exist (see Gschwandtner et al., 2016; Bors et al., 2020). Overall, we can conclude that visually mapping the data and deciding how to present them on the screen are the most important steps when creating visualizations.

So far, we have outlined the basic aspects (*what*, *why*, and *how*) that need to be considered when visualizing time and time-oriented data. In the next section, we will return to each of these aspects and show in more detail and by means of examples how the visualization design is influenced by them.

4.2 Visualization Design Examples

We introduced three basic questions that have to be taken into account when designing visual representations for time and time-oriented data:

1. Data level: *What* is presented?
2. Task level: *Why* is it presented?
3. Presentation level: *How* is it presented?

We will now demonstrate the close interrelation of the three levels. By means of examples, we will illustrate the necessity and importance of finding answers to each of these questions in order to arrive at visual representations that allow viewers to gain insight into the analyzed data.

4.2.1 Data Level

In the first place, the characteristics of time-oriented data strongly influence the design of appropriate visual representations. Two examples will be used to demonstrate this: one is related to the time axis itself, and the other will deal with the data. First, we point out how significantly different the expressiveness of a visual representation can be depending on whether the time domain is linear or cyclic. Secondly, we will illustrate that spatial time-oriented data[2] require a visualization design that is quite different from that of abstract time-oriented data, and that is usually more complex and involves making well-balanced design decisions.

Time Characteristics: Linear vs. Cyclic Representation of Time

Figure 4.7 shows three different visual representations of the same time-oriented dataset, which contains the daily number of cases of influenza that occurred in the northern part of Germany during a period of three years. The data exhibit a strong cyclic pattern. The leftmost image of Figure 4.7 uses a simple line plot ($\hookrightarrow$ p. 233) to visualize the data. Although peaks in time can be recognized easily in the plot, the cyclic behavior of the data, however, can only be guessed and it is hard to discern which cyclic temporal patterns in fact do exist. In contrast, the right part of Figure 4.7

---

[2] Commonly referred to as *spatio-temporal data*.

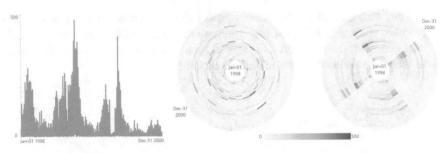

Fig. 4.7: Different insights can be gained from visual representations depending on whether the linear or cyclic character of the data is emphasized. ⓒⓘ *The authors.*

shows radial representations that emphasize cyclic characteristics of time-oriented data by using spiral-shaped time axes (see Weber et al. (2001) and ↪ p. 284). For the left spiral, the cyclic pattern is not visible. This is due to the fact that the cycle length has been set to 24 days, which does not match the pattern in the data. The right spiral in Figure 4.7 is adequately parametrized with a cycle length of 28 days, which immediately reveals the periodic pattern present in the data. The significant difference in the number of cases of influenza reported on Sundays and Mondays, respectively, is quite obvious. We would also see this weekly pattern if we set the cycle length to 7 or 14 days, or any (low) multiple of 7.

The example illustrates that in addition to using the right kind of representation of time (linear vs. cyclic), it is also necessary to find an appropriate parametrization of the visual representation. Interaction (see Chapter 5) usually enables users to reparametrize the visualization, but the difficulty is to find parameter settings suitable to discover patterns in unknown datasets. Automatically animating through possible parameter values – for the spiral's cycle length in our example – is one option to assist users in finding such patterns. During the course of the animation, visual patterns emerge as the spiral's cycle length is approaching cycles in the data that match in length. Upon the emergence of such patterns, the user stops the animation and can fine-tune the display as necessary. Analytical methods (see Chapter 6) can help in narrowing down the search space, which in our example means finding promising candidates with adequate cycle length (see Yang et al., 2000). Combining interactive exploration and analytical methods is helpful for guiding users to less sharp or uncommon patterns, which are hard to distill using either approach alone (see Ceneda et al., 2018).

In summary, we see that it is very important to take the specifics time into account. This applies not only to the question of whether the time axis is linear or cyclic, but to other properties of the time axis as well. However, it is difficult to consider the entire breadth of properties of the time domain, so most visualization approaches focus on only a few important ones. Consequently, we do the same and focus our overview on visualization techniques in Appendix A on two key characteristics: the arrangement of the time axis (linear vs. cyclic) and additionally the type of time primitives (points vs. intervals). Differentiating points and intervals makes sense because it reflects the distinction between states and events. Point-based data express events. In contrast, interval-based data describe states, in which the data remain stable. The visualization design needs to consider this.

Data Characteristics: Aabstract Data vs. Spatial Data

We used linear vs. cyclic time to demonstrate the impact of the characteristics of time on the visualization design. Let us now do likewise with abstract vs. spatial data to illustrate the impact of data characteristics.

Abstract data are not associated per se with a spatial visual mapping. Therefore, when designing a visual representation of such data, one can fully concentrate on aspects related to the characteristics of the dimension of time. Figure 4.8 shows the *ThemeRiver* technique (↪ p. 293) by Havre et al. (2000) as an example of an approach for which the focus is on the time aspect. The dimension of time is mapped to the horizontal display axis and multiple time-dependent variables are mapped to the thickness of individually colored currents, which form an overall visual stream of data values along the time axis. Because time is the only dimension of reference in abstract time-oriented data, the visual representation can make the best of the available screen space to convey the variables' dependency on time. The full-screen design, where the ThemeRiver occupies the entire screen, even makes it possible to display additional information, such as a time scale below the ThemeRiver, labels in the individual currents, or extra annotations for important events in the data.

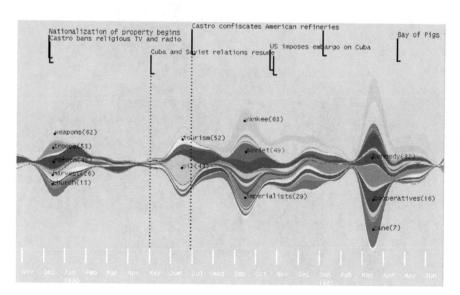

Fig. 4.8: The ThemeRiver technique is fully focused on communicating the temporal evolution of abstract time-oriented data. © *2002 IEEE. Reprinted, with permission, from Havre et al. (2002).*

When considering time-oriented data with spatial references, the visualization design has to address an additional requirement: Not only the temporal character of the data needs to be communicated, but also the spatial dependencies in the data must be revealed. Of course, this implies a conflict in which the communication of temporal aspects competes with the visualization of the spatial frame of reference for visual resources, such as screen space, visual encodings, and so forth. Providing too many resources to the visualization of aspects of time will most likely lead to a poorly represented spatial context – and vice versa. The goal is to find a well-balanced compromise. An example of such a compromise is given in Figure 4.9, where the data are visualized using ThemeRiver thumbnails superimposed on a two-dimensional map display (↪ p. 379). The position of a ThemeRiver thumbnail on the map is the visual anchor for the spatial context of the data. The ThemeRiver thumbnail itself encodes the temporal context of the data. The compromise that has been made implies that the map display is rather basic and avoids showing any geographic detail; just the borders of regions are visible. On the other hand, the ThemeRiver representation has to get along with much less screen space (compared to the full-screen counterpart). This is the reason why labels or annotations are no longer visible constantly, but instead are displayed only on demand.

On top of the compromises made, all visualization approaches that embed (time-representing) thumbnails (or glyphs or icons) into a map share two common prob-

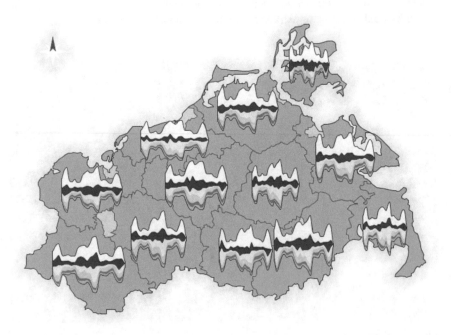

Fig. 4.9: Embedding ThemeRiver thumbnails on a map allows for communicating both temporal and spatial dependencies of spatial time-oriented data. ©① *The authors.*

lems: finding a good glyph design and finding a good layout for the embedding. The performance of glyphs for time-oriented data depends on tasks and data density (see Fuchs et al., 2013). In other words, a good glyph design is heavily application-dependent. The same applies to the layout, that is, a good glyph layout also depends on the application context. There is consensus that having an overlap-free layout is generally a good starting point. However, minimizing (i) occlusion between thumbnails and (ii) overlap between thumbnails and geographic features is a difficult problem. In fact, the problem is related to the general map labeling problem, which is NP-hard. Pursuing a globally optimized solution is computationally complex (see Petzold, 2003; Been et al., 2006), whereas locally optimizing approaches usually perform less expensive iterative adjustments that lead to suitable, but not necessarily optimal layouts (see Fuchs and Schumann, 2004a; Luboschik et al., 2008). We will not go into any details of possible solutions, but instead refer the interested reader to a more recent publication by McNabb and Laramee (2019), who introduce an algorithm for a guided glyph placement.

We have seen that the distinction between abstract and spatial data is essential to create meaningful visualizations. This was demonstrated by two examples. In terms of abstract data, we considered time series. In this case, the data values were given in a linear order with respect to a linear time axis. In addition to such a temporal dependence, there may exist further relations in the time-oriented data, for example, semantic relationships between data items. Just like the data itself, such additional relations might also change over time. These changes can be described by a dynamic graph. The visualization of dynamic graphs requires customized visualization strategies.

In terms of spatial data, we considered a 2-dimensional geo-spatial frame of reference. However, data values might also be embedded into a 3-dimensional spatial context that is not geo-related. Typical examples of such data are MRI data or other medical imaging data, which are often given on regular 3D grids. Such data are referred to as volume data, and they can be time-dependent as well.

Visualizing dynamic graphs and dynamic volume data are research topics on their own and will not be discussed in this book. For more details on these topics, we refer to works by Beck et al. (2017) on dynamic graphs as well as Reinders et al. (2001) and Bai et al. (2020) on dynamic volumes.

To conclude, the properties of time-oriented data strongly influence the design of visual representations. However, it is difficult to observe all properties in the same way. Therefore, common visualization techniques consider only some of them. Also in our overview of visualization techniques in Appendix A, we will not address every data aspect. Instead, we will focus there on the two key characteristics of time-oriented data: the frame of reference (abstract vs. spatial) and additionally the number of variables (single vs. multiple). It makes an essential difference if one time primitive is associated with only one single data value (univariate) or with multiple data values (multivariate). The visualization of multivariate data is much more complex. Thus, it requires sophisticated visualization methods.

4.2.2 Task Level

We introduced the user task as a second important visualization aspect. Incorporating the users' tasks into the visualization design process on a general level is a challenging endeavor. To illustrate what this means in practice, we introduce two concrete examples of how visualization design choices are driven by user tasks.

In the first example, we present a pragmatic solution for the specific case of *color-coding*. Earlier in this chapter, we indicated that in addition to the positional encoding of data values along a time axis, color-coding plays an important role when visualizing multiple time-dependent data variables. The design of the color scale employed for the visual encoding substantially influences the overall expressiveness of the visual representation. To obtain expressive visual results, flexible color-coding schemes are needed that can be adapted to the data as well as to the task at hand. In the following, we will explain how color scales can be generated in a task-dependent manner, and how they can be applied to visualize time-oriented data.

In the second example, we show how to choose axes scales to better support certain user tasks for visualizing multivariate developments over time for line plots (↪ p. 233). Depending on the choices of tasks, one might choose from different combinations of *superimposition*, i.e., arranging plots on top of each other, *juxtaposition*, i.e., arranging plots next to each other, or *indexing*, i.e., plotting values relative to a selected point in time.

Color-Coding

The general goal of color-coding is to find an expressive mapping of data to color. This can be modeled as a color-mapping function $f : D \rightarrow C$ that maps values of a dataset D to colors from a color scale C. A fundamental requirement for effective color-coding is that the color-mapping function f be injective, that is, every data value (or every well-defined group of data values) is associated with a unique color. This, in turn, allows users to mentally associate that unique color with a distinct data value (or group of values). On top of that, the color-mapping function f needs to be designed in such a way that different values will be mapped to different colors and similar colors will imply similar values. In this way, two quite different data values result in two colors that are easy to discern visually, and visually similar colors infer that they represent data values that are similar. Figure 4.10 demonstrates a basic mapping strategy where a data *value* between $[min, max]$ is normalized to t on a $[0, 1]$ range and then mapped to a *color* on a light-green to dark-green color scale.

Besides these fundamental requirements, color-coding depends on further factors (see Telea, 2014). In particular, the characteristics of data and tasks must be taken into account. In terms of the *data characteristics*, first and foremost the statistical features of the data and the time scale should be considered such as extreme values, the overall distribution of data values as well as data variation speeds and domain sampling frequencies. For example, using a linear color-mapping function on a skewed dataset will result in the majority of data values being compressed to a narrow range of colors,

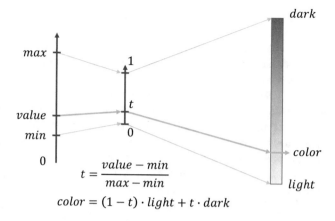

$$t = \frac{value - min}{max - min}$$

$$color = (1 - t) \cdot light + t \cdot dark$$

Fig. 4.10: Simple strategy for mapping data to color. ⓒ① *The authors.*

which is usually not desirable. With regard to the *characteristics of the tasks*, a main distinction is whether the task requires the comparison of exact quantitative values or the assessment of qualitative differences. Furthermore, certain tasks may lead to specific regions of interest in the data domain. These regions should be accentuated, for instance by using bright, warm, and fully saturated colors. In general, we can say that different tasks require different color-coding schemes.

On top of these two fundamental influencing factors, an effective color-coding also depends on the *characteristics of the user* and the *characteristics of the output device*. In terms of the user, the cultural and professional background, conventions of the application domain as well as individual color perception have to be considered. In terms of the output devices, we need to take into account different systems to define and display colors. A color-coding scheme that is appropriate for displaying data on a computer display might be inappropriate when showing the same data on other media.

We see that a variety of factors influence the encoding of data via colors. Effective color scales therefore must be designed with care. Comprehensive overviews on the topic of color scales are available in the literature (see Silva et al., 2011; Mittelstädt et al., 2015; Bernard et al., 2015; Zhou and Hansen, 2016; Nardini et al., 2021). In the following, we will discuss the design of color scales depending on the tasks at hand in more detail.

Task-Dependent Color-Coding

In order to define color scales in a task-specific manner, an adequate specification of tasks is required. In Section 4.1.2, we distinguished between *elementary* analytic questions, which refer to data elements individually, and *synoptic* analytic questions, which address groups of data elements. Color-coding individual data values requires unsegmented color scales. Unsegmented color scales associate unique colors with all individual data values, that is, every color of the color scale represents exactly one data value. In contrast to that, segmented color scales should be used to encode sets of data values. Each color of a segmented color scale stands for a data subset, usually a range of data values.

In Section 4.1.2, we also distinguished between the two basic analytic questions *identify* (what are the data?) and *locate* (where are the data?), which can be applied to both elementary and synoptic tasks. In order to facilitate identification tasks, it should be made easy for the user to mentally map a perceived color to a concrete data value or a set of data values. Moreover, perceived color distances should correspond to distances in the data, which requires color scales that take the capabilities of human perception into account. For example, Mittelstädt et al. (2014) optimizes color scales to reduce physiological color contrasts, which can considerably improve the identification of data values. To support location tasks, on the other hand, color scales should be designed so that data of interest can be located quickly and easily, ideally pre-attentively (see Healey and Enns, 2012). This can be achieved, for example, by accentuation and de-accentuation for which various options are possible, including highlighting with color (see Hall et al., 2014; Waldner et al., 2017; Mairena et al., 2022).

The specification of color scales for elementary and synoptic identification and location tasks is a well-investigated problem (see Bergman et al., 1995; Harrower and Brewer, 2003; Silva et al., 2007; Silva et al., 2011; Mittelstädt et al., 2015). Figure 4.11 shows examples of such color scales. The segmented color scale for identification represents five different colors, and thus allows us to identify five different sets of values. The unsegmented version can be used to identify individual values. The segmented color scale for location supports users in making a binary decision: Yellow encodes a match of some selection criteria; otherwise, there is no match. The unsegmented color scale represents a smooth interpretation of the selection criteria.

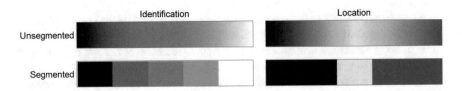

Fig. 4.11: Examples of unsegmented and segmented color scales for identification and location of data values in a visual representation. ©① *The authors.*

Figure 4.12 illustrates the difference between color scales for identification and location for the case of time-oriented data. The figure shows daily temperature values for about three and a half years mapped to a color-coded spiral display (↪ p. 274). While the color scale in Figure 4.12a supports identification, that is, one can easily associate a color with a particular range of values, the color scale in Figure 4.12b

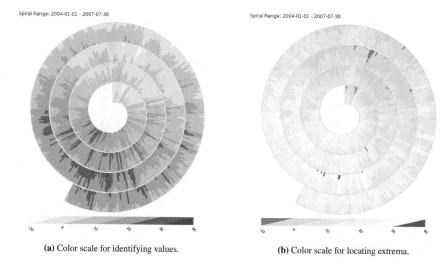

Spiral Range: 2004-01-01 - 2007-07-30

Spiral Range: 2004-01-01 - 2007-07-30

(a) Color scale for identifying values.

(b) Color scale for locating extrema.

Fig. 4.12: Daily temperature values visualized along a spiral time axis using different color scales for different tasks. ⓒⓘ *The authors.*

is most suited to locate specific data values in time. In our example, the highest and lowest values are accentuated using saturated red and blue, respectively. All other values are encoded to shades of gray, effectively attenuating these parts of the data. This way, it is easy to locate where in time high and low values occur.

Identification and location tasks have in common that they involve a form of lookup, either of particular values or of certain references in time and space. In the literature, *lookup* tasks are differentiated from *comparison* tasks (see Andrienko and Andrienko, 2006). Comparison tasks are concerned with relationships in the data. For example, we may ask is one value higher than another value (elementary task) or do values develop systematically to form a trend (synoptic task). The distinction between lookup and comparison tasks deserves a more detailed investigation. Supporting the lookup task basically requires color scales that allow for precise association of particular colors with concrete data values. In order to facilitate comparison tasks, all variables involved in the comparison must be represented by a common unified color scale, which can be problematic when variables exhibit quite different value ranges. The next paragraphs will provide more details on how efficient color scales for lookup and comparison tasks can be designed.

Color-coding for the lookup task As mentioned before, there are two kinds of lookup tasks: identification and location. Location tasks are basically a search for certain references in time and space that exhibit specific data characteristics. For this purpose, relevant data values (or subsets) are known beforehand and hence can be easily accentuated using a highlighting color. On the other hand, the design of color scales for identification is intricate because the whole range of data values is potentially relevant and must be easily identifiable. One way to facilitate lookup

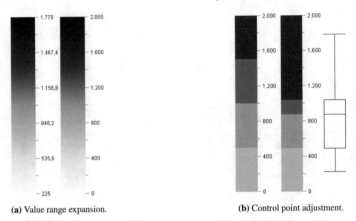

(a) Value range expansion.   **(b)** Control point adjustment.

Fig. 4.13: Value range expansion and control point adjustment help to make color legends more readable and to better adapt the color-coding to the underlying data distribution, which is depicted as a box-whisker plot. ⓒ① *The authors.*

tasks is to extract statistical metadata from the underlying dataset and utilize them to adjust predefined color scales (see Schulze-Wollgast et al., 2005; Tominski et al., 2008). Let us take a look at three possible ways of adaptation.

Expansion of the value range The labels displayed in a color scale legend are the key to an easy and correct interpretation of a color-coded visualization. Commonly a legend shows labels at uniformly sampled points between the data's minimum and maximum. As the left color scale in Figure 4.13a illustrates, this usually results in odd and difficult-to-interpret labels. Even if the user has a clear picture of the color, it takes considerable effort to mentally compute the corresponding value, or even the range of plausible values. The trick of value range expansion is to extend the data range that is mapped to the color scale. This is done in such a way so as to arrive at a color-mapping that is easier to interpret. The right color scale in Figure 4.13a demonstrates this positive effect.

Adjustment of control points A color-map is defined by several control points, each of which is associated with a specific color. Appropriate interpolation schemes are used to derive intermediate colors in between two control points. The left color scale in Figure 4.13b shows an example where control points are uniformly distributed (interpolation is not applied for this segmented color scale). While this is generally a good starting point, more information can be communicated when using an adapted control point distribution. This is demonstrated in the right color scale of Figure 4.13b, where control points have been shifted in accordance with the data distribution. The advantage is that users can easily associate colors with certain ranges of the data distribution.[3]

---

[3] The box-whisker plot or box plot used in the figures depict minimum, 1st quartile, median, 3rd quartile, and maximum value (horizontal ticks from bottom to top).

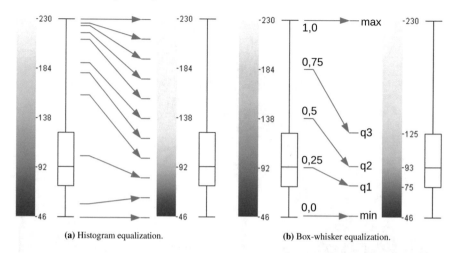

(a) Histogram equalization.   (b) Box-whisker equalization.

Fig. 4.14: Equalization schemas for adapting a color scale to the data distribution, which is depicted as box-whisker plots. ⓒⓘ *The authors.*

Skewing of the color-mapping function Uneven value distributions can be problematic because they lead to situations where the majority of data values are represented by only a narrow range of colors. This is unfavorable for the identification of individual data values. Logarithmic or exponential color-mapping functions are useful when visualizing data with skewed value distributions. In cases where the underlying data distribution cannot be described by an analytical function, equalization can be applied to generate adapted color scales. The net effect of equalization is that the scale of colors is in accord with the data's value distribution. Histogram equalization and box-whisker equalization are examples of this kind of adaptation:

- Histogram equalization works as follows. First, one subdivides the value range into n uniform bins and counts the number of data values falling into the bins. Secondly, the color scale is sampled at $n + 1$ points, where the points' locations are determined by the cumulative frequencies of the bins. Finally, the colors at these sample points are used to construct an adapted color scale as illustrated in Figure 4.14a. As a result, more colors are provided there where larger numbers of data values are located, making values in high-density regions easier to distinguish.
- Box-whisker equalization works similarly. Here, colors are sampled at points determined by quartiles. Quartiles partition the original data into four parts, each of which contains one-fourth of the data. The second quartile is defined as the median of the entire set of data (one half of the data lies below the second quartile, and the other half lies above it). The first and the third quartile are the medians of the lower and upper half of the data, respectively. The adapted color scale is constructed from the colors sampled at the quartiles (see Figure 4.14b).

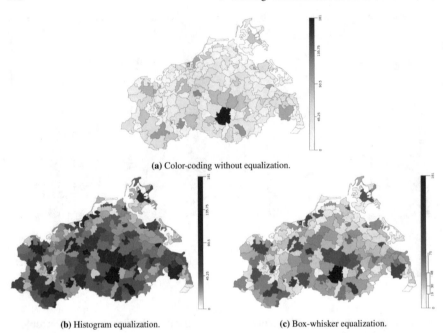

(a) Color-coding without equalization.

(b) Histogram equalization.   (c) Box-whisker equalization.

Fig. 4.15: Color scale equalization applied to the visualization of spatio-temporal health data.
©⊕ *The authors.*

How equalization affects the visualization of spatio-temporal data compared to using unadapted color scales is shown in Figure 4.15. It can be seen that colors are hard to distinguish in dense parts of the data unless histogram or box-whisker equalization is applied, which improves discriminability.

Color-coding for the comparison task The comparison of two or more time-dependent variables requires a global color scale that comprises the value ranges of all variables participating in the comparison. Particularly problematic are comparisons where the individual value ranges are quite different. For example, a variable with a small value range would be represented by only a small fraction of the global color scale, which makes it hard for viewers to differentiate colors in that range. An approach to alleviating this problem is to derive distinct intervals from the union of all value ranges and to create a separate encoding for each interval. To this end, a unique constant hue is assigned to each interval, while varying only brightness and saturation to encode data values. Finally, the separately specified color scales for the intervals are integrated into one global comparison color scale. To avoid discontinuities at the tieing points of two intervals, the brightness and saturation values of one interval have to correspond with the respective values of the adjacent interval. In other words, within one interval, the hue is constant while brightness and saturation vary, whereas at the boundary from one interval to the next, the hue is modified while brightness and saturation are kept constant. This way, even small

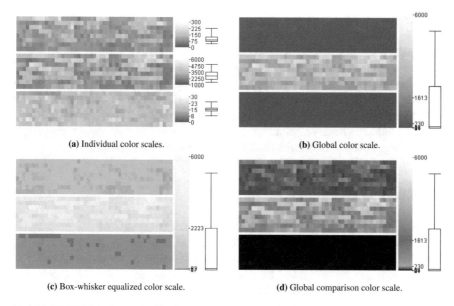

(a) Individual color scales.   **(b)** Global color scale.

(c) Box-whisker equalized color scale.   **(d)** Global comparison color scale.

Fig. 4.16: Different color scales for visual comparison of three time-dependent variables. ©① *The authors.*

value ranges will be represented by their own brightness-varying subscale of the global color scale and the differentiation of data values is improved.

Figure 4.16 shows how different color-coding schemes influence the task of comparing three time-dependent variables. Figure 4.16a uses individual color scales for each variable. A visual comparison is hardly possible because one and the same color stands for three different data values (one in each value range). A global color scale as shown in Figure 4.16b allows visual comparison, but data values of the first and third variables are no longer distinguishable because their value ranges are rather small compared to the one of the second variable. Figure 4.16c illustrates that adapting the color scale to the global value distribution is beneficial. Figure 4.16d shows the visualization outcome when applying the color scale construction as described above: the recognition of values has been improved significantly. However, these results cannot be guaranteed for all cases, in particular, then when the merging process generates too many or too few distinct value ranges.

After reflecting on different options to support lookup and comparison tasks for color-coding, we will now discuss task-dependent considerations for line plots. While color-coding is mainly applied for representing data values in the context of time-oriented data, line plots employ positional encoding for both data and time values. This opens a number of options for parametrization and transformation based on the tasks at hand.

Line Plots

Line plots connect successive data points with lines in order to emphasize the overall change over time. They are very well suited for visually representing time series. However, several difficulties arise with lookup and comparison tasks, particularly when it comes to multivariate time-oriented data. For example, if the developments of time series of different units or value ranges need to be compared, a straightforward overlay could be visually misleading. Yet, by using different options of arrangement and scaling, different user tasks can be supported more appropriately.

Task-Based Line Plots

In the following paragraphs, we describe particular challenges of line plots in the context of lookup and comparison tasks and how they can be mitigated. For the lookup task, the effectiveness with which data and time elements are identifiable can for example be influenced by the appropriate scaling of the plot's axes. For the comparison task, largely different value domains, the comparison of percent changes, as well as heterogeneous data, i.e., data measured in different units, pose problems that can be addressed by specific arrangements, dedicated axes scaling, and indexing.

Line plots for the lookup task The design of line plots for the lookup task must ensure that the whole range of data values is easily identifiable. One way to facilitate this is to extract statistical metadata from the underlying dataset and to scale the time axis accordingly. An example of this is a method called *banking to 45 degrees* which was originally introduced by Cleveland et al. (1988) and refined by Heer and Agrawala (2006) as well as Talbot et al. (2012). It is an optimization technique for computing the aspect ratio of a line plot such that the average orientation of line segments equals 45 degrees (see Figure 4.17).

Line Plots for the comparison task When considering comparison tasks, several difficulties may arise for multivariate time-oriented data. One main challenge in this regard is largely different value domains. In the following, we discuss methods that help us to solve such problems.

Arrangement The simplest case is to superimpose the different variables within a single coordinate system. This employs the major advantage that the individual lines are laid out close to each other and thus allow for an easy direct comparison. However, the superimposition approach stated above might be problematic if variables with largely different value domains are involved. Figure 4.18a illustrates this with superimposed line plots of the closing prices of the two stocks of Amazon (AMZN) and Twitter (TWTR) over the time range of January 3, 2022 to July 27, 2022. For this time interval, AMZN has a value domain in the range of 100–180 whereas TWTR has a value domain of 30–60. These largely different value domains lead to an under-representation of the dynamics of the smaller value domain and make relative

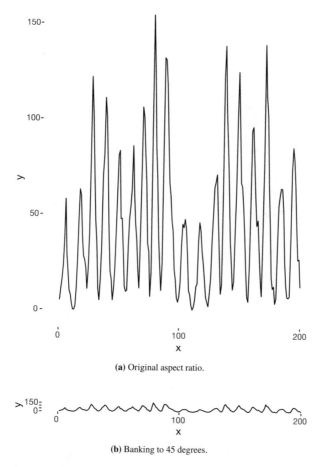

(a) Original aspect ratio.

(b) Banking to 45 degrees.

Fig. 4.17: Example of different width-to-height ratios for supporting lookup tasks. ⓒⓘ *The authors.*

comparisons prone to errors. A possible solution to this is a different *arrangement* of the data display. One option is juxtaposition, which displays the different plots next to each other while adjusting the scale dynamically to make relative changes and the overall shape of variable development better comparable. Figure 4.18b shows the same data as Figure 4.18a by presenting the second variable underneath the first one on a synchronized time scale. In doing so, the dynamics of the smaller value range are much better perceivable. Other layout arrangements are also possible, and in its generalized form, this approach is related to small multiples ($\hookrightarrow$ p. 359).

Axes scaling Not only largely different value domains pose a challenge to line plots, but also the representation and comparison of *percent changes*, i.e., looking at changes relative to the absolute data value. On linear scales, constant percentual changes are displayed as exponentially increasing lines. Furthermore, the same percentual changes are represented via lines of different slopes. For example, an increase

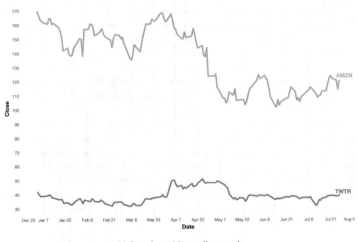

(a) Superimposition on linear scale.

(b) Juxtaposition on linear scale.

Fig. 4.18: Different layout configurations for multivariate time-series comparison of the closing prices of Amazon (AMZN, orange) and Twitter (TWTR, blue) on linear scales. ⓒⓘ *The authors.*

of 100% from a value of 10 to a value of 20 is represented by the same slope as an increase of only 10% from a value of 100 to a value of 110. A possible solution to mitigate this problem is to *scale the axes* of the line plot with respect to the distribution of the data, e.g., using logarithmic scales instead of linear ones. In this case, equal percentual changes are represented by equal slopes. This approach is shown in Figure 4.19a where percentage changes of AMZN and TWTR stock prices can be compared visually directly and also the largely different value domains problem can be overcome by using log scales.

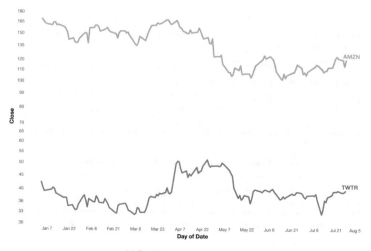

(a) Superimposition on log scale.

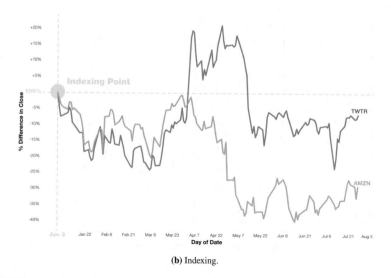

(b) Indexing.

Fig. 4.19: Different configurations for multivariate time-series comparison of the closing prices of Amazon (AMZN, orange) and Twitter (TWTR, blue) using log scales and indexing. ⓒⓘ *The authors.*

Indexing So far, we have focused on multivariate homogeneous data. In contrast to that, heterogeneous time series involve different kinds of data or units. The simplest solution is again to use juxtaposition as described earlier. A further, frequently applied approach is to use superimposition combined with multiple y-axes. However, this also introduces two main problems. First, it is limited to only very few heterogeneous variables (mostly not more than two). Secondly, and most important, the visual appearance and interrelationship of different variables are largely dependent on the selection of the scales for the individual y-axes. Thus, these relationships (especially line crossings and vertical position in relation to each other) are mostly arbitrary.

Bertin (1983) also dealt with this problem several decades ago and introduced *indexing* as a possible solution. The indexing method avoids the problems mentioned before by using a simple transformation of the original values for each time series. The result is a set of new values of a percent unit (see Figure 4.19b). The heterogeneous time series are converted into homogeneous data, which can easily be compared by superimposition. Bertin defines the indexing method with the following formula:

$$\text{index-value}_k = \frac{v_k}{v_{ip}} * 100 \ [\%] : 0 \leq k < n$$

The new indexed value is calculated for every element in the original time series. The point ip refers to the *indexing point*. This is a special point in time of the time series. It is the base point for all percent calculations. The index value for the point k is thus calculated via the formula described above. v_{ip} is the value of the indexing point and represents 100%. v_k is the original value of the time series. By using this method, all displayed time-series values use the same percent dimension, which makes heterogeneous time series far easier to compare. For example, the time series can be drawn in superimposition without any arbitrary choice of scales and ranges of the different axes dimensions. One of the two main benefits of indexing is the ability to superimpose any data by the transformation of values into a percent dimension. The other benefit is the user-defined setting of an indexing point. This makes comparisons more effective and precise. A study by Aigner et al. (2011) gathered empirical evidence showing that using indexing in general yields a higher correctness rate than the two other visualization types linear scale with juxtaposition and log scale with superimposition. With regard to task completion times, the results are less clear and only slight advantages for indexing were found.

Summary

In the previous paragraphs, we discussed the influence of the task at hand on the visualization of time-oriented data. The examples of color-coding, plot arrangement, axes scaling, and indexing served to demonstrate how the task can be taken into account in the visualization process. The figures in this section showed that visualizing the same data using different mapping strategies leads to visual representations that are quite different from each other. Hence, it is important to consider the tasks of users in the visualization design to come up with effective visual tools for supporting them. As we have seen, visualization results can be improved when distinguishing between the following major task categories:

- Elementary vs. synoptic analytic questions,
- Identification vs. location, and
- Lookup vs. comparison.

However, still more research is required to investigate new methods of task-orientation, especially with regard to the wide range of visualization options. An example of such work with a particular focus on time-series visualization is the research of Albers et al. (2014), where different color and positional encodings have been compared. In their study, the authors confirm that different designs support different tasks. In particular, they show that positional encodings are better suited for elementary location tasks (i.e., locating minima, maxima, and ranges) while color encodings are better suited for synoptic location tasks (i.e., involving visual aggregations such as average, spread, and outliers).

4.2.3 Presentation Level

Finally, there are design issues at the level of the visual representation. Communicating the time-dependence of data primarily requires a well-considered placement of the time axis. This will make it easier for users to associate data with a particular time, and vice versa. In Section 4.1.3, we have differentiated between 2D and 3D presentations of time-oriented data. Let us take up this distinction as an example of a design decision to be made at the level of the visual representation. Visualization approaches that use a 2D presentation space have to ensure that the time axis is emphasized because time and data dimensions often have to share the two available display dimensions. In the case of 3D representations, a third display dimension is allocatable. In fact, many techniques utilize it as a dedicated dimension for the time axis, clearly separating time from other (data) dimensions. In the following, we will illustrate the 2D and the 3D approach with two examples.

2D Presentation of Time-Oriented Data

We discuss the presentation of time-oriented data in 2D by the example of axes-based visualizations. Axes-based visualization techniques are a widely used approach to represent multidimensional datasets in 2D (see Claessen and van Wijk, 2011). The basic idea is to construct a visual axis for each variable of the n-variate data space and to scale the axes with respect to the corresponding value range. Then, a suitable layout of the visual axes on the display has to be found. Finally, the data representation is realized by placing additional visual objects along the visual axes and in accord with the data. In this way, a lossless projection of the n-variate data space onto the 2-dimensional screen space can be accomplished. *Parallel coordinates* by Inselberg and Dimsdale (1990) are a well-known example of this approach. As shown in Figure 4.20, parallel coordinates use equidistant and parallel axes to represent multiple variables, and each data tuple is represented by a polygonal line linking the corresponding variable values. In the case of time-oriented data, however, this means that the axis encoding time is considered as one of many, not taking into account the outstanding importance of this axis.

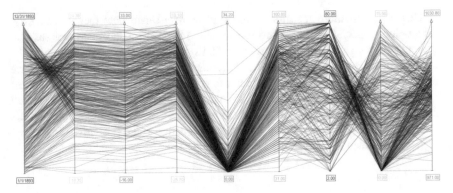

Fig. 4.20: In parallel coordinates, the time axis (leftmost) is just one of many axes. The importance of time is not particularly emphasized. ⓒ① *The authors.*

In contrast, Tominski et al. (2004) describe an axes-based visualization called *TimeWheel*, which focuses on one specific axis of interest, in our case the time axis (↪ p. 298). The basic idea of the TimeWheel technique is to distinguish between one independent variable, in our case time, and multiple dependent variables representing the time-oriented data. Figure 4.21 illustrates the design. The dimension of time is presented by the reference time axis in the center of the display and time-dependent

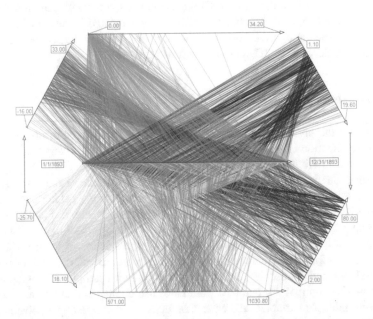

Fig. 4.21: The TimeWheel shows the reference time axis in a prominent central position and arranges data axes representing time-dependent variables around the time axis. Data are visualized by drawing lines between points at the time axis and values at the data axes. ⓒ① *The authors.*

variables are shown as data axes that are circularly arranged around the time axis, where each dependent variable has a specific color hue associated with it. For each time value on the time axis, colored lines are drawn that connect the time value with the corresponding data value at each of the data axes, effectively establishing a visual link between time and multivariate data. By doing so, the time dependency of all variables can be visualized. Note that the interrelation of time values and data values of a variable can be explored most efficiently when a data axis is parallel to the time axis. Interactive rotation of the TimeWheel can be used to move data axes of interest into such a parallel position.

Two additional visual cues support data interpretation and guide the viewer's attention: color fading and length adjustment. Color fading is applied to attenuate lines drawn from the time axis to axes that are almost perpendicular to the time axis. During rotation, lines gradually fade out and eventually become invisible when the associated data axis approaches an upright orientation. To provide more display space for the data variables of interest, the length of the data axes is adjusted according to their angle to the time axis. When the TimeWheel is rotated, data axes that are going to become parallel to the time axis are stretched to make them longer and data axes that head for an upright orientation are shrunk to make them shorter. Figure 4.21 shows a TimeWheel that visualizes eight time-dependent variables, where color fading and length adjustment have been applied to focus on the orange and the green data axes.

The TimeWheel is an example of a 2D visualization technique that acknowledges the important role of the time axis. The time axis' central position emphasizes the temporal character of the data and additional visual cues support interactive analysis and exploration of multiple time-dependent data variables.

3D Presentation of Time-Oriented Data

3D presentation spaces provide a third display dimension. This opens the door to additional possibilities of encoding time and time-oriented data. Particularly, the visualization of data that have further independent variables in addition to the dimension of time can benefit from the additional dimension of the display space.

Spatio-temporal data are an example where data variables do not only depend on time, but also on space (e.g., on points given by longitude and latitude or on geographic regions). When visualizing such data, the temporal frame of reference as well as the spatial frame of reference have to be represented. For this purpose, the *space-time cube* design (see Kraak (2003) and ↪ p. 377) can be applied: The z-axis of the display space exclusively encodes time, while the x- and y-axes represent spatial dimensions. Spatio-temporal data are then encoded by embedding visual objects into the space-time cube (e.g., visual markers or icons) and by mapping data to visual attributes (e.g., color or texture). Kristensson et al. (2009) provide evidence that space-time cube representations can facilitate intuitive recognition and interpretation of data in their spatio-temporal context.

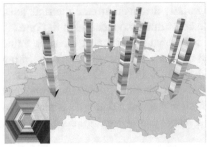

(a) Pencil icons for linear time.   (b) Helix icons for cyclic time.

Fig. 4.22: 3D visualization of spatio-temporal data using color-coded icons embedded into a map display. ⓒⓘ *The authors.*

Figure 4.22 shows two examples of this approach as described by Tominski et al. (2005b). Figure 4.22a represents multiple time-dependent variables by so-called pencil icons (↪ p. 386). The linear time axis is encoded along the pencil's faces starting from the tip. Each face of the pencil is associated with a specific data variable and a specific color hue and represents the corresponding data values by varying color saturation. Figure 4.22b uses so-called helix icons (↪ p. 389). Here, we assume a cyclic character of time and thus, a ribbon is constructed along a spiral helix. For each time step, the ribbon extends in angle and height, depending on the number of time elements per helix cycle and the number of cyclic passes. Again color-coding is used to encode the data values. To represent more than one data variable, the ribbon can be subdivided into narrower sub-ribbons.

The embedded 3D icons are suited for visualizing data that are anchored at certain points in space. When the goal is to understand spatio-temporal data along paths, one can use a different visualization. The *great wall of space-time* (↪ p. 369) by Tominski and Schulz (2012) provides a dedicated path-oriented 3D representation. An example is shown in Figure 4.23. The construction of the wall is based on (1) defining a topological path through the neighborhood graph of the map, (2) deriving a geometric path based on the map geometry, (3) extruding the geometric path to create a 3D wall above the map, and (4) projecting a visualization of the data associated with the defined path onto the wall. In the figure, the wall shows a color-coded matrix, where rows correspond to time steps and columns correspond to the different map areas along the path.

The 3D display space used in the previous examples is advantageous in terms of the prominent encoding of time, but it also exhibits two problems that need to be addressed: perspective distortions and occlusion (see Section 4.1.3). Perspective distortions are problematic because they could impair the interpretation of the visualized data. Therefore, the visual mapping should avoid or reduce the use of geometric visual attributes that are subject to perspective projections (e.g., shape, size, or orientation). This is the reason why the given examples apply color-coding instead of geometric encoding. The occlusion aspect has to be addressed by addi-

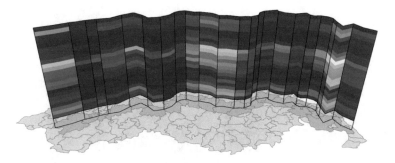

Fig. 4.23: The great wall of space-time visualizes spatio-temporal data along a path through the map. The color-coded matrix projected onto the wall represents health data. ©① *The authors.*

tional mechanisms. For example, users should be allowed to rotate the icons or the whole map in order to make back faces visible. Another option is to incorporate additional 2D views that do not suffer from occlusion. Such views are shown for a user-selected region of interest in the bottom-left corner of Figures 4.22a and 4.22b. Similar 2D views could be generated from the matrix-based wall representation in Figure 4.23. Again this approach is a compromise. On the one hand, the 2D views are occlusion-free, but on the other hand, one can show only a limited number of additional views, and moreover, one unlinks the data from their spatial point or path of reference.

Irrespective of whether one uses a 2D or 3D representation, the visualization design for time-oriented data requires a special handling of the time axis to effectively communicate the time-dependence of the data. Both approaches have to take care to emphasize the dimension of time among other data dimensions.

4.3 Summary

Solving the visualization problem primarily requires answering the three questions: (1) What is visualized? (2) Why is it visualized? (3) How is it visualized? The answers to the first two questions determine the answer to the third question.

In the case of visualizing time-oriented data, answering the what-question requires both specifying the characteristics of the time domain as well as specifying the characteristics of the data associated with time. In Chapter 3, we have shown that many different aspects characterize time and time-oriented data. It is virtually impossible to simultaneously cover all of them within a single visualization process. On top of this, there exists no visualization technique that is capable of handling all of the different aspects simultaneously and presenting all of them in an appropriate way. Here, the answer to "why are we visualizing the data?" comes into play. Those aspects of the data that are of specific interest with regard to the tasks at hand have to be communicated by the visual representation, while others can be diminished or

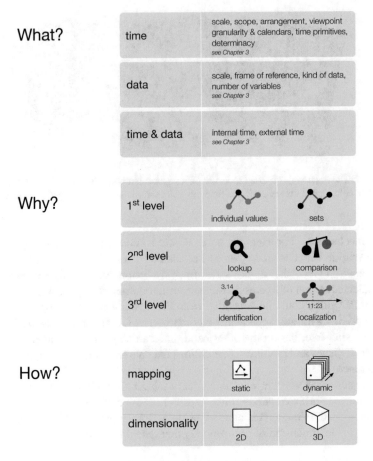

Fig. 4.24: Three key questions of the visualization problem.

even omitted. However, this is an intricate problem, since most visualization systems do not support the process of generating suitable task-specific visual representations. Thus, our primary aim can only be to communicate the problem, and also to demonstrate the necessity and potential of considering the interrelation between the what, why, and how aspects by example, as we have done in Section 4.2.

Figure 4.24 again summarizes the key characteristics of the three aspects. The *what* aspect addresses characteristics of time and data as detailed in Chapter 3. For describing the *why* aspect, we considered typical analytical questions as described in Section 4.1.2. The *how* aspect is mainly categorized by the differentiation of static and dynamic as well as 2D and 3D representations (see Section 4.1.3).

We will see that there are a variety of techniques for handling and accounting for these key characteristics. Accordingly, many different visual representations of time-oriented data can be generated. Appendix A attests to this statement.

References

Aigner, W., C. Kainz, R. Ma, and S. Miksch (2011). "Bertin Was Right: An Empirical Evaluation of Indexing to Compare Multivariate Time-Series Data Using Line Plots". In: *Computer Graphics Forum* 30.1, pp. 215–228. DOI: 10.1111/j.1467-8659.2010.01845.x.

Aigner, W., S. Miksch, W. Müller, H. Schumann, and C. Tominski (2007). "Visualizing Time-Oriented Data – A Systematic View". In: *Computers & Graphics* 31.3, pp. 401–409. DOI: 10.1016/j.cag.2007.01.030.

Aigner, W., S. Miksch, W. Müller, H. Schumann, and C. Tominski (2008). "Visual Methods for Analyzing Time-Oriented Data". In: *IEEE Transactions on Visualization and Computer Graphics* 14.1, pp. 47–60. DOI: 10.1109/TVCG.2007.70415.

Aigner, W., S. Miksch, B. Thurnher, and S. Biffl (2005). "PlanningLines: Novel Glyphs for Representing Temporal Uncertainties and Their Evaluation". In: *Proceedings of the International Conference Information Visualisation (IV)*. IEEE Computer Society, pp. 457–463. DOI: 10.1109/IV.2005.97.

Albers, D., M. Correll, and M. Gleicher (2014). "Task-Driven Evaluation of Aggregation in Time Series Visualization". In: *Proceedings of the SIGCHI Conference on Human Factors in Computing Systems (CHI)*. ACM Press, pp. 551–560. DOI: 10.1145/2556288.2557200.

Andrienko, N. and G. Andrienko (2006). *Exploratory Analysis of Spatial and Temporal Data*. Springer. DOI: 10.1007/3-540-31190-4.

Bach, B., P. Dragicevic, D. W. Archambault, C. Hurter, and S. Carpendale (2017). "A Descriptive Framework for Temporal Data Visualizations Based on Generalized Space-Time Cubes". In: *Computer Graphics Forum* 36.6, pp. 36–61. DOI: 10.1111/cgf.12804.

Bai, Z., Y. Tao, and H. Lin (2020). "Time-varying Volume Visualization: A Survey". In: *Journal of Visualization* 23.5, pp. 745–761. DOI: 10.1007/s12650-020-00654-x.

Beck, F., M. Burch, S. Diehl, and D. Weikopf (2017). "A Taxonomy and Survey of Dynamic Graph Visualization". In: *Computer Graphics Forum* 36.1, pp. 133–159. DOI: 10.1111/cgf.12791.

Been, K., E. Daiches, and C.-K. Yap (2006). "Dynamic Map Labeling". In: *IEEE Transactions on Visualization and Computer Graphics* 12.5, pp. 773–780. DOI: 10.1109/TVCG.2006.136.

Bergman, L., B. E. Rogowitz, and L. A. Treinish (1995). "A Rule-based Tool for Assisting Colormap Selection". In: *Proceedings of the IEEE Visualization Conference (Vis)*. IEEE Computer Society, pp. 118–125. DOI: 10.1109/VISUAL.1995.480803.

Bernard, J., M. Steiger, S. Mittelstädt, S. Thum, D. A. Keim, and J. Kohlhammer (2015). "A Survey and Task-based Quality Assessment of Static 2D Colormaps". In: *Proceedings of the Conference on Visualization and Data Analysis (VDA)*. Vol. 9397. SPIE Proceedings. SPIE. DOI: 10.1117/12.2079841.

Bertin, J. (1983). *Semiology of Graphics: Diagrams, Networks, Maps*. Translated by William J. Berg. University of Wisconsin Press.

Borland, D. and R. M. Taylor (2007). "Rainbow Color Map (Still) Considered Harmful". In: *IEEE Computer Graphics and Applications* 27.2, pp. 14–17. DOI: 10.1109/mcg.2007.323435.

Bors, C., C. Eichner, S. Miksch, C. Tominski, H. Schumann, and T. Gschwandtner (2020). "Exploring Time Series Segmentations Using Uncertainty and Focus+Context Techniques". In: *Proceedings of the Eurographics / IEEE Conference on Visualization (EuroVis) - Short Papers*. Eurographics Association, pp. 7–11. DOI: 10.2312/evs.20201040.

Card, S., J. Mackinlay, and B. Shneiderman (1999). *Readings in Information Visualization: Using Vision to Think*. Morgan Kaufmann Publishers.

Ceneda, D., T. Gschwandtner, S. Miksch, and C. Tominski (2018). "Guided Visual Exploration of Cyclical Patterns in Time-series". In: *Proceedings of the IEEE Symposium on Visualization in Data Science (VDS)*. IEEE Computer Society.

Claessen, J. H. T. and J. J. van Wijk (2011). "Flexible Linked Axes for Multivariate Data Visualization". In: *IEEE Transactions on Visualization and Computer Graphics* 17.12, pp. 2310–2316. DOI: 10.1109/TVCG.2011.201.

Cleveland, W. S., M. E. McGill, and R. McGill (1988). "The Shape Parameter of a Two-Variable Graph". In: *Journal of the American Statistical Association* 83.402, pp. 289–300. DOI: 10.1080/01621459.1988.10478598.

Constantine, L. L. (2003). "Canonical Abstract Prototypes for Abstract Visual and Interaction Design". In: *Interactive Systems: Design, Specification, and Verification*. Edited by Jorge, J., Nunes, N. J., and e Cunha, J. F. Vol. 2844. Lecture Notes in Computer Science. Springer, pp. 1–15. DOI: 10.1007/978-3-540-39929-2_1.

Courage, C. and K. Baxter (2005). *Understanding Your Users*. Morgan Kaufmann. DOI: 10.1016/B978-1-55860-935-8.X5029-5.

Daassi, C., L. Nigay, and M.-C. Fauvet (2005). "A Taxonomy of Temporal Data Visualization Techniques". In: *Interaction Information Intelligence* 5.2, pp. 41–63. URL: https://www.irit.fr/journal-i3/volume05/numero02/revue_i3_05_02_02.pdf.

Draper, G. M., Y. Livnat, and R. F. Riesenfeld (2009). "A Survey of Radial Methods for Information Visualization". In: *IEEE Transactions on Visualization and Computer Graphics* 15.5, pp. 759–776. DOI: 10.1109/TVCG.2009.23.

Dübel, S., M. Röhlig, H. Schumann, and M. Trapp (2014). "2D and 3D Presentation of Spatial Data: A Systematic Review". In: *Proceedings of the International Workshop on 3DVis (3DVis@IEEE VIS)*. IEEE Computer Society, pp. 11–18. DOI: 10.1109/3DVis.2014.7160094.

Elmqvist, N. and P. Tsigas (2007). "A Taxonomy of 3D Occlusion Management Techniques". In: *Proceedings of the IEEE Conference on Virtual Reality (VR)*. IEEE Computer Society, pp. 51–58. DOI: 10.1109/vr.2007.352463.

Fang, Y., H. Xu, and J. Jiang (2020). "A Survey of Time Series Data Visualization Research". In: *IOP Conference Series: Materials Science and Engineering* 782. DOI: 10.1088/1757-899x/782/2/022013.

Farquhar, A. B. and H. Farquhar (1891). *Economic and Industrial Solutions*. New York, NY: G. B. Putnam's Sons.

Fuchs, G. and H. Schumann (2004a). "Intelligent Icon Positioning for Interactive Map-Based Information Systems". In: *Proceedings of the International Conference of the Information Resources Management Association (IRMA)*. Idea Group Inc., pp. 261–264. URL: https://www.irma-international.org/proceeding-paper/intelligent-icon-positioning-interactive-map/32349/.

Fuchs, J., F. Fischer, F. Mansmann, E. Bertini, and P. Isenberg (2013). "Evaluation of Alternative Glyph Designs for Time Series Data in a Small Multiple Setting". In: *Proceedings of the SIGCHI Conference on Human Factors in Computing Systems (CHI)*. ACM Press, pp. 3237–3246. DOI: 10.1145/2470654.2466443.

Gapminder Foundation (2021). *Gapminder Tools*. URL: https://www.gapminder.org/tools.

Gschwandtner, T., M. Bögl, P. Federico, and S. Miksch (2016). "Visual Encodings of Temporal Uncertainty: A Comparative User Study". In: *IEEE Transactions on Visualization and Computer Graphics* 22.1, pp. 539–548. DOI: 10.1109/TVCG.2015.2467752.

Hackos, J. T. and J. C. Redish (1998). *User and Task Analysis for Interface Design*. John Wiley & Sons, Inc.

Hall, K. W., C. Perin, P. G. Kusalik, C. Gutwin, and S. Carpendale (2014). "Formalizing Emphasis in Information Visualization". In: *Computer Graphics Forum* 35.3, pp. 717–737. DOI: 10.1111/cgf.12936.

Harris, R. L. (1999). *Information Graphics: A Comprehensive Illustrated Reference*. Oxford University Press. URL: https://global.oup.com/academic/product/information-graphics-9780195135329.

Harrower, M. A. and C. A. Brewer (2003). "ColorBrewer.org: An Online Tool for Selecting Color Schemes for Maps". In: *The Cartographic Journal* 40.1, pp. 27–37. DOI: 10.4324/9781351191234-18.

Havre, S., E. Hetzler, and L. Nowell (2000). "ThemeRiver: Visualizing Theme Changes Over Time". In: *Proceedings of the IEEE Symposium Information Visualization (InfoVis)*. IEEE Computer Society, pp. 115–124. DOI: 10.1109/INFVIS.2000.885098.

Havre, S., E. Hetzler, P. Whitney, and L. Nowell (2002). "ThemeRiver: Visualizing Thematic Changes in Large Document Collections". In: *IEEE Transactions on Visualization and Computer Graphics* 8.1, pp. 9–20. DOI: 10.1109/2945.981848.

Healey, C. G. and J. T. Enns (2012). "Attention and Visual Memory in Visualization and Computer Graphics". In: *IEEE Transactions on Visualization and Computer Graphics* 18.7, pp. 1170–1188. DOI: 10.1109/TVCG.2011.127.

Heer, J. and M. Agrawala (2006). "Multi-Scale Banking to 45 Degrees". In: *IEEE Transactions on Visualization and Computer Graphics* 12.5, pp. 701–708. DOI: 10.1109/TVCG.2006.163.

Inselberg, A. and B. Dimsdale (1990). "Parallel Coordinates: A Tool for Visualizing Multi-Dimensional Geometry". In: *Proceedings of the IEEE Visualization Conference (Vis)*. IEEE Computer Society, pp. 361–378. DOI: 10.1109/VISUAL.1990.146402.

Jabbari, A., R. Blanch, and S. Dupuy-Chessa (2018). "Composite Visual Mapping for Time Series Visualization". In: *Proceedings of the IEEE Pacific Visualization Symposium (PacificVis)*. IEEE, pp. 116–124. DOI: 10.1109/PacificVis.2018.00023.

Knight, D., B. Knight, M. Pearson, and M. Quintana (2018). *Microsoft Power BI Quick Start Guide: Build Dashboards and Visualizations to Make Your Data Come to Life*. 1st edition. Packt Publishing. URL: https://www.packtpub.com/product/microsoft-power-bi-quick-start-guide/9781789138221.

Kraak, M.-J. (2003). "The Space-Time Cube Revisited from a Geovisualization Perspective". In: *Proceedings of the 21st International Cartographic Conference (ICC)*. The International Cartographic Association (ICA), pp. 1988–1996. URL: https://icaci.org/files/documents/ICC_proceedings/ICC2003/Papers/255.pdf.

Kraus, M., K. Klein, J. Fuchs, D. A. Keim, F. Schreiber, M. Sedlmair, and T.-M. Rhyne (2021). "The Value of Immersive Visualization". In: *IEEE Computer Graphics and Applications* 41.4, pp. 125–132. DOI: 10.1109/MCG.2021.3075258.

Kristensson, P. O., N. Dahlback, D. Anundi, M. Bjornstad, H. Gillberg, J. Haraldsson, I. Martensson, M. Nordvall, and J. Stahl (2009). "An Evaluation of Space Time Cube Representation of Spatiotemporal Patterns". In: *IEEE Transactions on Visualization and Computer Graphics* 15.4, pp. 696–702. DOI: 10.1109/TVCG.2008.194.

Loth, A. (2019). *Visual Analytics with Tableau*. Wiley. DOI: 10.1002/9781119561996.

Luboschik, M., H. Schumann, and H. Cords (2008). "Particle-Based Labeling: Fast Point-feature Labeling Without Obscuring Other Visual Features". In: *IEEE Transactions on Visualization and Computer Graphics* 14.6, pp. 1237–1244. DOI: 10.1109/tvcg.2008.152.

MacEachren, A. M. (1995). *How Maps Work: Representation, Visualization, and Design*. Guilford Press.

Mackinlay, J. (1986). "Automating the Design of Graphical Presentations of Relational Information". In: *ACM Transactions on Graphics* 5.2, pp. 110–141. DOI: 10.1145/22949.22950.

Mairena, A., C. Gutwin, and A. Cockburn (2022). "Which Emphasis Technique to Use? Perception of Emphasis Techniques with Varying Distractors, Backgrounds, and Visualization Types". In: *Information Visualization* 21.2, pp. 95–129. DOI: 10.1177/14738716211045354.

Marriott, K., F. Schreiber, T. Dwyer, K. Klein, N. H. Riche, T. Itoh, W. Stuerzlinger, and B. H. Thomas, eds. (2018). *Immersive Analytics*. Vol. 11190. Lecture Notes in Computer Science. Springer. DOI: 10.1007/978-3-030-01388-2.

McNabb, L. and R. S. Laramee (2019). "Multivariate Maps – A Glyph-Placement Algorithm to Support Multivariate Geospatial Visualization". In: *Information* 10.10. DOI: 10.3390/info10100302.

Mittelstädt, S., D. Jäckle, F. Stoffel, and D. A. Keim (2015). "ColorCAT: Guided Design of Colormaps for Combined Analysis Tasks". In: *Proceedings of the Eu-*

rographics / IEEE Conference on Visualization (EuroVis) - Short Papers. Eurographics Association, pp. 115–119. DOI: 10.2312/eurovisshort.20151135.

Mittelstädt, S., A. Stoffel, and D. A. Keim (2014). "Methods for Compensating Contrast Effects in Information Visualization". In: *Computer Graphics Forum* 33.3, pp. 231–240. DOI: 10.1111/cgf.12379.

Müller, W. and H. Schumann (2003). "Visualization Methods for Time-Dependent Data - An Overview". In: *Proceedings of Winter Simulation Conference (WSC)*. IEEE Computer Society, pp. 737–745. DOI: 10.1109/WSC.2003.1261490.

Munzner, T. (2009). "A Nested Process Model for Visualization Design and Validation". In: *IEEE Transactions on Visualization and Computer Graphics* 15.6, pp. 921–928. DOI: 10.1109/TVCG.2009.111.

Munzner, T. (2014). *Visualization Analysis and Design*. A K Peters/CRC Press. DOI: 10.1201/b17511.

Nardini, P., M. Chen, F. Samsel, R. Bujack, M. Böttinger, and G. Scheuermann (2021). "The Making of Continuous Colormaps". In: *IEEE Transactions on Visualization and Computer Graphics* 27.6, pp. 3048–3063. DOI: 10.1109/TVCG.2019.2961674.

Paternò, F., C. Mancini, and S. Meniconi (1997). "ConcurTaskTrees: A Diagrammatic Notation for Specifying Task Models". In: *Proceedings of IFIP TC13 International Conference on Human-Computer Interaction (INTERACT)*. Springer, pp. 362–369. DOI: 10.1007/978-0-387-35175-9_58.

Perin, C., T. Wun, R. Pusch, and S. Carpendale (2018). "Assessing the Graphical Perception of Time and Speed on 2D+Time Trajectories". In: *IEEE Transactions on Visualization and Computer Graphics* 24.1, pp. 698–708. DOI: 10.1109/TVCG.2017.2743918.

Petzold, I. (2003). "Beschriftung von Bildschirmkarten in Echtzeit". PhD thesis. Rheinische Friedrich-Wilhelms-Universität Bonn. URL: https://hdl.handle.net/20.500.11811/1870.

Reinders, F., F. H. Post, and H. J. W. Spoelder (2001). "Visualization of Time-Dependent Data with Feature Tracking and Event Detection". In: *The Visual Computer* 17.1, pp. 55–71. DOI: 10.1007/pl00013399.

Robertson, G., R. Fernandez, D. Fisher, B. Lee, and J. Stasko (2008). "Effectiveness of Animation in Trend Visualization". In: *IEEE Transactions on Visualization and Computer Graphics* 14.6, pp. 1325–1332. DOI: 10.1109/TVCG.2008.125.

Röhlig, M., M. Luboschik, and H. Schumann (2017). "Visibility Widgets for Unveiling Occluded Data in 3D Terrain Visualization". In: *Journal of Visual Languages & Computing* 42, pp. 86–98. DOI: 10.1016/j.jvlc.2017.08.008.

Schulze-Wollgast, P., C. Tominski, and H. Schumann (2005). "Enhancing Visual Exploration by Appropriate Color Coding". In: *Proceedings of the International Conference in Central Europe on Computer Graphics, Visualization and Computer Vision (WSCG)*. University of West Bohemia, pp. 203–210.

Silva, S., J. Madeira, and B. S. Santos (2007). "There is More to Color Scales than Meets the Eye: A Review on the Use of Color in Visualization". In: *Proceedings of the International Conference Information Visualisation (IV)*. IEEE Computer Society, pp. 943–950. DOI: 10.1109/iv.2007.113.

Silva, S., B. S. Santos, and J. Madeira (2011). "Using Color in Visualization: A Survey". In: *Computers & Graphics* 35.2, pp. 320–333. DOI: 10.1016/j.cag.2010.11.015.

Silva, S. F. and T. Catarci (2000). "Visualization of Linear Time-Oriented Data: A Survey". In: *Proceedings of the International Conference on Web Information Systems Engineering (WISE)*. IEEE Computer Society, pp. 310–319. DOI: 10.1109/WISE.2000.882407.

Simons, D. J. and R. A. Rensink (2005). "Change Blindness: Past, Present, and Future". In: *Trends in Cognitive Sciences* 9.1, pp. 16–20. DOI: 10.1016/j.tics.2004.11.006.

Talbot, J., J. Gerth, and P. Hanrahan (2012). "An Empirical Model of Slope Ratio Comparisons". In: *IEEE Transactions on Visualization and Computer Graphics* 18.12, pp. 2613–2620. DOI: 10.1109/TVCG.2012.196.

Telea, A. C. (2014). *Data Visualization: Principles and Practice*. 2nd edition. A K Peters/CRC Press. DOI: 10.1201/b17217.

Thompson, J. R., Z. Liu, W. Li, and J. Stasko (2020). "Understanding the Design Space and Authoring Paradigms for Animated Data Graphics". In: *Computer Graphics Forum* 39.3, pp. 207–218. DOI: 10.1111/cgf.13974.

Tominski, C., J. Abello, and H. Schumann (2004). "Axes-Based Visualizations with Radial Layouts". In: *Proceedings of the ACM Symposium on Applied Computing (SAC)*. ACM Press, pp. 1242–1247. DOI: 10.1145/967900.968153.

Tominski, C., G. Fuchs, and H. Schumann (2008). "Task-Driven Color Coding". In: *Proceedings of the International Conference Information Visualisation (IV)*. IEEE Computer Society, pp. 373–380. DOI: 10.1109/IV.2008.24.

Tominski, C. and H.-J. Schulz (2012). "The Great Wall of Space-Time". In: *Proceedings of the Workshop on Vision, Modeling & Visualization (VMV)*. Eurographics Association, pp. 199–206. DOI: 10.2312/PE/VMV/VMV12/199-206.

Tominski, C., P. Schulze-Wollgast, and H. Schumann (2005b). "3D Information Visualization for Time Dependent Data on Maps". In: *Proceedings of the International Conference Information Visualisation (IV)*. IEEE Computer Society, pp. 175–181. DOI: 10.1109/IV.2005.3.

Tominski, C. and H. Schumann (2020). *Interactive Visual Data Analysis*. AK Peters Visualization Series. CRC Press. DOI: 10.1201/9781315152707.

Tufte, E. R. (1983). *The Visual Display of Quantitative Information*. Graphics Press. URL: https://www.edwardtufte.com/tufte/books_vdqi.

Tversky, B., J. B. Morrison, and M. Betrancourt (2002). "Animation: Can It Facilitate?" In: *International Journal of Human-Computer Studies* 57.4, pp. 247–262. DOI: 10.1006/ijhc.2002.1017.

Unger, A. and H. Schumann (2009). "Visual Support for the Understanding of Simulation Processes". In: *Proceedings of the IEEE Pacific Visualization Symposium (PacificVis)*. IEEE Computer Society, pp. 57–64. DOI: 10.1109/PACIFICVIS.2009.4906838.

Vande Moere, A. (2004). "Time-Varying Data Visualization Using Information Flocking Boids". In: *Proceedings of the IEEE Symposium Information Visual-*

ization (InfoVis). IEEE Computer Society, pp. 97–104. DOI: 10.1109/INFVIS.2004.65.

Waldner, M., A. Karimov, and M. E. Gröller (2017). "Exploring Visual Prominence of Multi-channel Highlighting in Visualizations". In: *Proceedings of the Spring Conference on Computer Graphics (SCCG)*. ACM Press, 8:1–8:10. DOI: 10.1145/3154353.3154369.

Weber, M., M. Alexa, and W. Müller (2001). "Visualizing Time-Series on Spirals". In: *Proceedings of the IEEE Symposium Information Visualization (InfoVis)*. IEEE Computer Society, pp. 7–14. DOI: 10.1109/INFVIS.2001.963273.

Weiss, D. J., A. Nelson, H. S. Gibson, W. Temperley, S. Peedell, A. Lieber, M. Hancher, E. Poyart, S. Belchior, N. Fullman, B. Mappin, U. Dalrymple, J. Rozier, T. C. D. Lucas, R. E. Howes, L. S. Tusting, S. Y. Kang, E. Cameron, D. Bisanzio, K. E. Battle, S. Bhatt, and P. W. Gething (2018a). "A Global Map of Travel Time to Cities to Assess Inequalities in Accessibility in 2015". In: *Nature* 553.7688, pp. 333–336. DOI: 10.1038/nature25181.

Weiss, D., H. Gibson, U. Dalrymple, J. Rozier, T. Lucas, R. Howes, L. Tusting, S. Kang, E. Cameron, K. Battle, S. Bhatt, and P. Gething (2018b). *Accessibility to Cities*. https://malariaatlas.org/research-project/accessibility-to-cities/. URL: https://malariaatlas.org/wp-content/uploads/2017/12/MAP_Accessibility_To_Cities_Press_Release.zip.

Wills, G. (2012). *Visualizing Time - Designing Graphical Representations for Statistical Data*. Springer. DOI: 10.1007/978-0-387-77907-2.

Wolter, M., I. Assenmacher, B. Hentschel, M. Schirski, and T. Kuhlen (2009). "A Time Model for Time-Varying Visualization". In: *Computer Graphics Forum* 28.6, pp. 1561–1571. DOI: 10.1111/j.1467-8659.2008.01314.x.

Yang, J., W. Wang, and P. S. Yu (2000). "Mining Asynchronous Periodic Patterns in Time Series Data". In: *Proceedings of the ACM SIGKDD International Conference on Knowledge Discovery and Data Mining (KDD)*. ACM Press, pp. 275–279. DOI: 10.1145/347090.347150.

Zhou, L. and C. D. Hansen (2016). "A Survey of Colormaps in Visualization". In: *IEEE Transactions on Visualization and Computer Graphics* 22.8, pp. 2051–2069. DOI: 10.1109/TVCG.2015.2489649.

Chapter 5
Involving the Human via Interaction

> A graphic is not "drawn" once and for all; it is "constructed" and reconstructed until it reveals all the relationships constituted by the interplay of the data. The best graphic operations are those carried out by the decision-maker himself.
>
> Bertin (1981, p. 16)

The previous chapter discussed diverse options for designing visual representations that help people understand time and time-oriented data. 'Seeing' trends, correlations, and patterns in a visual representation is indeed a powerful way for people to extract knowledge from data. Yet, 'seeing' alone is not sufficient, or as Thomas and Cook (2005) put it:

> Visual representations alone cannot satisfy analytical needs. Interaction techniques are required to support the dialogue between the analyst and the data.
>
> Thomas and Cook (2005, p. 30)

From the previous chapter, we know that various aspects are involved when creating a visual representation: the characteristics of time and data, the user tasks, as well as the choice and the parametrization of visualization techniques. As a consequence, a generated visual representation might not yield the desired outcome, particularly when feeding unknown data into a visualization method. A related problem is that we sometimes do not know exactly what to expect from a visual representation or whether it is effective with regard to the task to be accomplished. One way to deal with this problem is to include the human user into the loop. So, visual exploration and analysis is not a one-way street where data are transformed into images, but it is in fact a human-in-the-loop process controlled and manipulated by one or more users.

Having said that, it becomes clear that in addition to visual methods, a high degree of interactivity and advanced interaction techniques for working with time-oriented data are important. Interaction helps users not only see the data but also understand them. By interacting, users can comprehend the visual mapping, realize the effect of visualization parameters, carve out hidden patterns, and become confident about

W. Aigner et al., *Visualization of Time-Oriented Data*, Human–Computer Interaction Series, https://doi.org/10.1007/978-1-4471-7527-8_5

the visualization and its underlying data. Users want and many times need to get their hands on their data – which is particularly true when engaging in exploratory data analyses. The importance of interaction is nicely summarized in the following statement:

> While visual representations may provoke curiosity,
> interaction provides the means to satisfy it.
>
> Tominski and Schumann (2020, p. 132)

While visualization research is naturally more focused on the visual output, the interactive operations involved in carrying out data analyses must also be considered. This chapter provides an overview of how interactivity can support the exploration of time and time-oriented data. For a deeper discussion of interaction for visualization in general, the interested reader is referred to Tominski (2015).

5.1 Motivation & User Intents

The constantly increasing size and complexity of today's datasets are major challenges for interactive visualization. Large datasets cannot simply be loaded to limited computer memory and then be mapped to an even smaller display. Users are only able to digest a fraction of the available information at a time. Complex data contain many different aspects and may stem from heterogeneous sources. As complexity increases, so does the number of questions that one might ask about the data and to which visual representations should help us find answers.

In our particular case, we need to account for the specific aspects of time and time-oriented data in the context of what, why, and how they are visualized (see Chapters 3 and 4). Any attempt to indiscriminately encode all facets of a complex time-oriented dataset into a single visual representation is condemned to failure, as this would lead to a confusing and overloaded display that users can hardly interpret.

Instead, the big problem has to be split into smaller pieces by focusing on relevant data aspects and particular tasks per visual representation. Several benefits can be gained: computational costs are reduced in a kind of divide-and-conquer way, the visual representations become more effective because they are tailored to emphasize a particular point, and users find it easier to explore and analyze the data since they can concentrate on important and task-relevant questions.

Dividing the visualization problem and separating different aspects into individual views raise the question of how users can visually access and mentally combine these. The answer is *interaction*. In an iterative process, the user will focus on different parts of the data, look at them from alternative perspectives, and actively construct answers to diverse questions. Typically, this process follows the *visual information seeking mantra*:

> "Overview first,
> zoom and filter,
> then details-on-demand."
>
> Shneiderman (1996, p. 2)

Starting with an overview, the user will first identify interesting parts of the time domain to focus on for a more detailed examination. From there, it might make sense to move on to data that are related or similar, or it might be better to return to the overview and investigate the data from a different point of view, or with regard to a different question. In other words, the user forms a mental model of the data by interactively navigating from one focus to the next, where the focus may be any part of the time domain, a certain data aspect, or a specific analysis task. While exploring data in this way, users develop understanding and insight.

The general motivation for interaction is clear now. But what specifically motivates a user to interact? An answer to this question is given in a study by Yi et al. (2007), who worked toward a deeper understanding of interaction in visualization. As already briefly mentioned in Section 1.1, they identified several user intents for interaction and introduced a list of categories that describe on a high level why users want to or need to interact. In the following, we make use of these categories and adapt them to the case of interacting with time and time-oriented data:

Select – Mark something as interesting When users spot something interesting in the visual representation, they want to mark and visually highlight it as such, be it to temporarily tag an intriguing finding or to permanently memorize important analysis results. The pieces to be marked can be manifold: interesting points in time, an entire time-dependent variable, a particular temporal pattern, or certain identified events.

Explore – Show me something else Time-oriented data are often large and can be visualized only partially. That is, only a subset of time and a subset of the time-dependent variables are visible at a time. To arrive at a full view of the data, users have to explore different subsets of the data. This includes interactively navigating the time domain to bring different parts of it to the display, and also constructing different subsets of variables to uncover multivariate temporal dependencies.

Reconfigure – Show me a different arrangement Different arrangements of time and associated data can communicate completely different aspects, a fact which becomes obvious when recalling the distinction between linear and cyclic representations of time. As users want to look at time from different angles, they need to be provided with facilities that allow them to interactively generate different spatial arrangements of time-oriented data.

Encode – Show me a different representation Similarly to what was said about the spatial arrangement, the visual encoding of data values has a major impact on what can be derived from a visual representation. Because data and tasks are versatile, users need to be able to adapt the visual encoding to suit their needs, be it to carry out location or comparison tasks, or to confirm a hypothesis generated from one visual encoding by checking it against an alternative one.

Abstract/Elaborate – Show me more or less detail During a visual analysis, users need to look at certain things in detail, while for other things schematic representations are sufficient. The hierarchically structured levels of granularity of time are a natural match to drive such an interactive information drill-down into time-oriented data. Higher levels contain abstractions and provide aggregated overviews, whereas lower levels hold the increasingly elaborate details.

Filter – Show me something conditionally When users search for particular information in the data or evaluate a certain hypothesis about the data, it makes sense to restrict the visualization to show only those data items that match the conditions imposed by the search criteria or the hypothesis' constraints. Interactively filtering out or attenuating irrelevant data items clears the view for users to focus on those parts of the data being relevant to the task at hand.

Connect – Show me related items When users make a potentially interesting finding for one part of the data, they usually ask whether similar or related discoveries can be made in other parts of the data as well. So, users intend to interactively find, compare, and evaluate such similarities or relations. For example, for a trend discovered in one season of a certain year, it could be interesting to investigate if the trend is repeated at the same time in subsequent years.

These seven intents apply to interactive visualization, and we linked them specifically to interacting with time-oriented data. On top of that, Yi et al. (2007) mention two general interaction intents that are also relevant when exploring time.

Undo/Redo – Let me go to where I have already been Users have to navigate in time and study it at different levels of granularity, they have to try different arrangements and visual encodings, and they have to experiment with filtering conditions and similarity thresholds. A history mechanism for undoing and redoing interactions enables users to try out new views on the data and to return effortlessly to a previous visual representation if new ones did not work out as expected.

Change configuration – Let me adjust the interface In addition to adapting the visual representation to the data and the tasks at hand, it is also often necessary to configure the overall system that provides the visualization. This includes configuring not only the user interface (e.g., the arrangement of windows or the items in toolbars), but also the general management of system resources (e.g., the amount of memory to be used for undo and redo).

Taken together, the discussed intents represent on a high conceptual level what interactions a visualization system for time-oriented data should provide. For specific types of time-oriented data, additional interactions may be worth considering, such as faceting and warping for multivariate longitudinal data (see Cheng et al., 2016).

Many of the visualization approaches we describe in Appendix A support interaction of one kind or another. While marking (or selecting) interesting data items and navigation in time are quasi-mandatory, facilities for other intents are often rudimentary or not considered at all. This is often due to the extra effort one has to expend for designing and implementing effective interaction techniques. But in fact, all of the outlined user intents are equally important and corresponding techniques should be provided in order to take full advantage of the synergy of the human's skills in creative problem-solving and the machine's computational capabilities.

5.2 Interaction Fundamentals

Now that we know about the general motivation and the specific user intents behind interaction, we can move on and take a look at how interaction is actually performed. We will next describe fundamental aspects of interaction, which naturally are more general and less specific to interacting with time-oriented data.

5.2.1 Conceptual Background

Let us first look at aspects that concern interaction on a conceptual level, including how interaction can be modeled as a loop, what costs are involved when interacting, how interaction can be performed in a discrete or continues manner, and what the role of interaction latency is.

The interaction loop When users interact they express their intent to change what they see on the display, and they expect the visual representation to reflect the intended change. Consequently, Norman (2013) models interaction as a loop of two phases: an execution phase and an evaluation phase. The first phase subsumes steps that are related to the execution of interaction, including the intention to interact, the mental construction of an interaction plan, and the physical actions (e.g., pressing a button) to actually execute the plan. The second phase is related to understanding the system-generated visual feedback and involves perceiving and interpreting the feedback as well as evaluating the success of the interaction. Figure 5.1 illustrates Norman's conceptual model.

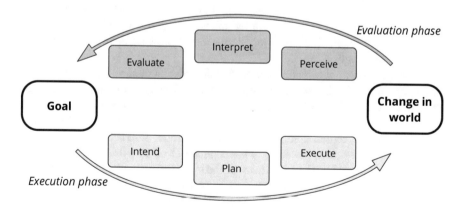

Fig. 5.1: Norman's model of interaction comprised of the execution and the evaluation phases. ⓒⓘ *The authors. Adapted from Norman (2013).*

Interaction costs The individual steps of both phases of the loop incur costs (see Lam, 2008). These costs can be physical or mental. Physical costs relate to flexing one's muscles, for example, when moving the hands to press a button or when moving the eyes to perceive the system response. Mental costs pertain to brain activities when thinking about how to achieve a goal or when interpreting the visual feedback. In a sense, the costs are associated with building bridges between the human and the system. Therefore, Norman (2013) calls the loop's phases the *gulf of execution* and the *gulf of evaluation*.

A primary goal of interaction design should be to narrow the gulfs by keeping the interaction costs low. On the execution side, this involves, for example, making interactions easy to discover and avoiding longer pointer movements through cascades of settings. On the evaluation side, it is important to let the visual response stand out clearly so that users can understand the effects of their actions easily.

Modes of interaction Technically, Jankun-Kelly et al. (2007) model the loop of user interaction as adjustments of visualization parameters, where concrete parameters can be manifold, e.g., the rotation angle of a 3D helix glyph, the focus point of a fisheye-transformed time axis, thresholds of a filter operation, or parameters that control a clustering algorithm.

Different modes of interaction can be identified depending on how parameter changes are performed. Spence (2007) distinguishes two modes of active user interaction:

- stepped interaction and
- continuous interaction.

For *stepped interaction*, a parameter change is executed in a discrete fashion. That is, the user performs one interaction step and evaluates the visual feedback. Much later, the user might perform another step of interaction. As an example, one can imagine a user looking at a visualization of the data at the granularity of years. If more details are required, the user might take an interaction step to switch to a finer granularity of months.

The term *continuous interaction* is used to describe interaction for which a visualization parameter is changed at a higher frequency. The user continuously performs an action and evaluates the generated feedback for a sustained period of time. This enables the user to quickly scan a larger range of parameter values and their corresponding visual representations. As such, continuous interaction is particularly useful in the context of exploratory 'what if' analyses of time-oriented data.

An example would be the adjustment of the cycle length for a spiral visualization in order to find out if and if so, which cyclic patterns exist in the data. For stepped interaction, the user has to explicitly specify different cycle lengths in a successive manner (e.g., by entering a numeric value). The stepped approach is quite time-consuming already when exploring only a moderate number of possible parameter values. Moreover, the discrete stepping does not allow cyclic patterns to emerge naturally as different cycle lengths are tried out. With continuous interaction (e.g., by dragging a slider), the user can explore any parameter range at any speed with a single continuous action. The risk of missing interesting patterns is reduced because cyclic patterns would crystallize gradually as suitable parameter values are approached. An important requirement though is to keep the interaction latency low.

Interaction latency For smooth and efficient interaction, the ensemble of visual and interaction methods has to generate feedback in a timely manner (within 50 - 100 ms according to Shneiderman (1994) and Spence (2007)). However, time-oriented data tend to be large and can pose considerable computational challenges. On the one hand, mapping and rendering the visual representation takes time, particularly if complex visual abstractions have to be displayed. On the other hand, computational methods (see Chapter 6) involved in the visualization process consume processing time before generating results. The adverse implication for interaction is that visual feedback might lag, disrupting the interaction loop (see Liu and Heer, 2014).

Another aspect adds to the time costs for presenting visual feedback. As interaction involves change, we want users to understand what is happening. However, abrupt changes in the visual display will hurt the mental model that users are developing while exploring unknown data. Pulo (2007) and Heer and Robertson (2007) provide evidence that smoothly interpolating the parameter change and applying animation to present the visual feedback can be a better solution. However, animation consumes time as well, not to mention the possibly costly calculations for interpolating parameter changes.

Thus, there are two conflicting requirements. On the one hand, interaction needs synchronicity. An interactive system has to be responsive at all times and should provide visual feedback immediately. From the interaction perspective, a system that is blocked and unresponsive while computing is the worst scenario. On the other hand, interaction needs asynchronicity – for both generating the feedback (i.e., computation) and presenting the feedback (i.e., animation). The difficulty is to integrate synchronicity and asynchronicity. One option to address this difficulty is to take a progressive approach.

Progressive visualization The goal of progressive visualization is to facilitate a smooth interaction cycle by generating visual feedback as quickly as possible (see Stolper et al., 2014; Angelini et al., 2018). This is achieved by a *divide & conquer* approach: Long-running computations are subdivided into smaller steps, and these operate on smaller data chunks rather than the whole dataset. For time-oriented data, data chunks can be obtained simply by sampling with respect to the dimension of time, by considering the semantics of time (e.g., workdays vs. weekends or day vs. night), or based on the increasingly detailed granularities of time (e.g., yearly, monthly, or daily data). The subdivision of computations into smaller steps depends very much on the concrete algorithms involved in the analytical and visual transformation of the data.

Working in smaller steps and on smaller data, progressive visualization generates a series of preliminary or partial results of increasing quality until a complete final image of the entire data is rendered. The quick and incremental generation of partial results has several advantages. First of all, the system is responsive at all times, and the interaction loop can run smoothly, even if there are still some computations running in the background. Second, users can observe the system computing the visualization. This makes otherwise hidden calculations more transparent and understandable. Third, as partial results arrive, users can early on develop an idea of the data and, if necessary, can steer the running computations to more fruitful results. For example, if partial results do not show the expected outcome, the computations can be canceled early to stop wasting time. If partial results already show promising features in the data, these parts can be prioritized to further crystallize interesting patterns early on.

Overall, we can see that interaction is a human-in-the-loop process during which a diverse set of user intents have to be satisfied. For the user, costs should be kept low, which requires interactions that are easy to carry out and visual feedback that is easy to understand. From a technical perspective, the execution and evaluation phases of the interaction loop must run smoothly, which can be supported by progressive visualization. What ultimately counts is that both user concerns and technical aspects are addressed under the umbrella of an effective and cost-efficient user interface.

5.2.2 User Interface

The user interface is the channel through which a human and a machine exchange information (i.e., interaction input and visual feedback). This interface is the linch-pin of interactive visual exploration and analysis of time-oriented data. Any visual representation is useless if the user interface fails to present it to the user in an appropriate way, and the diversity of available visualization techniques lies idle if the user interface fails to provide interactive access to them. In order to succeed, the user interface has to bridge the gap between the technical aspects of a visualization approach and the users' mental models of the problems to be solved. In this regard, Cooper et al. state:

> [...] user interfaces that are consistent with users' mental models are vastly superior to those that are merely reflections of the implementation model.
>
> Cooper et al. (2007, p. 30)

The user interface is responsible for numerous tasks. It has to provide visual access to time-oriented data and to information about the visualization process itself at different levels of graphical and semantic detail. Appropriate controls need to be integrated to allow users to steer exploration and analysis with regard to the interaction intents mentioned before, including marking interesting points in time, navigating in time at different levels of granularity, rearranging data items and elements of the visual representation, or filtering for relevant conditions. Moreover, the user interface has to support bookkeeping in terms of the annotation of findings, storage of results, and management of the working history (undo/redo).

In general, the user interface has to offer facilities to present information to the user and to accept interaction input from the user. This separation is reflected in the *model-view-controller* (MVC) architecture by Krasner and Pope (1988), where views provide visual representations of some model (in our case time, time-oriented data, and visualization parameters) and controllers serve for interactive (or automatic) manipulation of the model. Next, we look at visualization views and interaction controls in more detail.

Visualization views Especially the different temporal granularities make it necessary to present the data at different levels of graphical and semantic detail. Overview+detail, focus+context, and multiple coordinated views are key strategies to address this demand.

Overview+detail methods present overview and detail separately. The separation can be either spatial or temporal. Spatial separation means that separate views show detail and overview. For example, on the bottom of Figure 5.2, an overview shows the entire time domain at a high level of abstraction. On top of the overview, there is a separate detail view, which shows the data in full detail (i.e., detailed planning information), but only for a narrow time interval. Temporal separation means a view is capable of showing any level of detail, but only one at a time. This is usually referred to as *zooming*, where the user can interactively zoom into details or return to an overview. *Geometric zooming* operates solely in the presentation space to scale a visual representation, whereas *semantic zooming* denotes zooming that can go beyond purely geometrical scaling and may involve recoding the data in the presentation space as well as in the data space depending on the zoom level.

Contrary to overview+detail, *focus+context* methods smoothly integrate detail and overview. For the user-chosen focus, full detail is presented, and the focus is embedded into a less-detailed display of the context. Figure 5.3 shows the perspective wall technique ($\hookrightarrow$ p. 256) as a prominent example of the focus+context approach. Cockburn et al. (2009) provide a comprehensive survey of overview+detail, zooming, and focus+context and discuss the advantages and disadvantages of these concepts.

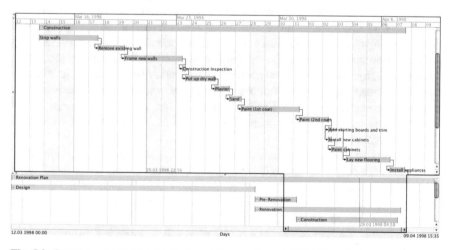

Fig. 5.2: Overview+detail. The detail view at the top shows individual steps of the construction phase of a renovation plan. In the overview at the bottom, the entire project is shown, including the design, pre-renovation, renovation, and construction phases. ©① *The authors.*

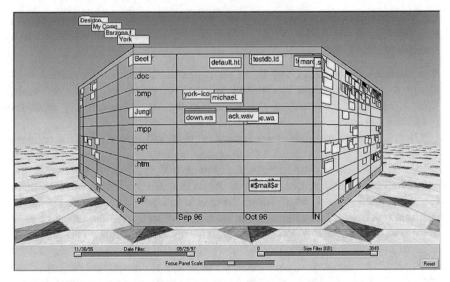

Fig. 5.3: Focus+context. The center of the perspective wall shows the focus in full detail. The focus is flanked on both sides by context regions. Due to perspective distortion, the context regions intentionally decrease in size and show less detail. © *Inxight Federal Systems. Used with permission.*

When visualizing time-oriented data, it is also often helpful to provide *multiple coordinated views*,[1] each of which is dedicated to particular aspects of time, certain data subsets, or specific visualization tasks. When there are multiple views, the user interface obviously needs a strategy for arranging them. One option is to use a fixed arrangement that has been designed by an expert and has proved to be efficient. It is also possible to provide users with the full flexibility of windowing systems, allowing them to move and resize views arbitrarily. Both options have their advantages and disadvantages and both are actually applied. An interesting third alternative is to maintain the flexibility of user-controlled arrangements, but to impose certain constraints in terms of what arrangements are possible (e.g., disallow partially overlapping views or enforce adjacency of related views). Irrespective of the strategy applied, the visualization should be *responsive* in the sense that it automatically adjusts itself to match the size and the aspect ratio of views (see Hoffswell et al., 2020).

In addition to arranging multiple views, coordinating the views plays an important role. Views are coordinated to help develop and maintain a consistent overall image of the visualized data. This means that an interaction which is initiated in one view is automatically propagated to all coordinated views, which in turn update themselves to reflect the change visually. A practical example is browsing in time. When the user navigates to a particular range of the time axis in one view, all other views (that are coordinated) follow the navigation automatically, which otherwise would be a cumbersome task to be manually accomplished by the user on a per-view

---

[1] Baldonado et al. (2000) provide general guidelines for when to use multiple coordinated views.

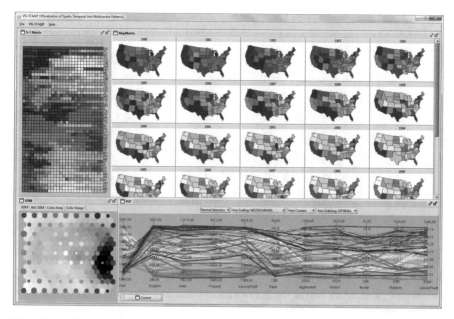

Fig. 5.4: Multiple coordinated views. Analysts can look at the data from different perspectives. The views are coordinated, which means selecting objects in one view will automatically highlight them in all other views as well. ⓒ① *The authors. Generated with the VIS-STAMP system by Guo et al. (2006).*

basis. Figure 5.4 shows an example where multiple coordinated views are applied to visualize spatio-temporal data in the VIS-STAMP system (↪ p. 380).

Interaction controls In addition to one or several visualization views, the user interface also consists of various interaction controls to enable users to tune the visualization process to the data and task at hand. Figure 5.5 shows a simple example with a single spiral view to its left (see Tominski and Schumann (2008) and ↪ p. 274). Already this single view depends on a number of parameters for which a corresponding number of controls must be provided in the control panel to the right. The control panel contains sliders for continuous adjustments of parameters such as *segments per cycle*, *spiral width*, and *center offset*. Buttons, drop boxes, and custom controls are provided for selecting different modes of encoding (e.g., adjusting individual colors or choosing different color scales).

In this example, user input (e.g., pressing a button or dragging a slider) is immediately committed to the system, which is a requirement for *continuous interaction*. However, this puts high demands on the system in terms of generating visual feedback quickly at interactive rates (see Piringer et al., 2009). Therefore, a commonly applied alternative is to allow users to perform a number of adjustments and to commit the adjustments as a single transaction only when the user presses an "Apply" button, which corresponds to *stepped interaction*.

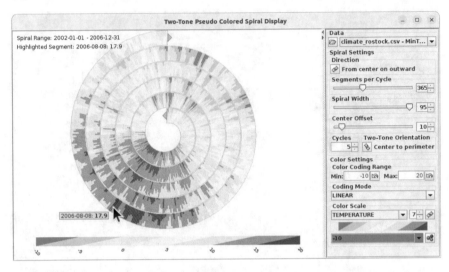

Fig. 5.5: User interface for a spiral visualization. The interface consists of one spiral view and one control panel, which in turn consists of various controls to adjust visualization parameters. ⓒⓘ *The authors.*

Certainly, there are visualization parameters that are adjusted more often than others during interactive visual exploration. Resources should preferably be spent on facilitating continuous interaction for important parameters. Moreover, Gajos et al. (2006) provide evidence that duplicating important functionality from an all-encompassing control panel to an exposed position is a useful way to drive adaptable user interfaces. For example, toolbars allow for interaction that is most frequently used, whereas rarely applied tools have to be selected from an otherwise collapsed menu structure.

5.3 Basic Interaction with Time-Oriented Data

It is clear now that we need visualization views on the one hand, and interaction controls on the other hand. Views are usually equipped with visual data representations, of which we described many examples for time and time-oriented data in the previous chapters. Let us now take a closer look at interactive means of controlling the visualization beyond standard graphical user interface controls. To this end, we briefly describe navigation in time, direct manipulation, brushing & linking, and dynamic queries as key methods for the interactive exploration and analysis of time-oriented data.

5.3.1 Navigation in Time

Time-oriented data typically contain very many time primitives, often too many to be displayed in a single visual representation. As a consequence, usually only a part of the time axis is visible at a time, and users have to navigate in time in order to develop a comprehensive understanding of the data. This navigation in time is essential.

Interactive sliders are control elements commonly found in user interfaces facilitating the exploration of data. For the case of time-oriented data, standard sliders are usually not enough for two reasons. First, standard sliders only have one handle to set a single value. For navigating in time, however, often two handles are required for defining the time interval to be visualized. One handle is for adjusting the interval's start, and the other handle sets the interval's end. Second, a standard slider cannot provide precise access to the time domain when the number of time primitives exceeds the interaction resolution. What is needed is a slider that can operate on different scales to facilitate quick and still precise access to all parts of time.

Figure 5.6 illustrates how such a slider may work for a time axis that extends from January 1, 2000 to December 31, 2010. In Figure 5.6b, the right handle has already been set to October 8, 2010. The figure further shows how the user can easily and accurately adjust the left handle to August 8, 2006. The interaction starts by horizontally dragging the handle roughly toward the desired date. Then the cursor is dragged in a downward movement to trigger the dynamic appearance of a higher-resolution on-demand slider. The interaction continues there horizontally, and thanks to the higher precision, the desired start date can be set exactly, which would not have been possible with the main slider alone.

Navigation in time via dedicated sliders is a widely applied approach. In the following, we will learn that interaction can also be performed directly on the visual representation of the data.

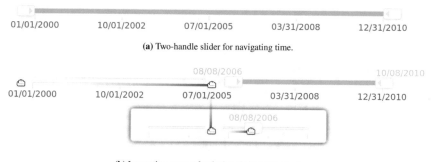

(a) Two-handle slider for navigating time.

(b) Interaction gesture for dual-scale interval adjustment.

Fig. 5.6: Navigation in time with a two-handle slider. (a) The slider's handles define the start and end of the time interval to be visualized. (b) Using a continuous interaction gesture, the interval start is adjusted coarsely on the main slider and fine-tuned precisely on a higher-resolution on-demand slider. ©① *The authors. Adapted from Tominski and Schumann (2020).*

5.3.2 Direct Manipulation

Graphical controls in user interfaces often have the advantage of being standardized components (e.g., buttons, single-handle sliders, and value spinners), which are easy to integrate and use. However, a disadvantage is that visual feedback usually does not appear where the interaction is performed. Recall the example from Figure 5.5 where the interaction takes place in the control panel to the right, whereas visual feedback is displayed in the visualization view to the left. Direct manipulation as introduced by Shneiderman (1983) is the classic means to address this disadvantage.

The goal is to enable users to manipulate the visual representation directly without a detour. To this end, a visualization view or its graphical elements are implemented so as to be responsive to user input. A visualization may for instance allow zooming into details under the mouse cursor simply by rotating the mouse wheel, or visiting different parts of the visual representation simply by dragging the view. Such functionality is often present in zoomable user interfaces (see Cockburn et al., 2009). Virtual trackballs (see Henriksen et al., 2004) are more object-centric in that they allow users to grab and rotate virtual objects to view them from different angles.

In terms of interacting with visual representations of time-oriented data, we just learned that navigating time is of particular importance. To support navigation, many tools rely on standard slider or calendar controls in the user interface. However, for direct manipulation, the interaction has to be tightly coupled with the display of the data. We explain what this means by two examples.

First, we take a look at DimpVis (↪ p. 305), which facilitates navigation to points in time (see Kondo and Collins, 2014). Figure 5.7 shows DimpVis in action on a basic point plot. The interaction works as follows. When the user grabs a dot, a path shows up indicating the selected data item's trajectory through time. In order to navigate, the user can now drag the dot along the path, where intermediate labels help the user find the desired moment in time. In a sense, DimpVis works like a slider, only the sliding operates on a curved path, rather than a straight line.

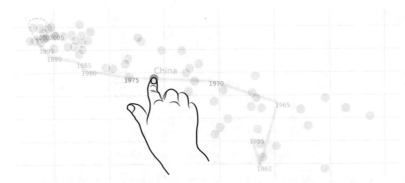

Fig. 5.7: Navigation in time via dragging a data item along its trajectory through time. ⓒ① *The authors. Generated with the DimpVis software by Kondo and Collins (2014).*

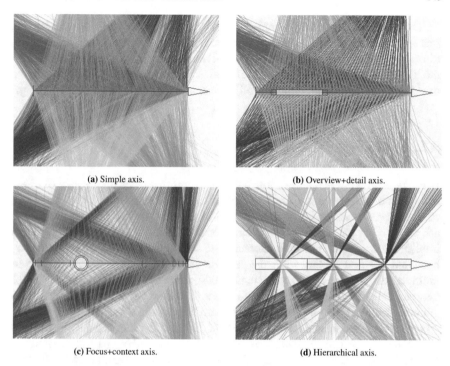

(a) Simple axis. **(b)** Overview+detail axis.

(c) Focus+context axis. **(d)** Hierarchical axis.

Fig. 5.8: The TimeWheel's mapping of time along the time axis can be manipulated directly in different ways. The simple axis uses a fixed linear mapping of time. The overview+detail axis allows users to select any particular range of the time domain to be mapped linearly to the time axis. The focus+context axis can be used to untangle dense parts of the time domain by applying a non-linear mapping. The hierarchical axis represents time at different levels of granularity, where individual axis segments can be expanded and collapsed. ⓒ① *The authors.*

For a second example of direct manipulation, we refer to the TimeWheel (↪ p. 298), in particular to its interactive axes (see Tominski et al., 2004). As Figure 5.8 illustrates, the TimeWheel provides (a) a simple non-interactive axis and three types of interactive axes: (b) overview+detail axis, (c) focus+context axis, and (d) hierarchical axis. Each of the axes displays time and the interactive ones offer different options for direct manipulation. The overview+detail axis basically extends the simple axis with three interactive handles to control the position and extent of the time interval to be displayed in the TimeWheel, effectively allowing users to zoom and scroll into any particular part of the data. The focus+context axis applies a non-linear distortion to the time axis in order to provide more drawing space for the user's focus and less space for the context. This allows users to untangle dense parts of the data. Finally, for the hierarchical axis, the display is hierarchically subdivided into segments according to the different granularities of time (e.g., years, quarters, months, and days). Users can expand or collapse these segments interactively to view the data at different levels of abstraction.

The advantage of directly manipulating the visual representation is, as indicated, that interaction and visual feedback take place at the very same location. However, direct manipulation always involves some learning and training of the interaction facilities provided (see Schwab et al., 2019a). This is necessary because most of the time the interaction is not standardized but custom-made to fit the visual mapping.

5.3.3 Brushing & Linking

Brushing & linking is a classic interaction concept that takes up the idea of direct manipulation. Becker and Cleveland (1987) describe brushing as a technique that enables users to select interesting data items directly from a data display. There are various options for selecting data items. We will often find brushing being implemented as point and click interaction to select individual data items. Rubber-band and lasso interaction serve the purpose of brushing subranges in the data or multiple data items at once. Hauser et al. (2002) introduce brushing based on angles between data items, and Doleisch and Hauser (2002) go beyond binary selection to allow for smooth brushing (i.e., data items can be partially selected).

After brushing, selected data items are highlighted in order to make them stand out against the rest of the data. The key benefit of the *linking* part of brushing & linking is that data brushed in one view are automatically highlighted in all other views. In this sense, brushing & linking is a form of coordination among multiple views. This is especially useful when visualizing the variables of a multivariate time-oriented dataset individually in separate views: Brushing a temporal interval of interest in one view will highlight the same interval and corresponding data values in all views. This makes it easy for users to compare how the individual variables develop during the brushed time period.

For complex data, using a single brush is often unsatisfactory. Instead, users need to perform multiple brushes on different time-dependent variables or in different views. Compound brushing as explained by Chen (2004) allows users to combine individual brushes into composite brushes by using various operators, including logical, analytical, data-centric, and visual operations. With such facilities, brushing is much like a visual query mechanism.

5.3.4 Dynamic Queries

Shneiderman (1994) describes dynamic queries as a concept for visual information seeking. It is strongly related to brushing & linking in that the goal is to focus on data of interest, which in the case of dynamic queries is often realized by filtering out irrelevant data. Because time-oriented data are often large, dynamically omitting data with respect to task-specific conditions can be very helpful.

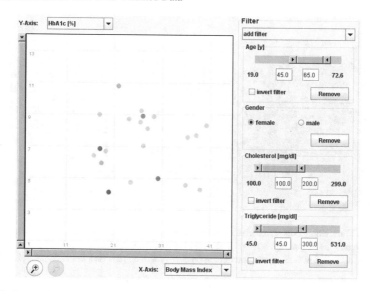

Fig. 5.9: Several filters can be adjusted in order to dynamically restrict the scatter plot visualization to data items that conform with the formulated conditions. ⓒⓘ *The authors.*

Depending on the view characteristics and visualization tasks, two alternatives can be applied to display filtering results: filtered objects can be displayed in less detail or they can be made invisible. Reducing detail is useful in views that maintain an overview, where all information needs to be displayed at all times, but filtered objects need only to be indicated. Making objects invisible is useful in views that notoriously suffer from cluttering.

Filter conditions are usually specified using dedicated mechanisms. Threshold or range sliders are effective for filtering time or any particular numerical variable; textual filters are useful for extracting objects with specific labels (e.g., data tagged by season). Similar to what has been said for brushing & linking, the next logical step is to combine filters to provide some form of multidimensional data reduction. For instance, a logical AND combination generates a filter that can be passed only if an object obeys all filter conditions; an object can pass a logical OR filter if it satisfies any of the involved filter conditions. Figure 5.9 shows an example of a dynamic query interface.

While some systems offer only fixed filter combinations or require users to enter syntactic constructs of some filter language, others implement a visual interface where the user can interactively specify filter conditions. Examples of querying time-oriented data that are visualized as line plot (↪ p. 233) are timeboxes and relaxed selection techniques.

Timeboxes (↪ p. 290) by Hochheiser and Shneiderman (2004) are used to filter out variables of a multivariate line plot. To this end, the user marks regions in the visual display by creating one or more elastic rectangles that specify particular value ranges and time intervals. The system then filters out all variables whose plots do not

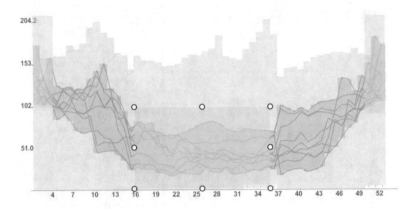

Fig. 5.10: Three timeboxes are used to dynamically query stock data. Only those stocks are displayed that are high at the beginning, but low in the middle, and again high at the end of the year. ⓒ① *The authors. Generated with the TimeSearcher software by Hochheiser and Shneiderman (2004).*

overlap with the rectangles, effectively performing multiple AND-combined range queries on the data. Figure 5.10 depicts a query that combines three timeboxes to restrict the display to stocks that performed well in the first and the last weeks of the year, but had a bad performance in the middle of the year.

The relaxed selection techniques by Holz and Feiner (2009) are useful for finding specific patterns in the data. For that purpose, the user specifies a query pattern by sketching it directly on the display. When the user is performing the sketching, either the distance of the sketch to the line plot or the user's sketching speed is taken into consideration in order to locally relax the query pattern. This relaxation is necessary to allow for a certain tolerance when matching the pattern in the data. An interactive display of the query sketch can be used to fine-tune the query pattern. Once the query pattern is specified, the system computes corresponding pattern matches and displays them in the line plot as depicted in Figure 5.11.

We should acknowledge that carrying out interactions directly on the visual representation as illustrated in this section is definitely useful, but the user can mark only those things that are already in the data and are actually displayed. Formulating queries with regard to potential but not yet existing patterns in the data beyond some tolerance requires additional formal query languages, and their utility hinges on the interface exposed to the user (see Monroe et al., 2013a).

Overall, navigating in time, direct manipulation, brushing & linking, and dynamic queries form an interaction vocabulary that any visualization of time-oriented data should support. Despite the advantages of being able to dynamically focus on data that are relevant to the task at hand, this vocabulary has still not yet become standard. While virtually all visualization tools for time-oriented data offer navigation in time, many do so using only rudimentary means that require users to take discrete steps rather than allowing them to browse the data in a continuous manner. Brushing the data directly in the visual representation and constructing more complex dynamic

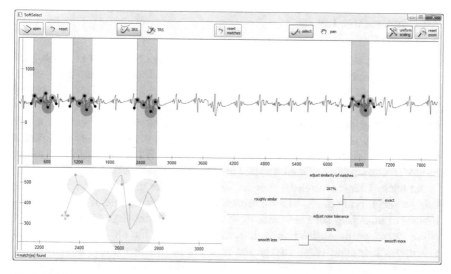

Fig. 5.11: The user can sketch a query pattern directly in the line plot and optionally refine it locally in a dedicated query view. The line plot then shows where in time the query matches with a certain tolerance. © *Courtesy of Christian Holz.*

queries are typically reserved for professional visualization systems. Again one can find a reason for that in the higher development costs for designing and implementing efficient interaction methods, particularly when direct manipulation and sketching are involved (see Mannino and Abouzied, 2018). Moreover, because visualization and interaction must be coupled tightly, it is typically difficult to develop interaction components that can be interchanged among the different visualization techniques for time-oriented data. One rare exception is the EazyPZ library (see Schwab et al., 2019b) whose zoom and pan functionality can be used as a basis for flexible navigation in time. Finding generally applicable solutions to other interaction problems is an open research question.

5.4 Advanced Interaction Methods

The previous section was concerned with basic interaction methods. In this section, we shed some light on advanced ways of interacting with time-oriented data. We start with interactive lenses as versatile tools for data exploration. When interesting data portions have been spotted, it is often necessary to compare them. This section will illustrate how visual comparison can be supported with naturally inspired interaction techniques. In order to help users make analytical progress, further advanced support can be offered in the form of guidance or by integrating automatic event-based methods. Finally, this section will consider advanced interaction using modern interaction modalities beyond mouse and keyboard interaction.

5.4.1 Interactive Lenses

Interactive lenses, originally introduced as magic lenses by Bier et al. (1993), are related to the focus+context concept discussed on p. 137. Tominski et al. (2017) define interactive lenses as lightweight tools that provide alternative visual representations for selected parts of the data on demand. Once activated, working with a lens is as easy as moving it across the visualization to specify where the lens is to take effect. The lens effect is automatically computed and merged with the base visualization to generate a locally enhanced visual representation. When the lens is no longer needed, it can simply be dismissed and the original visualization is restored.

As such, interactive lenses support scrutinizing the visualized data similar to using a magnifying glass. The difference though is that an interactive lens is not limited to enlarging selected parts of the visual representation. Conceptually, the effect generated by an interactive lens can include (i) the alternation of existing visualization content (e.g., change the coloring of selected time points), (ii) the omission of content (e.g., filter out less relevant data), or (iii) the addition of new content (e.g., add textual labels for clarification).

According to Tominski et al. (2017), more than 50 lens techniques for different data analysis scenarios are known in the literature, and eight of them are suited for time-oriented data. An additional example is the *regression lens* by Shao et al. (2017) shown in Figure 5.12. It is particularly useful for analyzing temporal trends. The lens' primary purpose is to enhance point plots ($\hookrightarrow$ p. 232) by adding locally computed regression curves for the data points within the perimeter of the lens. Our example shows two regression curves calculated by different algorithms. Additionally, the left and top borders of the lens are enhanced with histograms of the selected data. By

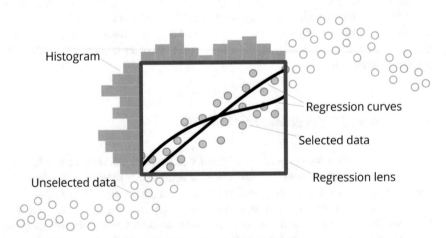

Fig. 5.12: The regression lens computes regression curves of its underlying data points and shows them as line plots on top of the base visualization. Additional histograms indicate the data distribution at the lens borders. ⊜⓪ *The authors. Adapted from Shao et al. (2017).*

moving and resizing the lens, the user can quickly explore the regression in local parts of the data without changing the original visualization globally.

While our example of the regression lens is focused on time-oriented data, inter-active lenses are highly versatile tools in general. The swiftness and naturalness with which lenses can be operated are their key advantages. How natural interaction can also benefit the comparison of time-oriented data will be discussed next.

5.4.2 Interactive Visual Comparison

Comparing data is a ubiquitous data analysis activity (see Gleicher et al., 2011; Gleicher, 2018; L'Yi et al., 2021). It is particularly relevant in the context of time-oriented data. For example, the detection of temporal trends requires the comparison of individual data values along the time axis in the first place. Once promising trends have been identified, it is usually also of interest to compare them with each other: Which trend has the steeper slope or which trend peaks at the global maximum?

Without dedicated support, visual comparison can be a demanding task. In Chapter 4, we already discussed visual color-coding specifically to support visual comparison tasks. But still it may be necessary to move the eyes back and forth between the data to be compared, which is costly and error-prone. In the following, we discuss interaction techniques that allow users to dynamically re-arrange parts of a visual representation to facilitate visual comparison.

The interaction techniques to be presented are inspired by natural human behavior (see Tominski et al., 2012a). When people compare information printed on paper they usually carry out three steps:

1. Select comparison candidates
2. Arrange candidates for comparison
3. Carry out the actual comparison

In the first step, people specify *what* they want to compare. The comparison candidates can be individual data values or data items at different points in time or sub-ranges of the time axis showing interesting behavior such as trends or recurring patterns. In the second step, the comparison candidates are arranged so as to enable or ease their comparison. Finally, the actual comparison is conducted to figure out what relationships might exist between the compared data. Two requirements should be fulfilled in this regard. First, the properties of the individual data being compared should be clearly visible. Second, the similarities and differences between the data need to be communicated as well. The degree to which both requirements are met depends largely on the arrangement generated in step two, so let us look at this aspect in more detail.

Assume two comparison candidates *A* and *B* have been selected. When *A* and *B* are printed on paper, people would naturally arrange them as juxtaposition or superposition. For juxtaposition, *A* and *B* are arranged side by side. This allows us to see the individual data properties of *A* and *B* clearly, but in order to detect

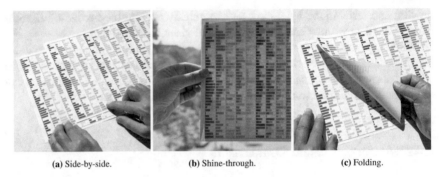

(a) Side-by-side.   (b) Shine-through.   (c) Folding.

Fig. 5.13: Natural comparison behavior when comparing information printed on paper. ⓒ① *Tominski and Schumann (2020).*

similarities or differences, the eyes have to switch between both sides frequently. For superposition, A and B are stacked on top of each other. As A and B are now co-located, similarities and differences are potentially easier to see, but either A occludes B or the other way around, which hinders the comparison and also deteriorates the visibility of either A or B. For real-world comparison on paper, the occlusion can be resolved in two ways. Either the stacked A and B are held against the light to let the occluded information shine through and generate a merged representation of A and B. Or the occlusion is resolved by folding the occluding piece of paper back and forth to reveal A and B in quick succession. Figure 5.13 illustrates these natural comparison behaviors: side-by-side, shine-through, and folding.

On the computer, this natural comparison between A and B can be replicated via advanced interaction techniques, as schematically depicted in Figure 5.14. Via simple drag gestures, side-by-side and overlapping arrangements can be created. For resolving occlusions, shine-through comparison can be implemented via alpha-blending, where the occluding view is made partially transparent. The folding technique makes it possible to peel off the occluding view very much like for real paper. To keep the interaction costs low, the folding can simply be triggered by clicking at the location where the occlusion between the views is to be resolved. Based on a heuristic, a natural fold is calculated and presented via a smooth animation.

Let us take a closer look at Figure 5.14 to understand the advantages and drawbacks of the different interactions. In the side-by-side variant, the user drags comparison candidate B next to A. This shows both subsets of the data clearly, however, determining which trend is steeper might not be so easy to figure out. The shine-through technique makes the direct comparison of the trends easier by superimposing A and B and allowing the user to manipulate the degree of occlusion via a vertical drag gesture or slider. Yet it is no longer clear which line plot belongs to which subset. The folding variant is a compromise. It clearly separates the superimposed line plots, and by quickly folding back and forth, the peaks can be compared reasonably well. Yet, the collateral occlusion caused by the folding need to be dealt with.

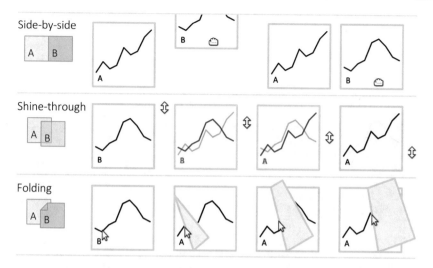

Fig. 5.14: Side-by-side, shine-through, and folding interaction. ⓒ⊕ *The authors.*

In summary, this section illustrated different interaction techniques for supporting visual comparison tasks, which are so common when analyzing time-oriented data. The naturalness of the interactions makes them easy to learn and carry out. Moreover, the outlined techniques are not limited to comparing line plots, but are generally applicable to any visual representation.

5.4.3 Guiding the User

The interaction techniques described in this chapter so far provide many degrees of freedom to enable users to study time-oriented data from different perspectives and to develop a comprehensive understanding. However, the many degrees of freedom can also be a challenge. During the data exploration, many questions arise: Where should I move the lens to identify a local cluster? Which partial trends should I select for comparison? Where should I navigate to find interesting patterns? These questions become problematic when there are too many of them and when the user has too many difficulties answering them. If this is the case, the analytical progress stalls and the interactive exploration comes to a halt.

To ensure steady progress and to keep the data exploration going, it makes sense to provide users with guidance. *Guidance* has been defined as a means to help users resolve problems they may encounter during interactive data exploration (see Schulz et al., 2013b; Ceneda et al., 2017; Collins et al., 2018). The important aspect here is that guidance is there to help and to assist. It is not a means to provide answers to analytic questions, but to enable and support users to arrive at answers on their own, that is, the human remains in the loop.

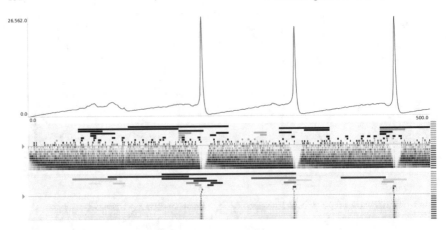

Fig. 5.15: Overview plot of a time series with 3.6 million time points (top) and color-coded difference bands (center: slope sign difference; bottom: absolute value difference) indicating where potentially interesting observations could be made. ©① *Martin Luboschik. Also see Luboschik et al. (2012).*

In the following, we will demonstrate how large time-oriented data can be explored at multiple scales with the help of an appropriate guidance strategy. The starting point is a large time series with millions of data points from a simulation of the cell division cycle in fission yeast (see Luboschik et al., 2012). We are going to visualize these data as a classic line plot (↪ p. 233). The problem though is that about 3.6 million time points usually do not fit in a line plot. Therefore, the time series has been aggregated at several levels of granularity, leading to a multi-scale representation of the data. Such a representation lends itself to being explored via zooming. When the zoom level changes, the visualization shows the level of granularity that matches the resolution of the display.

An overview of the whole time series is depicted at the top of Figure 5.15. At this level of granularity, one can easily see three peaks. But what we are seeing is only a coarse representation, in fact, the coarsest of our multi-scale time series. We do not know what is going on at the finer scales on the slopes or at the top of the peaks. Zooming and panning will allow us to access the details we seek. However, where in time and at what temporal scale can we make interesting observations? The guidance approach we are about to demonstrate uses the data themselves as an input to compute visual cues that provide users with orientation to narrow down their search on promising parts of the data.

The assumption is that differences between adjacent scales might serve as an indication for users to look more closely into particular parts of the data. Various measures can be employed to calculate the differences. Luboschik et al. (2012) consider absolute value differences and slope sign differences. These measures are calculated for all pairs of adjacent scales. Aggregating the measures and color-coding them leads to so-called difference bands that can be attached below our line plot on demand as shown in Figure 5.15.

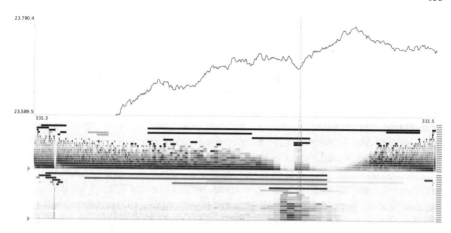

Fig. 5.16: Zoomed view of the tip of the second peak from Figure 5.15. The difference bands are magnified by means of a focus+context distortion. ©① *Martin Luboschik. Also see Luboschik et al. (2012).*

Interestingly, in the bluish bands for slope sign difference (center), we can see three notches exactly where the three peaks are in the line plot. There are also three greenish spikes in the absolute value difference bands (bottom). So, both bands *guide* the user to the peaks for more detailed inspection. And in fact, some interesting behavior can be observed. Looking at the notches for slope sign difference in Figure 5.15 more closely, one can see thin spikes.

To understand what is going on, we study the second notch in more detail. We magnify the second notch and the tip of its associated peak as shown in Figure 5.16. From the magnified difference bands, we can see that greater differences, indicated by darker colors, exist between the temporal scales of finer granularity. The zoomed line plot confirms that the tip of the peak is not a smooth curve as we might have thought. There is in fact a rather rough up and down of the curve.

This example of multi-scale exploration of time-oriented data illustrates the benefit of providing guidance. The additional difference bands provide on-demand support to help users decide which parts of the data are promising to study in detail. Other examples of guidance exist, where the focus is less on navigation, but on guiding the configuration of visualization techniques, for example, to suggest suitable cycle lengths of spiral representations (↪ p. 274) to help users find cyclic patterns in time-oriented data (see Ceneda et al., 2018). For a broader view on guidance and more examples, the interested reader is referred to the survey by Ceneda et al. (2019).

5.4.4 Integrating Interaction and Automation via Events

With the increasing complexity of data and visualization methods alike, it is not always easy for users to set visualization parameters appropriately for the analysis task at hand. Particularly if parameters are not self-explanatory, they are not easily set manually. Guidance can provide a form of support to assist users in the parametrization process.

Another possible solution is to employ the concept of *event-based visualization*, which combines visualization with event methodology (see Reinders et al., 2001; Tominski, 2011). In diverse application fields, including active databases, software engineering, and modeling and simulation, events are considered happenings of interest that trigger some automatic actions. In the context of visualization, such an event-action-scheme is useful for complementing manual interaction with automatic parametrization of visual representations.

The basic idea of event-based visualization is (1) to let users specify their interests, (2) to detect if and where these interests match in the data, and (3) to consider detected matches when generating the visual representation. This general procedure requires three main components: (1) *event specification*, (2) *event detection*, and (3) *event representation*. Figure 5.17 illustrates how they are attached to the visualization pipeline. Next, we will look at each of these components in more detail.

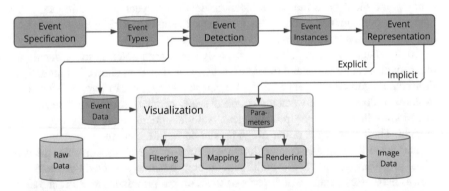

Fig. 5.17: The main ingredients of event-based visualization – event specification, event detection, and event representation – attached to the visualization pipeline. ©① *The authors.*

Describing User Interests

The event specification is an interactive step where users describe their interests as *event types*. To be able to find actual matches of user interests in the data, the event specification must be based on formal descriptions. Tominski (2011) uses elements of predicate logic to create well-defined event formulas that express interests with respect to relational datasets (e.g., data records whose values exceed a threshold or attribute with the highest average value). For an analysis of time-oriented data, sequence-related notations (for instance as introduced by Sadri et al. (2004)) enable users to specify conditions of interest regarding temporally ordered sequences (e.g., sequence of days with rising stock prices). A combination of event types to composite event types is possible via set operators.

As a simple example, we formulate our interest in *"Three successive days where the number of people diagnosed with influenza increases by more than 15% each day"* as the following event type:

$$\{(x, y, z)_{date} \mid z.flu \geq y.flu \cdot 1.15 \wedge y.flu \geq x.flu \cdot 1.15\}$$

The first part of the formula defines three variables $(x, y, z)_{date}$ that are sequenced by date. To express the condition of interest, these three variables are set into relation using predicates, functions, and logical connectors.

Certainly, casual users may find it difficult to describe their interests via event formulas. Therefore, sufficient specification support should consider dedicated means for experts, regular users, and visualization novices. In this regard, one can think of three levels of specification: *(i) direct specification, (ii) specification by parametrization*, and *(iii) specification by selection*. All levels are based on the aforementioned formalism, but the complete functionality is available only to expert users at the level of direct specification. The second level works with parametrizable templates that hide the complexity of event formulas from the user. Non-expert users can adjust the templates via easy-to-set parameters, but otherwise do not need to fiddle with the internals of event formulas. For example, exposing the increase rate (15% in our previous example) as a template parameter would be reasonable. At the third level, users simply select from a predefined collection of event types that are particularly tailored to the application context.

Finding Relevant Data Portions

The event detection is an automatic step that determines whether the interests defined interactively are present in the data. The outcome of the event detection is a set of *event instances*. They describe where in the data interesting information is located. That is, entities that match user interests are marked as event instances. For event detection, the variables used in event formulas are substituted with concrete data entities. In the second step, predicates, functions, and logical connections are evaluated, so that the event formula as a whole can be evaluated as either true or false. Because this procedure can be quite costly in terms of computation time, efficient methods must be utilized for the event detection. A combination of the capabilities of relational database management systems and efficient algorithms (e.g., the OPS algorithm by Sadri et al. (2004)) is useful for static data. When dynamic data (i.e., data that change over time, see Section 3.3) have to be considered, detection efficiency becomes even more crucial. Here, incremental detection methods can help. Such methods operate on a differential dataset, rather than on the whole data. However, incremental methods also impose restrictions on possible event types, because they do not have access to the entire dataset.

Considering User Interests in Visual Representations

The last important step of event-based visualization is the event representation. The goal of this step is to incorporate detected event instances, which reflect the interests of the user, into visual representations. The three requirements that have to be considered are as follows:

1. Communicate the fact that something interesting has been found.
2. Emphasize interesting data among the rest of the data.
3. Convey what makes the data interesting.

Most importantly, the visual representation must clearly express that something interesting is contained in the data. To meet this requirement, easy-to-perceive visual cues (e.g., a red frame around the visual representation, exclamation marks, or annotations) can be used. Alpha-blending can be applied to fade out past events. The second requirement aims at emphasizing those parts of the visual representation that are of interest. Additionally, the visualization should communicate what makes the highlighted parts interesting (i.e., what the particular event type is). However, when facing arbitrarily definable event formulas, this last requirement is difficult to fulfill.

We can distinguish two basic options for representing events: *explicit* and *implicit* event representation. For the explicit case, the focus is set exclusively on event instances, neglecting the raw data. Since the number of events is usually smaller than the number of data items, explicit event representation can grant insight even into very large datasets. For implicit event representation, the goal is to automatically adjust visualization parameters so as to highlight the points of interest detected in the data. Assuming that user interests relate to user tasks and vice versa, implicit event representation can help us obtain better-targeted visual representations. The big challenge though is to meet the aforesaid requirements solely by adapting visualization parameters. Apparently, the availability of adequate visualization parameters is a prerequisite for implicit event representation.

Let us illustrate the potential of event-based visualization with an example. Assume a user has to analyze multivariate time-dependent human health data for uncommonly high numbers of cases of influenza. The task at hand is to find out if and where in time these situations have occurred. A possible way to accomplish this task is to use the TimeWheel technique (↪ p. 298).

Figure 5.18a shows a TimeWheel that uses the standard parametrization, where time is encoded along the central axis and multiple diagnoses are mapped to the axes surrounding the time axis. In particular, influenza happens to be the diagnosis that is mapped to the upper right axis (light green). Alpha-blending is applied by default to reduce visual clutter. Looking at this TimeWheel, the user can only guess from the labels of the axis showing influenza that there are higher numbers of cases

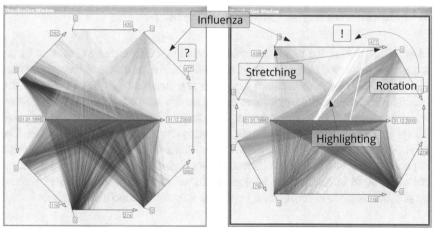

(a) Default parametrization.   (b) Targeted parametrization.

Fig. 5.18: Default vs. targeted parametrization of a TimeWheel. (a) TimeWheel representing a time-dependent health dataset using the default configuration, which aims at showing main trends, but does not consider the interests of the user. (b) TimeWheel representing the same data, but matches with the user's interests have been detected and corresponding data are emphasized via highlighted lines and automatic rotation and stretching; the presentation is better targeted to the user's task at hand. ©⊕ *The authors.*

because the alpha-blending made the particular lines almost invisible (see question mark). Several interaction steps are necessary to re-parametrize the TimeWheel to accomplish the task at hand.

In contrast to this, in an event-based visualization environment, the user can specify the interest in *"Days with a high number of cases of influenza"* as the event type ($\{x \mid x.flu \geq 300\}$). If the event detection step confirms the existence of such events in the data, visualization parameters are altered automatically so as to provide an individually adjusted TimeWheel that reflects the special situation. In our particular example in Figure 5.18b, we change the color and transparency of line segments representing event instances: Days with high numbers of influenza cases are excluded from alpha-blending and are drawn in white (see exclamation mark). Additionally, rotation and stretching are applied such that the axis representing influenza is moved gradually to an exposed position and is provided with more display space. The application of a gradual process is important in this case to support users in maintaining their mental map of the visual representation. In this automatically adjusted TimeWheel, the identification of days with higher numbers of influenza infections is easy.

5.4.5 Interaction Beyond Mouse and Keyboard

Most of the interaction techniques discussed in this chapter, and also most of the techniques described in the literature, are designed for the classic desktop computer workplace where the mouse and keyboard are the dominant input devices. Yet, technological advances have brought us to a point where new interaction modalities are becoming more and more commonplace. Interaction beyond mouse and keyboard brings new possibilities for exploring and analyzing data in various ways (see Lee et al., 2012; Keefe and Isenberg, 2013). In this section, we briefly look at what is possible in terms of modern interaction for time-oriented data. In particular, we consider touch interaction for exploring time-oriented data visualized as stacked graphs and tangible interaction for exploring space-time cube visualizations.

Touching Stacked Graphs

Touch interaction has become the primary input modality for mobile devices. It can also be found on laptop computers and larger display surfaces (see Voida et al., 2009). Touch interaction has the advantage that the action takes place directly on the display, exactly where the operation is to take effect. Yet, a difficulty with touch is that the input devices, our fingers, are rather imprecise making it harder to point at fine details in a visualization. Using the fingers for interaction can also cause the hand to occlude relevant information on the display. Nonetheless, the directness and intuitiveness of touch interaction are the key motivation for using it in the context of visualization.

The example we are looking at here is TouchWave by Baur et al. (2012). Touch-Wave is specifically designed for direct and fluid interaction with time-oriented data visualized as stacked graphs ($\hookrightarrow$ p. 286). For improving the legibility, comparability, and scalability of stacked graphs, several concrete touch interactions and corresponding visual feedback are offered. Legibility can be improved by touching the visualization background, which triggers the display of an on-demand vertical ruler showing the exact value distribution for the time point corresponding to the finger position. By using more than one finger, which is called multi-touch interaction, additional rulers can be activated to facilitate the visual comparison of several points in time.

As the order of individual streams in a stacked graph is important, reordering the streams is an essential operation. By long-pressing the stacked graph, its streams can be sorted so that the stream with the highest value for the time point being touched is at the top. Double tapping a stream will make it the baseline stream on top of which all other streams are stacked. Moreover, individual streams can be pulled out of the stacked graph via simple drag gestures. These interactive rearrangements are particularly useful for comparison, as we have already seen in Section 5.4.2.

To support multi-scale data exploration, the TouchWave utilizes pinch gestures. Pinching horizontally will create a focus+context distortion of the time line revealing details in the focus, while compressing the context. Vertical pinching can be used to

Fig. 5.19: Using a pinch gesture for scaling a stacked graph visualization vertically. © *Courtesy of Dominikus Baur.* `https://do.minik.us/projects/touchwave`

perform a hierarchical zoom with respect to the streams in a stacked graph. Such a vertical pinch gesture is illustrated in Figure 5.19.

TouchWave is designed particularly for stacked graphs. Yet, touch-based interaction also works for other visualizations of time-oriented data. For example, Riehmann et al. (2018) describe dedicated touch interactions for multiple time series depicted as horizon graphs (↪ p. 277). What all touch techniques have in common is that they facilitate the direct interaction *on* the display. Next, we will see how tangible interaction can support interaction *with* the display.

Exploring Space-Time Cubes with Tangible Interaction

Tangible interaction is a style of interaction where users interact by manipulating physical objects, so-called *tangibles* (see Shaer and Hornecker, 2009). This requires appropriate tracking equipment so that the system knows where the tangibles are located and how they are oriented in space. The spatial awareness can be utilized to define whole new interaction vocabularies. Basic interactions include horizontal and vertical translation and rotation, which in turn can be combined to gestures such as tilting, flipping, or shaking a tangible. These interactions can then be utilized to design new data exploration experiences.

In the context of exploring time-oriented data, tangible interaction opens up new possibilities for navigating the time axis and also for adjusting the visual representation depending on the user's tasks. To illustrate the usefulness of tangible interaction, we present two examples: tangible views and the Uplift system. In both cases, spatio-temporal data are visualized as a space-time cube (↪ p. 377) on a horizontal tabletop display. The cube's base plane resides in the horizontal x-y plane of the tabletop and the dimension of time extends from the base plane along the vertical z-axis. It is important to realize that the space-time cube is a virtual one, meaning that the space above the horizontal tabletop defines the space-time cube, but its content is not yet visible. Initially, there is only a map on the tabletop, but via tangible interaction, one can access the space-time cube and make different parts of time and space visible.

Tangible views The two terms *tangible* and *views* already hint at a duality between display and interaction: The views serve to show the visualization, and at the same time, the views are tangible and serve as an input device for interacting with the visualization. Conceptually, tangible views are spatially-aware lightweight displays. Spindler et al. (2010) describe an implementation where tangible views are made of cardboard onto which visual representations can be projected.

(a) Flipping the color-coding.   (b) Side-by-side comparison.

Fig. 5.20: Using tangible views for exploring spatio-temporal data in a virtual space-time cube. ⓒⓘ *The authors.*

In order to interactively explore a virtual space-time cube and adjust its visualization, one or more tangible views can be held in the space above the base map as illustrated in Figure 5.20. Different parts of the map can be accessed by moving a tangible view horizontally (i.e., navigation in space). The tangible view's partial map is then updated according to the horizontal position above the base map. Similarly, by raising and lowering the tangible view along the vertical axis, one can select particular time points to be displayed (i.e., navigation in time). By flipping the tangible view, it is possible to switch between two different color-coding strategies, for example, for identification and location tasks as described in Section 4.2.2. Tangible views can also facilitate visual comparison. To this end, two tangible views are used in combination. First, each view is moved individually to select two map regions and two time points to be compared. Then a lock operation is performed, which makes both tangible views insensitive to further motion. This in turn allows the user to bring the two tangible views together forming a side-by-side arrangement for comparison.

Uplift Our second example of tangible interaction with a space-time cube visualization is the Uplift system by Ens et al. (2021). In this example, the space-time cube is also located in the space above a tabletop display, but it is displayed virtually as an augmented-reality representation. This allows several persons to look at the data simultaneously as shown in Figure 5.21. Several tangibles are used in concert to interact with the system in various ways. Particularly interesting is the navigation through time and the unfolding of the space-time cube. By placing a tangible token on the tabletop, slider widgets with different temporal granularity can be activated. A physical slider widget can then be used to select a particular point in time. By using a hinge of the physical widget, the space-time cube can be unfolded to show several data layers for comparing multiple time steps.

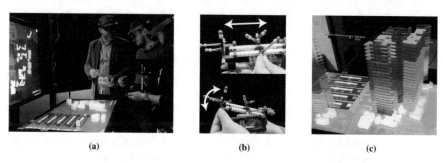

(a)	(b)	(c)

Fig. 5.21: Uplift: tangible and immersive tabletop system. (a) Collaborative exploration around a tabletop display using tangible objects. (b) Physical widget for navigating in time. (c) Unfolded space-time cube visualization above the tabletop surface. © *2021 IEEE. Reprinted, with permission, from Ens et al. (2021).*

What we can learn from tangible views, physical widgets, and TouchWave before is that there is more to interaction than just mouse and keyboard. Touch and tangible interaction are but two examples of modern ways of interacting with data. Further examples are gaze-based interaction (see Duchowski, 2018), where the eyes perform actions, and proxemic interaction (see Jakobsen et al., 2013), where the distance of the user to the display is considered. Natural language is another channel to be utilized for interaction, where combining language with other input modalities seems to be a quite promising approach (see Srinivasan and Stasko, 2018). Yet, further research needs to be conducted to take full advantage of these new interaction modalities and their combination for the particular case of visually exploring and analyzing time-oriented data.

5.5 Summary

The focus of this chapter was on interaction. We started with a brief overview of intents that motivate users to interact with the visualization. The most notable intent in the context of time-oriented data is the intent to navigate in time in order to visit different parts of the data. Users also need to view time-oriented data at different levels of detail, because the data are often given at multiple granularities. Further intents are related to interactively adjusting the visual mapping according to data and tasks at hand, and to managing the exploration process.

We explained that interactive visualization is an iterative loop where the user plans and carries out an interaction, and the computer generates feedback in order to visually reflect the change that resulted from the user's actions. This human-in-the-loop process brings together the computational power of the machine and the intellectual power of human beings. In order to take full advantage of this synergy, we need an efficient user interface that bridges the gap between the algorithmic structures being used for visualizing time and time-oriented data, and the mental models and analytic workflows of users. This also includes tackling technical challenges to guarantee the smooth execution of the interaction loop.

This chapter also presented basic interaction concepts, including temporal navigation, direct manipulation, brushing & linking, and dynamic queries. These concepts are vital for data exploration tasks where the user performs an undirected search for potentially interesting data features. Going beyond basic interaction, we considered interactive lenses, natural visual comparison, guidance, event-based visualization, and interaction beyond mouse and keyboard. These advanced concepts can further enhance the visual exploration of time-oriented data. But still, the potential of advanced interaction methods has not been fully exploited by current visualization techniques. There is room for future work to better adapt existing interaction methods or to develop new ones according to the specific needs of time-oriented data. Moreover, the examples of guidance and event-based visualization indicate that a combination of visual, interactive, and automatic methods can be quite useful. In the next chapter, we will take a closer look at computational analysis methods for supporting the visual analysis of time-oriented data.

References

Angelini, M., G. Santucci, H. Schumann, and H.-J. Schulz (2018). "A Review and Characterization of Progressive Visual Analytics". In: *Informatics* 5.3, p. 31. DOI: 10.3390/informatics5030031.

Baldonado, M. Q. W., A. Woodruff, and A. Kuchinsky (2000). "Guidelines for Using Multiple Views in Information Visualization". In: *Proceedings of the Conference on Advanced Visual Interfaces (AVI)*. ACM Press, pp. 110–119. DOI: 10.1145/345513.345271.

Baur, D., B. Lee, and S. Carpendale (2012). "TouchWave: Kinetic Multi-touch Manipulation for Hierarchical Stacked Graphs". In: *Proceedings of the International Conference on Interactive Tabletops and Surfaces (ITS)*. ACM Press, pp. 255–264. DOI: 10.1145/2396636.2396675.

Becker, R. A. and W. S. Cleveland (1987). "Brushing Scatterplots". In: *Technometrics* 29.2, pp. 127–142. DOI: 10.2307/1269768.

Bertin, J. (1981). *Graphics and Graphic Information-Processing*. Translated by William J. Berg and Paul Scott. de Gruyter.

Bier, E. A., M. C. Stone, K. Pier, W. Buxton, and T. D. DeRose (1993). "Toolglass and Magic Lenses: The See-Through Interface". In: *Proceedings of the Annual Conference on Computer Graphics and Interactive Techniques (SIGGRAPH)*. ACM Press, pp. 73–80. DOI: 10.1145/166117.166126.

Ceneda, D., T. Gschwandtner, T. May, S. Miksch, H.-J. Schulz, M. Streit, and C. Tominski (2017). "Characterizing Guidance in Visual Analytics". In: *IEEE Transactions on Visualization and Computer Graphics* 23.1, pp. 111–120. DOI: 10.1109/TVCG.2016.2598468.

Ceneda, D., T. Gschwandtner, and S. Miksch (2019). "A Review of Guidance Approaches in Visual Data Analysis: A Multifocal Perspective". In: *Computer Graphics Forum* 38.3, pp. 861–879. DOI: 10.1111/cgf.13730.

Ceneda, D., T. Gschwandtner, S. Miksch, and C. Tominski (2018). "Guided Visual Exploration of Cyclical Patterns in Time-series". In: *Proceedings of the IEEE Symposium on Visualization in Data Science (VDS)*. IEEE Computer Society.

Chen, H. (2004). "Compound Brushing Explained". In: *Information Visualization* 3.2, pp. 96–108. DOI: 10.1057/palgrave.ivs.9500068.

Cheng, X., D. Cook, and H. Hofmann (2016). "Enabling Interactivity on Displays of Multivariate Time Series and Longitudinal Data". In: *Journal of Computational and Graphical Statistics* 25.4, pp. 1057–1076. DOI: 10.1080/10618600.2015.1105749.

Cockburn, A., A. Karlson, and B. B. Bederson (2009). "A Review of Overview+Detail, Zooming, and Focus+Context Interfaces". In: *ACM Computing Surveys* 41.1, 2:1–2:31. DOI: 10.1145/1456650.1456652.

Collins, C., N. Andrienko, T. Schreck, J. Yang, J. Choo, U. Engelke, A. Jena, and T. Dwyer (2018). "Guidance in the Human-Machine Analytics Process". In: *Visual Informatics* 2.3. DOI: 10.1016/j.visinf.2018.09.003.

Cooper, A., R. Reimann, and D. Cronin (2007). *About Face 3: The Essentials of Interaction Design*. Wiley Publishing, Inc.

Doleisch, H. and H. Hauser (2002). "Smooth Brushing for Focus+Context Visualization of Simulation Data in 3D". In: *Proceedings of the International Conference in Central Europe on Computer Graphics, Visualization and Computer Vision (WSCG)*. University of West Bohemia, pp. 147–154. URL: http://wscg.zcu.cz/wscg2002/Papers_2002/E71.pdf.

Duchowski, A. T. (2018). "Gaze-based Interaction: A 30 Year Retrospective". In: *Computers & Graphics* 73, pp. 59–69. DOI: 10.1016/j.cag.2018.04.002.

Ens, B., S. Goodwin, A. Prouzeau, F. Anderson, F. Y. Wang, S. Gratzl, Z. Lucarelli, B. Moyle, J. Smiley, and T. Dwyer (2021). "Uplift: A Tangible and Immersive Tabletop System for Casual Collaborative Visual Analytics". In: *IEEE Transactions on Visualization and Computer Graphics* 27.2, pp. 1193–1203. DOI: 10.1109/TVCG.2020.3030334.

Gajos, K. Z., M. Czerwinski, D. S. Tan, and D. S. Weld (2006). "Exploring the Design Space for Adaptive Graphical User Interfaces". In: *Proceedings of the Conference on Advanced Visual Interfaces (AVI)*. ACM Press, pp. 201–208. DOI: 10.1145/1133265.1133306.

Gleicher, M. (2018). "Considerations for Visualizing Comparison". In: *IEEE Transactions on Visualization and Computer Graphics* 24.1, pp. 413–423. DOI: 10.1109/TVCG.2017.2744199.

Gleicher, M., D. Albers, R. Walker, I. Jusufi, C. D. Hansen, and J. C. Roberts (2011). "Visual Comparison for Information Visualization". In: *Information Visualization* 10.4, pp. 289–309. DOI: 10.1177/1473871611416549.

Guo, D., J. Chen, A. M. MacEachren, and K. Liao (2006). "A Visualization System for Space-Time and Multivariate Patterns (VIS-STAMP)". In: *IEEE Transactions on Visualization and Computer Graphics* 12.6, pp. 1461–1474. DOI: 10.1109/TVCG.2006.84.

Hauser, H., F. Ledermann, and H. Doleisch (2002). "Angular Brushing of Extended Parallel Coordinates". In: *Proceedings of the IEEE Symposium Information Visualization (InfoVis)*. IEEE Computer Society, pp. 127–130. DOI: 10.1109/INFVIS.2002.1173157.

Heer, J. and G. Robertson (2007). "Animated Transitions in Statistical Data Graphics". In: *IEEE Transactions on Visualization and Computer Graphics* 13.6, pp. 1240–1247. DOI: 10.1109/tvcg.2007.70539.

Henriksen, K., J. Sporring, and K. Hornbæk (2004). "Virtual Trackballs Revisited". In: *IEEE Transactions on Visualization and Computer Graphics* 10.2, pp. 206–216. DOI: 10.1109/tvcg.2004.1260772.

Hochheiser, H. and B. Shneiderman (2004). "Dynamic Query Tools for Time Series Data Sets: Timebox Widgets for Interactive Exploration". In: *Information Visualization* 3.1, pp. 1–18. DOI: 10.1057/palgrave.ivs.9500061.

Hoffswell, J., W. Li, and Z. Liu (2020). "Techniques for Flexible Responsive Visualization Design". In: *Proceedings of the SIGCHI Conference on Human Factors in Computing Systems (CHI)*. ACM Press, pp. 1–13. DOI: 10.1145/3313831.3376777.

Holz, C. and S. Feiner (2009). "Relaxed Selection Techniques for Querying Time-Series Graphs". In: *Proceedings of the ACM Symposium on User Interface Software and Technology (UIST)*. ACM Press, pp. 213–222. DOI: 10.1145/1622176.1622217.

Jakobsen, M. R., Y. S. Haile, S. Knudsen, and K. Hornbæk (2013). "Information Visualization and Proxemics: Design Opportunities and Empirical Findings". In: *IEEE Transactions on Visualization and Computer Graphics* 19.12, pp. 2386–2395. DOI: 10.1109/TVCG.2013.166.

Jankun-Kelly, T. J., K.-L. Ma, and M. Gertz (2007). "A Model and Framework for Visualization Exploration". In: *IEEE Transactions on Visualization and Computer Graphics* 13.2, pp. 357–369. DOI: 10.1109/tvcg.2007.28.

Keefe, D. F. and T. Isenberg (2013). "Reimagining the Scientific Visualization Interaction Paradigm". In: *Computer* 46.5, pp. 51–57. DOI: 10.1109/MC.2013.178.

Kondo, B. and C. Collins (2014). "DimpVis: Exploring Time-varying Information Visualizations by Direct Manipulation". In: *IEEE Transactions on Visualization and Computer Graphics* 20.12, pp. 2003–2012. DOI: 10.1109/TVCG.2014.2346250.

Krasner, G. E. and S. T. Pope (1988). "A Cookbook for Using the Model-View-Controller User Interface Paradigm in Smalltalk-80". In: *Journal of Object-Oriented Programming* 1.3, pp. 26–49.

L'Yi, S., J. Jo, and J. Seo (2021). "Comparative Layouts Revisited: Design Space, Guidelines, and Future Directions". In: *IEEE Transactions on Visualization and Computer Graphics* 27.2, pp. 1525–1535. DOI: 10.1109/TVCG.2020.3030419.

Lam, H. (2008). "A Framework of Interaction Costs in Information Visualization". In: *IEEE Transactions on Visualization and Computer Graphics* 14.6, pp. 1149–1156. DOI: 10.1109/TVCG.2008.109.

Lee, B., P. Isenberg, N. H. Riche, and S. Carpendale (2012). "Beyond Mouse and Keyboard: Expanding Design Considerations for Information Visualization Interactions". In: *IEEE Transactions on Visualization and Computer Graphics* 18.12, pp. 2689–2698. DOI: 10.1109/TVCG.2012.204.

Liu, Z. and J. Heer (2014). "The Effects of Interactive Latency on Exploratory Visual Analysis". In: *IEEE Transactions on Visualization and Computer Graphics* 20.12, pp. 2122–2131. DOI: 10.1109/TVCG.2014.2346452.

Luboschik, M., C. Maus, H.-J. Schulz, H. Schumann, and A. Uhrmacher (2012). "Heterogeneity-Based Guidance for Exploring Multiscale Data in Systems Biology". In: *Proceedings of the IEEE Symposium on Biological Data Visualization (BioVis)*. IEEE Computer Society, pp. 33–40. DOI: 10.1109/BioVis.2012.6378590.

Mannino, M. and A. Abouzied (2018). "Expressive Time Series Querying with Hand-Drawn Scale-Free Sketches". In: *Proceedings of the SIGCHI Conference on Human Factors in Computing Systems (CHI)*. ACM Press. DOI: 10.1145/3173574.3173962.

Monroe, M., R. Lan, J. M. del Olmo, B. Shneiderman, C. Plaisant, and J. Millstein (2013a). "The Challenges of Specifying Intervals and Absences in Temporal Queries: A Graphical Language Approach". In: *Proceedings of the SIGCHI Conference on Human Factors in Computing Systems (CHI)*. ACM Press, pp. 2349–2358. DOI: 10.1145/2470654.2481325.

Norman, D. A. (2013). *The Design of Everyday Things*. Revised and expanded edition. Basic Books.

Piringer, H., C. Tominski, P. Muigg, and W. Berger (2009). "A Multi-Threading Architecture to Support Interactive Visual Exploration". In: *IEEE Transactions on Visualization and Computer Graphics* 15.6, pp. 1113–1120. DOI: 10.1109/TVCG.2009.110.

Pulo, K. (2007). "Navani: Navigating Large-Scale Visualisations with Animated Transitions". In: *Proceedings of the International Conference Information Visualisation (IV)*. IEEE Computer Society, pp. 271–276. DOI: 10.1109/iv.2007.82.

Reinders, F., F. H. Post, and H. J. W. Spoelder (2001). "Visualization of Time-Dependent Data with Feature Tracking and Event Detection". In: *The Visual Computer* 17.1, pp. 55–71. DOI: 10.1007/pl00013399.

Riehmann, P., J. Reibert, J. Opolka, and B. Fröhlich (2018). "Touch the Time: Touch-Centered Interaction Paradigms for Time-Oriented Data". In: *Proceedings of the Eurographics / IEEE Conference on Visualization (EuroVis) - Short Papers*. Eurographics Association, pp. 113–117. DOI: 10.2312/eurovisshort.20181088.

Sadri, R., C. Zaniolo, A. Zarkesh, and J. Adibi (2004). "Expressing and Optimizing Sequence Queries in Database Systems". In: *ACM Transactions on Database Systems* 29.2, pp. 282–318. DOI: 10.1145/1005566.1005568.

Schulz, H.-J., M. Streit, T. May, and C. Tominski (2013b). *Towards a Characterization of Guidance in Visualization*. Poster at IEEE Conference on Information Visualization (InfoVis). Atlanta, USA.

Schwab, M., S. Hao, O. Vitek, J. Tompkin, J. Huang, and M. A. Borkin (2019a). "Evaluating Pan and Zoom Timelines and Sliders". In: *Proceedings of the SIGCHI Conference on Human Factors in Computing Systems (CHI)*. ACM Press, pp. 1–12. DOI: 10.1145/3290605.3300786.

Schwab, M., J. Tompkin, J. Huang, and M. A. Borkin (2019b). "EasyPZ.js: Interaction Binding for Pan and Zoom Visualizations". In: *IEEE Visualization Conference, IEEE VIS 2019 - Short Papers*. IEEE Computer Society, pp. 31–35. DOI: 10.1109/VISUAL.2019.8933747.

Shaer, O. and E. Hornecker (2009). "Tangible User Interfaces: Past, Present and Future Directions". In: *Foundations and Trends in Human-Computer Interaction* 3.1-2, pp. 1–137. DOI: 10.1561/1100000026.

Shao, L., A. Mahajan, T. Schreck, and D. J. Lehmann (2017). "Interactive Regression Lens for Exploring Scatter Plots". In: *Computer Graphics Forum* 36.3, pp. 157–166. DOI: 10.1111/cgf.13176.

Shneiderman, B. (1983). "Direct Manipulation: A Step Beyond Programming Languages". In: *IEEE Computer* 16.8, pp. 57–69. DOI: 10.1109/mc.1983.1654471.

Shneiderman, B. (1994). "Dynamic Queries for Visual Information Seeking". In: *IEEE Software* 11.6, pp. 70–77. DOI: 10.1109/52.329404.

Shneiderman, B. (1996). "The Eyes Have It: A Task by Data Type Taxonomy for Information Visualizations". In: *Proceedings of the IEEE Symposium on Visual Languages*. IEEE Computer Society, pp. 336–343. DOI: 10.1109/VL.1996.545307.

Spence, R. (2007). *Information Visualization: Design for Interaction*. 2nd edition. Prentice-Hall.

Spindler, M., C. Tominski, H. Schumann, and R. Dachselt (2010). "Tangible Views for Information Visualization". In: *Proceedings of the International Conference on Interactive Tabletops and Surfaces (ITS)*. ACM Press, pp. 157–166. DOI: 10.1145/1936652.1936684.

Srinivasan, A. and J. T. Stasko (2018). "Orko: Facilitating Multimodal Interaction for Visual Exploration and Analysis of Networks". In: *IEEE Transactions on Visualization and Computer Graphics* 24.1, pp. 511–521. DOI: 10.1109/TVCG.2017.2745219.

Stolper, C. D., A. Perer, and D. Gotz (2014). "Progressive Visual Analytics: User-Driven Visual Exploration of In-Progress Analytics". In: *IEEE Transactions on Visualization and Computer Graphics* 20.12, pp. 1653–1662. DOI: 10.1109/TVCG.2014.2346574.

Thomas, J. J. and K. A. Cook (2005). *Illuminating the Path: The Research and Development Agenda for Visual Analytics*. IEEE Computer Society.

Tominski, C. (2011). "Event-Based Concepts for User-Driven Visualization". In: *Information Visualization* 10.1, pp. 65–81. DOI: 10.1057/ivs.2009.32.

Tominski, C. (2015). *Interaction for Visualization*. Synthesis Lectures on Visualization 3. Morgan & Claypool. DOI: 10.2200/S00651ED1V01Y201506VIS003.

Tominski, C., J. Abello, and H. Schumann (2004). "Axes-Based Visualizations with Radial Layouts". In: *Proceedings of the ACM Symposium on Applied Computing (SAC)*. ACM Press, pp. 1242–1247. DOI: 10.1145/967900.968153.

Tominski, C., C. Forsell, and J. Johansson (2012a). "Interaction Support for Visual Comparison Inspired by Natural Behavior". In: *IEEE Transactions on Visualization and Computer Graphics* 18.12, pp. 2719–2728. DOI: 10.1109/TVCG.2012.237.

Tominski, C., S. Gladisch, U. Kister, R. Dachselt, and H. Schumann (2017). "Interactive Lenses for Visualization: An Extended Survey". In: *Computer Graphics Forum* 36.6, pp. 173–200. DOI: 10.1111/cgf.12871.

Tominski, C. and H. Schumann (2008). "Enhanced Interactive Spiral Display". In: *Proceedings of the Annual Conference of the Swedish Computer Graphics Association (SIGRAD)*. Linköping University Electronic Press, pp. 53–56. URL: https://www.ep.liu.se/ecp/034/013/ecp083413.pdf.

Tominski, C. and H. Schumann (2020). *Interactive Visual Data Analysis*. AK Peters Visualization Series. CRC Press. DOI: 10.1201/9781315152707.

Voida, S., M. Tobiasz, J. Stromer, P. Isenberg, and S. Carpendale (2009). "Getting Practical with Interactive Tabletop Displays: Designing for Dense Data, Fat Fingers, Diverse Interactions, and Face-to-face Collaboration". In: *Proceedings of the International Conference on Interactive Tabletops and Surfaces (ITS)*. ACM Press, pp. 109–116. DOI: 10.1145/1731903.1731926.

Yi, J. S., Y. ah Kang, J. T. Stasko, and J. A. Jacko (2007). "Toward a Deeper Understanding of the Role of Interaction in Information Visualization". In: *IEEE Transactions on Visualization and Computer Graphics* 13.6, pp. 1224–1231. DOI: 10.1109/TVCG.2007.70515.

Chapter 6
Computational Analysis Support

> It is useful to think of the human and the computer together as a single cognitive entity, with the computer functioning as a kind of *cognitive coprocessor* to the human brain. [...] Each part of the system is doing what it does best. The computer can pre-process vast amounts of information. The human can do rapid pattern analysis and flexible decision making.

Ware (2008, p. 175)

Visualization and interaction as described in the previous Chapters 4 and 5 help users to visually analyze time-oriented data. Following Shneiderman's *information seeking mantra* (see p. 130), analysts can look at the data, explore them, and in this way understand them. This is possible thanks to human visual perception and the fact that humans are quite good at recognizing patterns, finding interesting and unexpected solutions, combining knowledge from different sources, and being creative in general.[1] Purely interactive and visual data analysis works well unless the problem to be solved exceeds a certain size. With massive, heterogeneous, dynamic, and ambiguous data, it becomes increasingly difficult to create overview visualizations without losing interesting patterns, and human observers have a hard time interpreting and understanding the data. Therefore, Keim et al. (2006a) revised and expanded Shneiderman's mantra in order to indicate that it is not sufficient to just retrieve and display the data using a visual and interactive approach. In fact, it is necessary to computationally analyze the data according to aspects of interest, to show the most relevant features of the data, and at the same time to provide interaction methods that allow the user to get details of the data on demand:

Analyze First -
Show the Important -
Zoom, Filter and Analyse Further -
Details on Demand.

Keim et al. (2006a, p. 6)

---

[1] Wegner (1997) makes some interesting statements about why interaction is better than algorithms.

© The Author(s) 2023
W. Aigner et al., *Visualization of Time-Oriented Data*, Human–Computer Interaction Series, https://doi.org/10.1007/978-1-4471-7527-8_6

Following this mantra, we can utilize the proficiency of computing systems to assist the knowledge crystallization from time-oriented data. Apparently, if the problem size is sufficiently large, computers are better (i.e., faster and more accurate) than humans at numeric and symbolic calculations, logical deduction, and searching. In general, *data mining* and *knowledge discovery* are commonly defined as the application of algorithms to extract useful structures from large volumes of data, where knowledge discovery explicitly demands that knowledge be the end product of the analytical calculations (see Fayyad et al., 1996; Fayyad et al., 2001; Han et al., 2012). A variety of concepts and methods are involved in achieving this goal, including databases, statistics, artificial intelligence, neural networks, machine learning, information retrieval, pattern recognition, data visualization, and high-performance computing.

This chapter will illustrate how automatic analytical calculations can be utilized to facilitate the exploration and analysis of larger and more complex time-oriented data. To this end, we will give a brief overview of typical temporal analysis tasks and ground these methods as *temporal data abstraction*. For selected tasks, we will present examples that demonstrate how visualization can benefit from considering analytical support. Our descriptions will intentionally be kept at a basic level. For details on the sometimes quite complex matter of temporal data analysis, we refer interested readers to the relevant literature.

6.1 Temporal Analysis Tasks

Temporal analysis and temporal data mining are concerned with extracting useful information from time-oriented data (see Brockwell and Davis, 1991; Antunes and Oliveira, 2001; Mitsa, 2010; Ali et al., 2019). More specifically, Laxman and Sastry (2006) characterize the following categories of temporal data analysis tasks:

Classification Given a predefined set of classes, the goal of classification is to determine which class a dataset, sequence, or subsequence belongs to. As a specific instance of classification, *segmentation and labeling* applies algorithms to divide multivariate time-oriented data into smaller segments and to assign to these segments class labels accordingly. Applications such as speech recognition and gesture recognition apply classification to identify spoken words or performed interactions. The analysis of sensor data or spatio-temporal movement data often requires segmentation and labeling to make the enormous volumes of data to be handled manageable.

Clustering Clustering is concerned with grouping data into clusters based on similarity, where the similarity measure used is a key aspect of the clustering process. In the context of time-oriented data, it makes sense to cluster similar time series or subsequences of them. For example, in the analysis of financial data, one may be interested in stocks that exhibit similar behavior over time. In contrast to classification, where the classes are known beforehand, clusters are not defined upfront but crystallize during the computational analysis.

Search & retrieval This task encompasses searching for a-priori specified queries in possibly large volumes of data. This is often referred to as *query-by-example*. Search & retrieval can be applied to locate exact matches for an example query or approximate matches. In the latter case, similarity measures are needed that define the degree of exactness or fuzziness of the search (e.g., to find customers whose spending patterns over time are similar but not necessarily equal to a given spending profile).

Pattern discovery While search & retrieval requires a predefined query, pattern discovery is concerned with *automatically* discovering interesting patterns in the data (without any a-priori assumptions). The term *pattern*[2] usually covers a variety of meanings, including sequential pattern and periodic pattern, but also temporal association rules. In a sense, a pattern can be understood as a local structure in the data or combinations thereof. Often, frequently occurring patterns are of interest, for example when analyzing whether a TV commercial actually leads to an increase in sales. But patterns that occur very rarely can also be interesting because they might indicate failures or malicious behavior.

Prediction An important task in analyzing time-oriented data is the prediction of likely future behavior. The goal is to infer from data collected in the past and present how the data will evolve in the future. To achieve this goal, one has to build a predictive model for the data first. Examples of such models are autoregressive models, non-stationary and stationary models, or rule-based models.

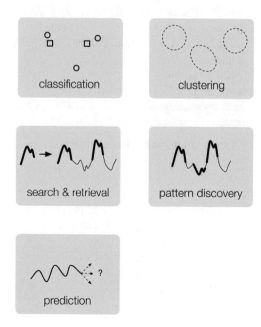

Fig. 6.1 Overview of temporal analysis tasks. ⓒ⒤ *The authors.*

---

[2] Andrienko et al. (2021) provide a deeper theoretical discussion of patterns.

The five fundamental temporal data analysis tasks are summarized in Figure 6.1. A variety of methods have to play in concert in order to accomplish these tasks. Statistical aggregation operators (e.g., sum, average, minimum, and maximum), methods from time-series analysis, and dedicated temporal data mining techniques are needed. For more details on the involved models and algorithms, the interested reader is referred to Laxman and Sastry (2006).

In the context of visualizing time-oriented data, these tasks share the common goal of temporal *data abstraction* in order to reduce the workload when computing visual representations and to keep the perceptual efforts required to interpret them low. For classification and clustering, we abstract from the raw data and work with classes and clusters. For search & retrieval and pattern discovery, we are primarily interested in relevant patterns and de-emphasize irrelevant data. For prediction, we focus on the future. In the following, we clarify the idea of temporal data abstraction and give a couple of examples afterward.

6.2 Principles of Temporal Data Abstraction

In practice, time-oriented datasets are often large and complex and originate from heterogeneous sources. The challenging question is how huge volumes of possibly continuously measured data can be analyzed to support decision-making. On the one hand, the data are too large to be interpreted all at once. On the other hand, the data are more erroneous than usually expected and some data are missing as well, a problem that we discussed in the context of data quality in Section 3.4. What is needed is a way to abstract the data in order to make them eligible for subsequent visualization.

The term *data abstraction* was originally introduced by Clancey (1985) in his classic proposal on heuristic classification (as the paper calls it). In the context of visual data analysis, Thomas and Cook (2005) describe what data abstraction is about:

> The objective is "to create an abstraction that conveys key ideas while suppressing irrelevant details."
>
> > Thomas and Cook (2005, p. 86) using, in quotation
> > marks, the words of Foley (2000, p. 67)

The basic idea is to use qualitative values, classes, or concepts, rather than raw data, for further analysis or visualization processes (see Lin et al., 2007; Combi et al., 2010). This helps in coping with data size and data complexity. To arrive at suitable data abstractions, several tasks must be conducted, including selecting relevant information, filtering out unneeded information, performing calculations, sorting, and grouping.

Let us now illustrate the concept of *temporal data abstraction* in medical contexts with a simple example. Figure 6.2 shows time-oriented data as generated when monitoring newborn infants that have to be ventilated artificially. The figure visualizes three variables plotted as points against a horizontal time axis: S_aO_2 (arterial oxygen saturation), $P_{tc}CO_2$ (transcutaneous partial pressure of carbon dioxide), and P_aCO_2 (arterial partial pressure of carbon dioxide). S_aO_2 and $P_{tc}CO_2$ are measured continuously at a regular rate, but with different frequencies. New values for P_aCO_2 arrive irregularly and some values for $P_{tc}CO_2$ are missing.

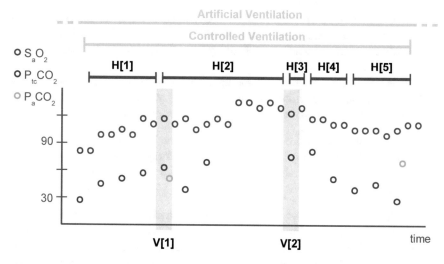

Fig. 6.2: Temporal data abstraction in the context of artificial ventilation. Vertical temporal abstractions are illustrated as V[1] and V[2] and horizontal temporal abstractions are illustrated as H[1]–H[5]. The context is given as "artificial ventilation" and its sub-context "controlled ventilation". ©① *The authors.*

The aim of temporal data abstraction is to arrive at qualitative values or patterns over time intervals. *Vertical* temporal abstraction (illustrated in V[1] and V[2]) considers multiple variables over a particular time point and combines them into a qualitative value or pattern. *Horizontal* temporal abstraction (illustrated as H[1]–H[5]) infers a qualitative value or pattern from one or more variables and a corresponding time interval. Usually, the abstraction process is context-dependent. In Figure 6.2, the abstraction is done in the context of artificial ventilation and in the sub-context of controlled ventilation.

In medical applications, there are different types of abstraction methods, ranging from rather simple to quite complicated ones. However, as pointed out by Combi et al. (2010), no exhaustive schema exists to categorize the available methods. Nevertheless, the common understanding is that even in very simple cases the process is knowledge-driven. The use of knowledge is the main characteristic that distinguishes data abstraction from statistical data analysis (e.g., trend detection using time-series analysis).

Simple methods involve single data values and usually do not need to consider time specifically. They generate vertical abstractions. The knowledge used are concept associations or concept taxonomies. Combi et al. (2010) distinguish three types of simple methods:

- *Qualitative abstraction* means converting numeric expressions to qualitative expressions. For example, the numeric value of 34.8°C of body temperature can be abstracted to the qualitative value "hypothermia".
- *Generalization abstraction* involves a mapping of instances into groups. For example, "hand-bagging is administered" is abstracted to "manual intervention is administered", where "hand-bagging" is an instance of the concept group "manual intervention".
- *Definitional abstraction* is a mapping across different concept categories. The movement here is not within the same concept taxonomy, as for the generalization abstraction, but across two different concept taxonomies.

More complex methods consider one or more variables jointly and specifically integrate the dimension of time in a kind of temporal reasoning. These methods generate horizontal temporal abstractions. According to Combi et al. (2010), four types of complex methods exist:

- *Merge (or state) abstraction* is the process of deriving maximal time intervals for which some constraints of interest hold. For example, several consecutive days with high fever and increased blood values can be mapped to "bed-ridden".
- *Persistence abstraction* means applying persistence rules to project maximal intervals for some property, both backward and forward in time. For example, "headache in the morning" can be abstracted to "headache in the evening before" or "headache in the afternoon afterward".
- *Trend (or gradient or rate) abstraction* is concerned with deriving significant changes and rates of change in the progression of some variable. For example, $P_{tc}CO_2$ has decreased from 130 to 90 in the last 20 minutes would result in "$P_{tc}CO_2$ is decreasing too fast".
- *Periodic abstraction* aims to derive repetitive occurrence, with some regularity in the pattern of repetition. For example, "headache every morning, but not during the day" would result in "repetitive headache in the morning".

In what follows, we demonstrate the applicability of temporal data abstraction methods for the analysis of time-oriented data using three examples: classification, clustering, and principal component analysis. Classification reduces data complexity by deriving qualitative statements, which are much easier to understand. Clustering decreases the number of data items to be represented and supports discerning similarities and unexpected behavior. Principal component analysis decreases the number of time-dependent variables by switching the focus to major trends in the data.

6.3 Classification via Segmentation and Labeling

Given a set of classes, segmentation and labeling splits long time series into segments (segmentation) and assigns to each segment a class (labeling). By doing so, the complexity of time-oriented data can be reduced substantially. The segmented and labeled data correspond to qualitative abstractions, which are simpler than the raw data, can be visualized more compactly, and hence, are easier to comprehend.

Segmentation algorithms can be distinguished by the type of data they process into algorithms for discrete domains (e.g., Cohen et al., 2002) and algorithms for continuous domains (e.g., Lin et al., 2002). Labeling algorithms can be divided into *supervised* methods, which use already labeled training data, and *unsupervised* methods, which seek hidden structures in unlabeled data autonomously. Supervised methods apply a model to partially or completely labeled segments (see Xing et al., 2010). Unsupervised methods calculate a grouping that can be used for further aggregation and analysis (see Warren Liao, 2005).

Segmentation and labeling as outlined above can be applied in various ways. In the following, we will describe several examples illustrating the wide range of involved methods and visual representations.

6.3.1 Data Classification in Medical Contexts

A specific area in medicine where time-oriented data play a crucial role is in monitoring the health of patients based on sensor data. In particular, the health of artificially ventilated infants is of great concern to medical personnel and parents alike. Addressing this challenging application domain, Miksch et al. (1996) developed VIE-VENT as an open-loop knowledge-based monitoring and therapy planning system.

In order to derive qualitative abstractions for different kinds of temporal trends (i.e., very-short, short, medium, and long-term trends) from continuously arriving quantitative data, the system utilizes context-sensitive and expectation-guided methods and incorporates background knowledge about data points, data intervals, and expected qualitative trend patterns. Smoothing and adjustment mechanisms help to keep qualitative abstractions stable in case of shifting contexts or data oscillating near thresholds. Context-aware schemata for data point transformation and curve fitting are used to express the dynamics of and the reaction to different data abnormalities. For example, during intermittent positive pressure ventilation (ippv), the transformation of the quantitative value $P_{tc}CO_2 = 56mmHg$ results in the qualitative abstraction "$P_{tc}CO_2$ substantially above target range". During intermittent mandatory ventilation (imv) however, $56mmHg$ represents the "target value". Qualitative abstractions and schemata of curve fitting are subsequently used to decide if the value progression happens too fast, at a normal rate, or too slow.

Figure 6.3 shows the user interface of VIE-VENT. In the top-left corner, the system displays exact values of the quantitative blood gas measurements CO_2, O_2, SaO_2. Arrows depict trends and qualitative abstractions are indicated by different

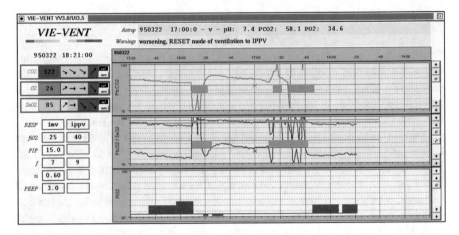

Fig. 6.3: VIE-VENT displays measured quantitative values as line plots. Qualitative abstractions and trends are represented by different colors and arrows in the top three boxes on the left. © *1996 Elsevier. Reprinted, with permission, from Miksch et al. (1996).*

colors (e.g., deep pink represents "extremely above target range"). The left panel further shows current and recommended ventilator settings in blue and red boxes, respectively. The right-hand side shows line plots of the most important variables for the last four hours.

In the context of medical data, strongly oscillating sensor signals pose a particular challenge for segmentation and labeling. The problem is that the derived data abstractions could change too quickly as to be interpretable by the observer. Therefore, Miksch et al. (1999) developed the Spread approach, a method for deriving steady qualitative abstractions from oscillating high-frequency data. It performs the following steps to classify the data:

1. *Eliminate data errors.* Sometimes up to 40% of the input data are obviously erroneous, i.e., exceed the limits of plausible values.
2. *Clarify the curve.* Transform the still noisy data into the *spread*, which is a steady curve with some additional information about the distribution of the data along that curve.
3. *Qualify the curve.* Abstract from quantitative values to qualitative values (i.e., the classes) like "normal" or "high". Concatenate contiguous segments labeled with the same class.

Figure 6.4 illustrates how the Spread approach can enhance the visual analysis. The Spread (in red) smooths out the strongly oscillating raw data (black line plot). Even the increased oscillation in the center of the display is dealt with gracefully: it leads to an increased width of the spread, but not to a change of the qualitative abstraction (in blue). With these abilities, the Spread can support physicians in making better qualitative assessments of otherwise difficult-to-interpret data.

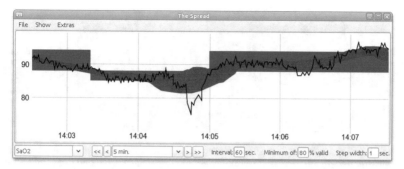

Fig. 6.4: The thin line shows the raw data. The red area depicts the *spread* and the blue rectangles represent the derived temporal intervals of steady qualitative values. The lower part of the figure shows the parameter settings. ©① *The authors. Adapted from Miksch et al. (1999).*

The above examples can only indicate the possible benefits that temporal data abstraction methods and their integration with the visualization can have in medical applications. We know of quite positive feedback from medical experts who found it easy to capture the health conditions of their patients. Moreover, these qualitative abstractions can be used for further reasoning or in guideline-based care for a simplified representation of treatment plans. For more medical examples, we refer to the survey of segmentation and labeling in clinical data analysis by Stacey and McGregor (2007).

6.3.2 Segmentation and Labeling of Multivariate Time Series

Segmenting and labeling multivariate time-oriented data is a problem that is difficult to solve automatically. Therefore, it makes sense to involve human expertise in the process. To this end, Bernard et al. (2018) propose a pipeline that consists of several steps operating on several data artifacts (see Figure 6.5). The key idea of this pipeline is to combine the different algorithms (A), their adequate parametrization (B), and the visual exploration of the parametrizations, the results, and their uncertainty (C).

While these concerns are usually handled separately, the pipeline-based approach tightly interconnects them to generate better results and also to support humans to develop a better understanding of the data and the data abstraction process. In particular, the explicit consideration of parametrizations and uncertainty makes the process transparent in terms of how segmentation results are generated and how (un)certain they are. Moreover, the pipeline is general and can be applied to various use cases and application domains, which might require the definition of dedicated algorithmic routines for specific time-oriented data.

By instantiating the pipeline, one can build upon the great variety of available segmentation and labeling algorithms to derive meaningful abstractions from multivariate time series. To make this possible, the unifiable characteristics of the involved

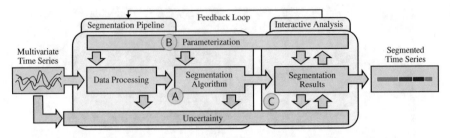

Fig. 6.5: Pipeline for the segmentation of multivariate time series. Data processing routines and segmentation algorithms process the multivariate time series (A), which requires setting several parameters (B). The segmentation results and information about involved uncertainties, which is collected throughout the pipeline, can be explored visually (C). ⓒⓘ *The authors. Adapted from Bernard et al. (2018).*

algorithms need to be combined into an appropriate software interface. Based on that, visual interfaces can be built for steering the algorithms, running them with different parametrizations, and visualizing parameters, results, and uncertainties. The benefit for the users is that they can experiment with different algorithms and their parametrizations to find the ones that yield the most meaningful results for a particular dataset or application problem.

Figure 6.6 shows an example interface from the VISSECT project (see Bernard et al., 2018) for a radiation observation dataset. On the left side (A), we can see the main steps of the general segmentation workflow. It starts with the selection of the data source and moves on to the parameter visualization and uncertainty analysis, all the way to the user feedback module. The screenshot shows three different

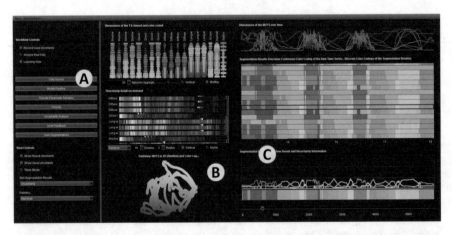

Fig. 6.6: Segmentation and labeling of radiation observations. (A) General segmentation workflow. (B) Data perspective including raw data, characteristics of individual variables, and dimensionality-reduced plot for patterns. (C) Overview and details of the multivariate time series and the segmentation results. ⓒⓘ *Courtesy of Jürgen Bernard.*

visualizations of multivariate time series, highlighting different characteristics (B). On the right side (C), we see the multivariate time series as line plots (top) and the juxtaposed segmentation results for different parametrizations for the selected algorithms (center). The user can also look at the details of one particular segmentation (bottom).

Eichner et al. (2020) provide a detailed discussion of how the combined visualization of parametrizations and segmentation results can facilitate understanding the influence of different algorithmic configurations on the segmentation and labeling pipeline. The different kinds of uncertainty (e.g., value uncertainty, result uncertainty, aggregation uncertainty, and cause & effect uncertainty) that stem from the selection of algorithms, parameters, and the calculation of multiple competing results are discussed by Bors et al. (2019). In VISSECT, uncertainty is consequently considered along all steps of the segmentation process. Through adequate visual representations of the uncertainty information (see Gschwandtner et al., 2016), users can better quantify and evaluate the various sources of uncertainty and so better understand the quality of the data abstraction obtained.

6.3.3 Linking Temporal and Visual Abstraction

The previous examples indicate that dedicated visual representations are applied to convey data abstractions derived from classification procedures. It is worth mentioning that in interactive environments, the visualization of time-oriented data and abstractions thereof can change dynamically due to user interaction, typically during navigation and zooming (see Chapter 5).

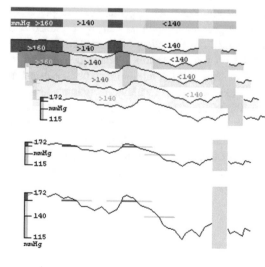

Fig. 6.7 Different steps of semantic zooming of a time-series visualization from a broad overview with qualitative values (top) to a detailed view with fine structures and quantitative details (bottom). Gray areas indicate missing data. ⓒⓘ *The authors.*

In such scenarios, the visualization must be able to capture as much temporal information as possible without losing overview and details, even if the available display space is very limited. Figure 6.7 demonstrates that this is possible by means of *semantic zooming*, which was introduced in Section 5.2.2. The idea is to combine temporal data abstractions with an appropriate set of visual abstractions for different levels of detail. For this purpose, Bade et al. (2004) propose reducing the graphical details in the visual representation when the available display space becomes smaller (↪ p. 340). Instead of showing a full-detail line plot, only colored segments are represented when reaching the highest level of visual abstraction.

Depending on the available display space (or the current zoom level), a suitable temporal abstraction is selected automatically and its corresponding visual abstraction is displayed. The advantage of this procedure is that it relieves the user of managing the levels of abstraction by hand. Moreover, the semantic zoom corresponds much better with the interactive nature of flexible and dynamic visual analysis scenarios.

In summary, the classification of time-oriented data via segmentation and labeling involves various algorithms and benefits from visualizations of parameter dependencies and uncertainty, allowing users to interactively steer the analysis toward promising and meaningful results. In the next section, we discuss clustering, which, in contrast to using predefined classes, aims to abstract time-oriented data into groups based on similarity.

6.4 Clustering Time Series

In general, grouping data into clusters and concentrating on the clusters rather than on individual data values makes it possible to analyze much larger datasets. Appropriate *distance* or *similarity measures* lay the groundwork for clustering. Distance and similarity measures are profoundly application-dependent and range from average geometric distance to measures based on longest common subsequences and to measures based on probabilistic models. Based on computed distances, clustering methods create groups of data, where the number of available techniques is large, including hierarchical clustering, partitional clustering, and sequential clustering.

Due to the diversity of methods, selecting appropriate algorithms is typically difficult. Careful adjustment of parameters and regular validation of the results are therefore essential steps in the process of clustering. More details on clustering methods and distance measures can be found in the work by Jain et al. (1999), Gan et al. (2007), and Xu and Wunsch II (2009).

Clustering and calendar-based visualization

A prominent example of how clustering can assist the visualization of time-oriented data is the work by van Wijk and van Selow (1999). The goal is to identify common and uncommon behavior in data with very many time series and to understand their distribution over time. The problem is that simply drawing line plots for all time series is not a satisfactory solution due to the overwhelmingly large number of time points and line plots. In order to tackle this problem, clustering methods and a calendar-based visualization are used.

In particular, the approach by van Wijk and van Selow (1999) works as follows. As the starting point, k daily time series describe some observed variable over the course of a day. The clustering process starts with the k daily series as the initial clusters. Figure 6.8 shows them at the bottom for an example with $k = 7$. The next step is to compute the differences between all pairs of clusters and to merge the two most similar clusters into a new cluster (i.e., an aggregated representative of the two clusters). This step runs repeatedly and results in a *clustering hierarchy* with $2k - 1$ clusters, where the root of the hierarchy represents the entire dataset as an aggregated abstraction.

Given the clustering hierarchy, we may now engage in two analysis tasks: (1) assess similarities among daily behavior and (2) locate common and uncommon days in time. A corresponding visualization of the clustered daily time series can

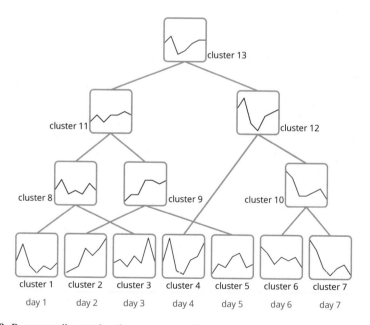

Fig. 6.8: By repeatedly merging the two most similar time series into new clusters, a clustering hierarchy is generated. The root cluster is an aggregated representative time series of the entire dataset. ⓒⓘ *The authors.*

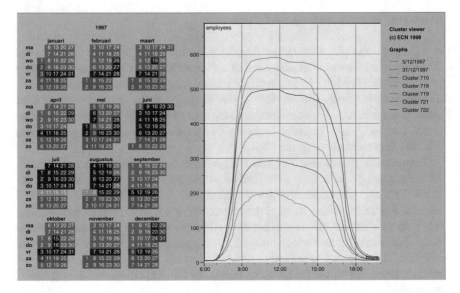

Fig. 6.9: Visual analysis of the number of employees at work. Day patterns for selected days and clusters are visualized as line plots (right). Individual days in a calendar display (left) are colored according to cluster affiliation. © *1999 IEEE. Reprinted, with permission, from van Wijk and van Selow (1999).*

then use two different views to support the two different tasks as shown in Figure 6.9. The first task is facilitated by a basic line plot (↪ p. 233) that shows a selected number of clusters, where each plot uses a unique color. To accomplish the second task, a calendar display is used where individual days are color-coded according to cluster affiliation. This way, analysts can see the daily behavior and at the same time understand when during a year this behavior occurs. Various interaction methods allow adjustments of the visual representation and data exploration. In terms of assessing similarities, the user can select a day from the calendar, and with the help of the clustering hierarchy, similar days (and clusters) can be retrieved automatically.

Figure 6.9 shows an example where the data contain the number of employees at work over the course of a day for all days in 1997. The line plot currently shows the concrete number of employees of two days (5/12/1997 and 31/12/1997) and the aggregated number of employees of five clusters (710, 718, 719, 721, and 722). van Wijk and van Selow (1999) demonstrate that several conclusions can be drawn from the visual representation. To give only a few examples:

- Employees follow office hours quite strictly and work between 8:30 am and 5:00 pm in most cases.
- Fewer people work on Fridays during summer (cluster 718).
- During weekends and holidays only very few people are at work (cluster 710).
- It is common practice to take a day off after a holiday (cluster 721).

These and similar statements were more difficult or even impossible to derive without the integration of clustering. For the visual analysis of time-oriented data, van Wijk and van Selow (1999) most convincingly demonstrate the advantages of clustering. While here the benefit lies in the abstraction from raw data to aggregated clusters, we will see in the next section that other kinds of analytical methods are needed if the number of variables gets larger.

6.5 Principal Component Analysis for Time-Oriented Data

Time-oriented data are often multivariate, that is, they contain several time-dependent variables. Visualizing very many variables can be prohibitively challenging. This challenge can be tackled by applying principal component analysis (PCA), which offers an excellent basis for data abstraction (see Jolliffe, 2002; Jackson, 2003; Jeong et al., 2009).

In the following, we will take a brief look at the basics of principal component analysis and illustrate by means of examples the benefit that this analytical concept has for the visual analysis of time-oriented data.

6.5.1 Basic Method

The key principle of PCA is a transformation of the original data space into the principal component space (see Figure 6.10). In the principal component space, the first coordinate, that is, the first principal component represents most of the original dataset's variance; the second principal component, which is orthogonal to the first one, represents most of the remaining variance; and so on. Visualizing the data in the new principal component space shows us how closely individual data records are related to the major trends, and thus PCA helps us to reveal the internal structure of the data. Moreover, since principal components are ordered by their significance, we can focus on fewer principal components than we have variables in our data.

The principal component space with its corresponding principal components can be computed as follows. Assume that we have modeled our multivariate dataset as a matrix:

$$
\mathbf{X} = \left(\mathbf{x_1 x_2} \cdots \mathbf{x_m} \right) = \begin{pmatrix} x_{1,1} & \cdots & x_{1,m} \\ x_{2,1} & \cdots & x_{2,m} \\ \vdots & & \vdots \\ \vdots & & \vdots \\ x_{n,1} & \cdots & x_{n,m} \end{pmatrix}
$$

where the columns of $\mathbf{X}$ correspond to the m variables $\mathbf{x_1}, \mathbf{x_2}, \ldots, \mathbf{x_m}$ of the dataset, and the rows represent n records of data (e.g., m sensor values measured n times).

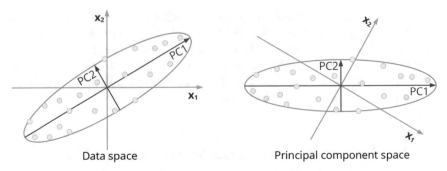

Data space   Principal component space

Fig. 6.10: Principal component analysis transforms multivariate data (with variables x_1 and x_2 in this case) into a new space, the so-called principal component space, which is spanned by the principal components (here PC1 and PC2). ⓒⓘ *The authors.*

For the case of time-oriented data, we would usually assume that one of the x_i is the dimension of time. However, it is important to mention that PCA does not distinguish between independent and dependent variables. In particular, the dimension of time would be processed indiscriminately from time-dependent variables, which would sacrifice the temporal dependencies in the data. Therefore, it is often preferable to exclude time from the analysis and to rejoin time and computed principal components afterward to restore the temporal context.

Moreover, depending on the application it can make sense to prepare the data such that they are mean-centered and normalized (by subtracting off the mean of each variable and scaling each variable according to its variance). Now our goal is to transform the data into the principal component space that is spanned by $r \leq m$ principal components.

For the purpose of explanation, we resort to *singular value decomposition (SVD)* according to which any matrix $\mathbf{X}$ can be decomposed as:

$$\mathbf{X} = \mathbf{W} \cdot \mathbf{\Sigma} \cdot \mathbf{C}^T$$

where $\mathbf{W}$ is an $n \times r$ matrix, $\mathbf{\Sigma}$ is an $r \times r$ diagonal matrix, and $\mathbf{C}^T$ is an $r \times m$ matrix:

$$\mathbf{X} = \begin{pmatrix} w_{1,1} & \cdots & w_{1,r} \\ w_{2,1} & \cdots & w_{2,r} \\ \vdots & & \vdots \\ \vdots & & \vdots \\ w_{n,1} & \cdots & w_{n,r} \end{pmatrix} \cdot \begin{pmatrix} \sigma_1 & \cdots & 0 \\ \vdots & \ddots & \vdots \\ 0 & \cdots & \sigma_r \end{pmatrix} \cdot \begin{pmatrix} c_{1,1} & c_{1,2} & \cdots & c_{1,m} \\ c_{2,1} & c_{2,2} & \cdots & c_{2,m} \\ \vdots & \vdots & & \vdots \\ c_{r,1} & c_{2,2} & \cdots & c_{r,m} \end{pmatrix}$$

The matrix $\mathbf{C}^T$ has in its rows the transposed eigenvectors $\mathbf{c_1}^T, \ldots, \mathbf{c_r}^T$ of the matrix $\mathbf{X}^T\mathbf{X}$, which corresponds to the *covariance matrix* of the original dataset. The $\mathbf{c_i}$ form the orthonormal basis of the principal component space; they are the principal components. Each $\mathbf{c_i}$ is the result of a linear combination of the original variables where the factors (or *loadings*) of the linear combination determine how much the original variables contribute to a principal component. The first principal component c_1 is chosen so as to be the one that captures most of the original data's variance, the second principal component most of the remaining variance, and so forth. The significance values $\sigma_1, \ldots, \sigma_r$ in Σ are determined by the likewise ranked square roots of the eigenvalues $\sqrt{\lambda_1}, \ldots, \sqrt{\lambda_r}$ of the eigenvectors (i.e., the principal components) $c_1, \ldots, c_r$. Finally, the ith row of the matrix W contains the coordinates of the ith data record in the new principal component space. The individual coordinates are often referred to as the *scores*.

This brief formal explanation provides a number of key take-aways. Let us summarize the ones that are most relevant for visualization:

- the significance values determine the ranking of principal components,
- the ranking is the basis for data abstraction, where principal components that bear little information can be omitted,
- the loadings describe the relationship of the original data variables and the principal components, and
- the scores describe the location of the original data records in the principal component space.

We will next demonstrate how PCA can be applied to enhance the visual analysis of time-oriented data. Our general goal is to uncover structure in the data and to reduce the analysis complexity by focusing on significant trends. In the first example, we will see that even a single principal component can bear sufficient information for discerning the main trends in the data. A second example will illustrate how one can determine the principal components to be retained for the visualization as well as the ones that can be omitted due to their low significance.

6.5.2 Gaining Insight into Climate Data with PCA

We consider the visual analysis of a climate dataset that contains daily meteorological observations of temperature (T_{min}, T_{avg}, and T_{max}) for a period of 105 years, which amounts to approximately 38,000 data records (see Nocke et al., 2004). We are only interested in the yearly summer season conditions. Therefore, the daily raw data are first aggregated into yearly data, for which five new variables are calculated for each year:

- *total heat (p1)*: the sum of the maximum temperatures for days with $T_{max} \geq 20°C$
- *summer days (p2)*: the number of days with $T_{max} \geq 25°C$
- *hot days (p3)*: the number of days with $T_{max} \geq 30°C$
- *mean of average (p4)*: the mean of the daily average temperatures T_{avg}
- *mean of extreme (p5)*: the mean of the daily maximum temperatures T_{max}

Apparently, these five quantitative variables are strongly correlated. The generated year-based dataset can be visualized as a centered layer area graph (↪ p. 289), as

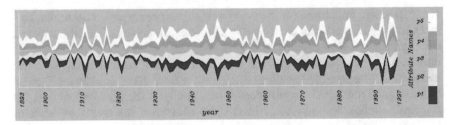

Fig. 6.11: Summer conditions (*p1–p5*) visualized as a centered layer area graph. © *Courtesy of Thomas Nocke.*

illustrated in Figure 6.11. This visual representation is quite useful to get an overview of the data. We can clearly distinguish valleys and peaks in the graph, which indicate particularly cold and hot summers, respectively. The general trend in the data is communicated quite well.

As we will see next, we can confirm our previous findings and gain further insight with the help of PCA and a simple bar graph ($\hookrightarrow$ p. 234). But instead of visualizing all five newly computed variables, our visual analysis will be based on just a single principal component. So, we apply PCA to the five variables derived from the raw data, where the dimension of time is excluded from the PCA, as indicated before. The computed PCA space is then fed to the visualization. In order to restore the temporal context, the bar graph in Figure 6.12 shows time along the horizontal axis, and the first principal component (PC1), to which all variables contribute because of their strong correlation, at the vertical axis. For each year, a bar is constructed

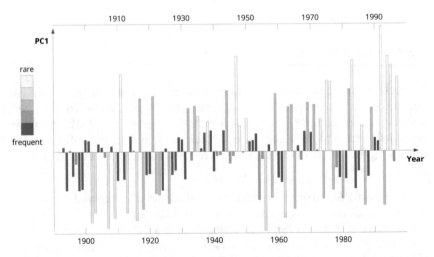

Fig. 6.12: The bar graph encodes years along the horizontal axis and the scores of the first principal component (PC1) along the vertical axis. Color indicates the frequency of score values. ©① *The authors.*

that connects the baseline with the year's PC1 coordinate (i.e., the year's score in principal component space). This effectively means upward bars encode a positive deviation from the major trend, that is, they stand for warmer summers, where long bars indicate summers with extreme conditions. In contrast, downward bars represent colder-than-normal summers. As an additional visual cue, frequencies of score values are mapped onto color to further distinguish typical (saturated green) and outlier (bright yellowish green) years. This visual representation allows us to discern the following interesting facts:

- The first third of the time axis is dominated by moderately warm summers mixed with the coldest summers.
- The hot summers in the 1910s and 1920s are immediately followed by cold summers.
- There were relatively nice summer seasons between 1930 and 1950.
- In general, outlier summers with extreme conditions accumulate at the end of the time axis.

Although the visualization in Figure 6.12 shows only the first principal component, rather than the five derived variables, it depicts corresponding trends very well. Nonetheless, one should recall that our data represent a special case where all five variables are strongly correlated. This correlation is the reason why PC1 separates warm and cold summers so well. When analyzing arbitrary time-oriented datasets, further principal components might be required to capture major structural relationships. The following example will illustrate how users can be assisted in making informed decisions about which principal component's scores to display.

6.5.3 Determining Relevant Components in Census Data

We now deal with a census dataset with multiple variables, including population, gross domestic product, literacy, and life expectancy. As before, the independent dimensions (i.e., time and space) are excluded to maintain the data's frame of reference, leaving ten variables to be processed analytically by the PCA. Accordingly, the analysis yields ten principal components, which correspond to the major trends in the data. The principal components' significance-weighted loadings indicate how individual variables contribute to these trends.

The significance-weighted loadings of our example are depicted in Figure 6.13, where longer bars stand for stronger contribution, and blue and yellow colors are used for positive and negative contribution, respectively. By definition, the principal components are ranked according to their significance from left to right. The figure indicates that the data's major trends (PC1-PC4) are largely influenced by the eight variables from literacy to life expectancy. But we can also see that if we look only at these first four principal components, we certainly lose reference to the two variables of population and population density, which do not contribute to the top four trends. Therefore, at least the principal components up to PC5, which is proportional to

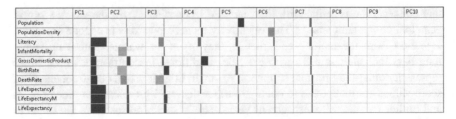

Fig. 6.13: The bars in the table cells visualize the loadings of principal components weighted by their significance, which clearly shows the left-to-right ranking of the principal components. ⓒⓘ *The authors.*

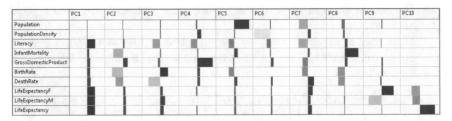

Fig. 6.14: The bars in the table cells visualize the unweighted loadings of principal components, that is, they indicate how much the individual variables contribute to any particular principal component. ⓒⓘ *The authors.*

population, and PC6, which is indirectly proportional to population density, should be retained. In turn, if we are interested in the main trends only, we can safely omit the remaining principal components (PC7-PC10).

If we are interested in outlier trends as well, we should be less generous with dropping principal components. This can be illustrated by a visualization of the plain (i.e., unweighted) loadings of the principal components as shown in Figure 6.14. The figure clearly reveals contradictory contributions of the variables to the lower-ranked trends. In particular, we can see a contradiction between life expectancy of females and males in the ninth principal component (PC9).

The visualization of the loadings helped us in identifying the top-ranked principal components and those that might bear potentially interesting outlier information. The knowledge that we derived about the principal components can also be interpreted in terms of the variables of the original data space. A number of findings can be gained, including the following:

- All the positive loadings in the main trend (PC1) indicate a direct proportional relationship for the literacy, infant mortality, gross domestic product, birth rate, death rate, and life expectancy.
- The second trend (PC2) is constituted by the gross domestic product, life expectancy as well as infant mortality, birth rate, and death rate, where the latter three variables are indirectly proportional to this trend.

- The major trends in the data (PC1-PC3) are largely independent of population and population density.
- An outlier trend is present in PC9, where the contradictory loadings of life expectancy of females and males might hint at an interesting aspect.

In summary, we have seen in this section that PCA is a useful tool for crystallizing major structural relationships in the data and for identifying possible candidates for data abstraction.

6.6 Summary

In this chapter, we provided a brief overview of how computational analysis methods can support the visual analysis of time-oriented data. We gave a list of typical temporal analysis tasks and illustrated the utility of analysis methods with three examples: segmentation and labeling (as a specific instance of classification), clustering, and principal component analysis. All of these examples perform a particular kind of temporal data abstraction. While our examples were simple, we still believe that they demonstrate the benefits of analytical methods quite well.

In fact, when confronted with really large datasets, a single analytical method alone will most certainly not suffice. Instead, a number of computational methods must play in concert to cope with the size and complexity of time-oriented data. Moreover, analytical methods are not solely a preprocessing step to support the visualization of data. The full potential of analytical methods unfolds only if they are considered at all stages of interactive exploration and visual analysis processes in an integrated fashion depending on the data, users, and tasks.

We will pick up this issue in the last chapter of this book, where we outline some ideas to arrive at an intertwined integration of visual, interactive, and analytical methods for the bigger goal of gaining insight into large and complex time-oriented data. Next in Chapter 7, we will return to visualization-related topics and discuss how data analysis practitioners can be supported in selecting visualization techniques appropriate for their needs.

References

Ali, M., A. Alqahtani, M. W. Jones, and X. Xie (2019). "Clustering and Classification for Time Series Data in Visual Analytics: A Survey". In: *IEEE Access* 7, pp. 181314–181338. DOI: 10.1109/ACCESS.2019.2958551.

Andrienko, N., G. Andrienko, S. Miksch, H. Schumann, and S. Wrobel (2021). "A Theoretical Model for Pattern Discovery in Visual Analytics". In: *Visual Informatics* 5.1, pp. 23–42. DOI: 10.1016/j.visinf.2020.12.002.

Antunes, C. M. and A. L. Oliveira (2001). *Temporal Data Mining: An Overview.* Workshop on Temporal Data Mining at the ACM SIGKDD International Conference on Knowledge Discovery and Data Mining (KDD). URL: https://www.researchgate.net/publication/284602094.

Bade, R., S. Schlechtweg, and S. Miksch (2004). "Connecting Time-oriented Data and Information to a Coherent Interactive Visualization". In: *Proceedings of the SIGCHI Conference on Human Factors in Computing Systems (CHI).* ACM Press, pp. 105–112. DOI: 10.1145/985692.985706.

Bernard, J., C. Bors, M. Bögl, C. Eichner, T. Gschwandtner, S. Miksch, H. Schumann, and J. Kohlhammer (2018). "Combining the Automated Segmentation and Visual Analysis of Multivariate Time Series". In: *Proceedings of the EuroVis Workshop on Visual Analytics (EuroVA).* Eurographics Association, pp. 49–53. DOI: 10.2312/eurova.20181112.

Bors, C., J. Bernard, M. Bögl, T. Gschwandtner, J. Kohlhammer, and S. Miksch (2019). "Quantifying Uncertainty in Multivariate Time Series Pre-Processing". In: Eurographics Association. DOI: 10.2312/eurova.20191121.

Brockwell, P. J. and R. A. Davis (1991). *Time Series: Theory and Methods.* 2nd edition. 2009 reprint. Springer. DOI: 10.1007/978-1-4419-0320-4.

Clancey, W. J. (1985). "Heuristic Classification". In: *Artificial Intelligence* 27.3, pp. 289–350. DOI: 10.1016/0004-3702(85)90016-5.

Cohen, P., B. Heeringa, and N. Adams (2002). "Unsupervised Segmentation of Categorical Time Series into Episodes". In: *Proceedings of the International Conference on Data Mining (ICDM).* IEEE Computer Society, pp. 99–106. DOI: 10.1109/ICDM.2002.1183891.

Combi, C., E. Keravnou-Papailiou, and Y. Shahar (2010). *Temporal Information Systems in Medicine.* Springer. DOI: 10.1007/978-1-4419-6543-1.

Eichner, C., H. Schumann, and C. Tominski (2020). "Making Parameter Dependencies of Time-Series Segmentation Visually Understandable". In: *Computer Graphics Forum* 39.1, pp. 607–622. DOI: 10.1111/cgf.13894.

Fayyad, U., G. G. Grinstein, and A. Wierse, eds. (2001). *Information Visualization in Data Mining and Knowledge Discovery.* Morgan Kaufmann.

Fayyad, U., G. Piatetsky-Shapiro, and P. Smyth (1996). "From Data Mining to Knowledge Discovery in Databases". In: *AI Magazine* 17.3, pp. 37–54. DOI: 10.1609/aimag.v17i3.1230.

Foley, J. D. (2000). "Getting There: The Ten Top Problems Left". In: *IEEE Computer Graphics and Applications* 20.1, pp. 66–68. DOI: 10.1109/38.814569.

Gan, G., C. Ma, and J. Wu (2007). *Data Clustering: Theory, Algorithms, and Applications.* ASA-SIAM Series on Statistics and Applied Probability. Society for Industrial and Applied Mathematics. DOI: 10.1137/1.9780898718348.

Gschwandtner, T., M. Bögl, P. Federico, and S. Miksch (2016). "Visual Encodings of Temporal Uncertainty: A Comparative User Study". In: *IEEE Transactions on Visualization and Computer Graphics* 22.1, pp. 539–548. DOI: 10.1109/TVCG.2015.2467752.

Han, J., M. Kamber, and J. Pei (2012). *Data Mining: Concepts and Techniques.* 3rd edition. Morgan Kaufmann. DOI: 10.1016/C2009-0-61819-5.

Jackson, J. E. (2003). *A User's Guide to Principal Components*. John Wiley & Sons.

Jain, A. K., M. N. Murty, and P. J. Flynn (1999). "Data Clustering: A Review". In: *ACM Computing Surveys* 31.3, pp. 264–323. DOI: 10.1145/331499.331504.

Jeong, D. H., C. Ziemkiewicz, B. Fisher, W. Ribarsky, and R. Chang (2009). "iPCA: An Interactive System for PCA-based Visual Analytics". In: *Computer Graphics Forum* 28.3, pp. 767–774. DOI: 10.1111/j.1467-8659.2009.01475.x.

Jolliffe, I. T. (2002). *Principal Component Analysis*. 2nd edition. Springer. DOI: 10.1007/b98835.

Keim, D. A., F. Mansmann, J. Schneidewind, and H. Ziegler (2006a). "Challenges in Visual Data Analysis". In: *Proceedings of the International Conference Information Visualisation (IV)*. IEEE Computer Society, pp. 9–16. DOI: 10.1109/IV.2006.31.

Laxman, S. and P. S. Sastry (2006). "A Survey of Temporal Data Mining". In: *Sādhanā* 31, pp. 173–198. DOI: 10.1007/bf02719780.

Lin, J., E. Keogh, S. Lonardi, and P. Patel (2002). "Finding Motifs in Time Series". In: *Proceedings of the SIGKDD Workshop on Temporal Data Mining*, pp. 53–68. URL: https://cs.gmu.edu/~jessica/Lin_motif.pdf.

Lin, J., E. J. Keogh, L. Wei, and S. Lonardi (2007). "Experiencing SAX: A Novel Symbolic Representation of Time Series". In: *Data Mining and Knowledge Discovery* 15.2, pp. 107–144. DOI: 10.1007/s10618-007-0064-z.

Miksch, S., W. Horn, C. Popow, and F. Paky (1996). "Utilizing Temporal Data Abstraction for Data Validation and Therapy Planning for Artificially Ventilated Newborn Infants". In: *Artificial Intelligence in Medicine* 8.6, pp. 543–576. DOI: 10.1016/s0933-3657(96)00355-7.

Miksch, S., A. Seyfang, W. Horn, and C. Popow (1999). "Abstracting Steady Qualitative Descriptions over Time from Noisy, High-Frequency Data". In: *Proceedings of the Joint European Conference on Artificial Intelligence in Medicine and Medical Decision Making (AIMDM)*. Springer, pp. 281–290. DOI: 10.1007/3-540-48720-4_31.

Mitsa, T. (2010). *Temporal Data Mining*. Chapman & Hall/CRC. DOI: 10.1201/9781420089776.

Nocke, T., H. Schumann, and U. Böhm (2004). "Methods for the Visualization of Clustered Climate Data". In: *Computational Statistics* 19.1, pp. 75–94. DOI: 10.1007/bf02915277.

Stacey, M. and C. McGregor (2007). "Temporal Abstraction in Intelligent Clinical Data Analysis: A Survey". In: *Artificial Intelligence in Medicine* 39.1, pp. 1–24. DOI: 10.1016/j.artmed.2006.08.002.

Thomas, J. J. and K. A. Cook (2005). *Illuminating the Path: The Research and Development Agenda for Visual Analytics*. IEEE Computer Society.

Van Wijk, J. J. and E. R. van Selow (1999). "Cluster and Calendar Based Visualization of Time Series Data". In: *Proceedings of the IEEE Symposium Information Visualization (InfoVis)*. IEEE Computer Society, pp. 4–9. DOI: 10.1109/INFVIS.1999.801851.

Ware, C. (2008). *Visual Thinking for Design*. Morgan Kaufmann. DOI: 10.1016/B978-0-12-370896-0.X0001-7.

Warren Liao, T. (2005). "Clustering of Time Series Data – A Survey". In: *Pattern Recognition* 38.11, pp. 1857–1874. DOI: 10.1016/j.patcog.2005.01.025.

Wegner, P. (1997). "Why Interaction Is More Powerful Than Algorithms". In: *Communications of the ACM* 40.5, pp. 80–91. DOI: 10.1145/253769.253801.

Xing, Z., J. Pei, and E. Keogh (2010). "A Brief Survey on Sequence Classification". In: *SIGKDD Explorations Newsletter* 12.1, pp. 40–48. DOI: 10.1145/1882471.1882478.

Xu, R. and D. C. Wunsch II (2009). *Clustering*. John Wiley & Sons. DOI: 10.1002/9780470382776.

Chapter 7
Guiding the Selection of Visualization Techniques

> Today we live in an information-based technological world. The problem is that this is an invisible technology. Knowledge and information are invisible. They have no natural form. It is up to the conveyor of the information and knowledge to provide shape, substance, and organization [...]
>
> Norman (1993, p. 104)

So far, this book approached the visualization of time-oriented data from a conceptual perspective. We looked at the characteristics of time and time-oriented data, considered aspects of visualization design, and learned how interaction and analysis methods can be employed to enhance the visualization. In this chapter, we switch to a more practically oriented perspective and address the question of how to select visualization techniques that are appropriate for an application problem at hand.

Such a practical perspective is immensely important because searching for suitable visualization techniques for time-oriented data can be a daunting tasks. There are so many options and so many solutions to be found in so many scientific publications, books, and online resources that it is hardly possible to check them all. With the examples of visualization techniques included in the previous chapters, we just scratched the surface of a rich body of existing work. In fact, there is an abundance of valuable techniques and tools. So, a person who has some time-oriented data to be visualized does not really know where to start and where to end. This calls for mechanisms to support researchers and practitioners in finding suitable techniques that fit their needs. This chapter describes how this call can be answered.

7.1 Structuring the Space of Solutions

When it comes to searching a large space, in our case the space of visualization techniques, it is very helpful to have a structure that helps us reduce the complexity of the search. Much like in *divide & conquer* approaches, we want to have a structure

W. Aigner et al., *Visualization of Time-Oriented Data*, Human–Computer Interaction Series, https://doi.org/10.1007/978-1-4471-7527-8_7

that subdivides the space into smaller partitions. Our search can then focus on relevant subspaces while ignoring the ones that are not of interest for a particular application problem.

The conceptual considerations from the earlier chapters of this book are promising starting points for finding such a structure. In particular, we looked at three key questions: What is visualized, why is it visualized, and how is it visualized? These three questions relate to the different characteristics of time and time-oriented data (what), the different analytical tasks one seeks to solve (why), and the different options for designing the visual representation (how). Yet, while the discussion of these relevant aspects provides us with valuable information, the involved details can be a too big hurdle for an easy entry into the field. For example, people who do not regularly work with visualization or other data analysis tools might not be able to fully grasp the diversity of aspects to be considered.

Thus, it makes sense to insert a new simpler layer of abstraction. This new layer should focus on aspects that are easy to decide, while leaving aspects being more subtle for later inclusion. Moreover, the simpler layer should focus on those aspects that are typically addressed by current visualization techniques. Therefore, we now consider only three key criteria: (1) time and (2) data, the what, and (3) visual representation, the how, while neglecting the why aspect. Each key criterion has two sub-criteria with two corresponding categories each, which gives us a much simpler overall structure compared to the full theoretical categorization from Chapter 4. The simplified schema is now more suitable for practical use:

- **Time**

  - *Primitives* – points vs. intervals
  - *Arrangement* – linear vs. cyclic

- **Data**

  - *Number of variables* – single vs. multiple
  - *Frame of reference* – abstract vs. spatial

- **Visual representation**

  - *Mapping of time* – static vs. dynamic
  - *Dimensionality* – 2D vs. 3D

The time and data criteria pick up four selected aspects (i.e., type of time primitives, arrangement of time, number of variables, and frame of reference) from Chapter 3, while the visual representation criterion refers to two aspects (i.e., mapping of time and dimensionality) discussed in Section 3.2. As such, this simplified categorization schema concentrates on aspects that are relatively easy to decide. For example, if data are abstract or have a spatial frame of reference is clear from the description of the data. If a visualization is 2D or 3D is also obvious and hence easy to categorize. On the other hand, the new schema does not include the analysis tasks, that is, the why aspect. There are two reasons for that. First, for many applications, there is not just one task to be tackled. Instead, tasks may be constantly in flux where working on one task naturally leads to other tasks. Moreover, deciding whether a particular visualization technique is suitable for solving a certain task is not easy. In fact, objective assessments of task suitability are rarely reported in the literature, and if so, mostly at the level of the visual encoding only (as we did for color-coding in Section 4.2.2), but not at the level of visualization techniques.

But still, even without the task level, we now have a simplified categorization schema that allows us to subdivide the space of solutions into smaller subspaces to restrict the search for suitable visualization techniques. How the search can actually be carried out will be described next.

7.2 The TimeViz Browser

As we said before, the problem is to find visualization techniques that match a user's particular needs. A tool that can help mitigate this problem is the *TimeViz Browser*. Designed as an interactive website, the TimeViz Browser enables practitioners and researchers to explore, investigate, and compare visualization techniques for time-oriented data. The key advantages of the TimeViz Browser are threefold. First, it brings together in a single place the available visualization techniques, which are otherwise scattered across a variety of sources. Second, it employs the simplified categorization schema to tag and structure the available techniques, which makes it easier to interactively explore the space of solutions. And third, the TimeViz Browser is freely available at `https://browser.timeviz.net` so that anyone can use it.

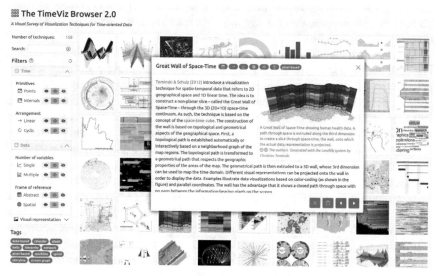

Fig. 7.1: The TimeViz Browser provides a visual overview of visualization techniques for time-oriented data and a filter interface for narrowing down the listed techniques according to six different criteria. ⓒⓘ *The authors.*

Fig. 7.2: Six filters grouped by data, time, and visual representation allow users to select visualization techniques that do or do not exhibit twelve different traits from six different categories. ⓒⓘ *The authors.*

Let us now take a closer look at how the TimeViz Browser works. Overall, it provides an overview of what is possible when visualizing time-oriented data. The diversity of visualization techniques is communicated visually by means of a collection of thumbnails, each showing a representative image of a visualization technique. Each technique can also be explored in greater detail. Selecting a technique opens up a detail view. This view offers a brief description of the technique, a larger figure, and a list of relevant publications. Small icons indicate the technique's place in the categorization schema (e.g., frame of reference: abstract vs. spatial or number of variables: single vs. multiple). Figure 7.1 shows a screenshot of the TimeViz Browser with the thumbnails in the background and the detail view on top of them. To the left of the figure, one can see the filter interface that allows users to narrow down the list of thumbnails.

This is where the TimeViz Browser utilizes the simplified categorization schema just introduced. There are six filters, one for each sub-criterion of the schema: primitives and arrangement (time), number of variables and frame of reference (data), and finally mapping of time and dimensionality (visual representation). As there are two categories per criterion, there are overall twelve filter controls in the interface (see Figure 7.2). Each filter control has three states *want, indifferent,* and *hide* based on which the list of thumbnails is filtered to allow users to express different interests. The want state of a filter signals the user's interest in techniques that belong to a certain category. So, by selecting *want*, the user asks the TimeViz Browser to show techniques that have a particular trait. Selecting indifferent literally means that the user is not concerned about a category. The hide state, on the other hand, is there to tell the TimeViz Browser to filter out techniques falling into a certain category. In short, for a visualization technique to be included as a thumbnail in the main view, it must satisfy all categories set to *want*, must not satisfy any category set to *hide*, and may or may not belong to categories set to *indifferent*.

Before we illustrate how the filters can be utilized to guide the search for suitable techniques in Section 7.4, we first provide a summarizing overview of the TimeViz Browser's corpus of available techniques in the next Section 7.3.

7.3 Overview of Visualization Techniques

The TimeVizBrowser covers overall 158 visualization techniques for time-oriented data. Table 7.1, which spans several pages, provides an overview of all techniques along with their categorization. Appendix A describes all listed techniques on a page-by-page basis, including a brief summary, an illustrative figure, and associated references. In a sense, Appendix A is like a static print version of the otherwise interactive TimeViz Browser.

Before we look at Table 7.1 in more detail, it is worth mentioning that it was not easy to decide on a good order for the techniques in the table. If we sorted by the name of a technique or by the year of its first publication, we would lose the semantic relationships of techniques, and similar techniques would be scattered across the table just because they have different names. Therefore, we looked for criteria in our categorization schema that would partition the corpus of techniques reasonably. A good first separation is possible with respect to the *frame of reference*. Accordingly, the table first lists abstract techniques and then techniques for data with a spatial frame of reference. At the next level of sorting, the techniques are ordered by the *number of variables* being visualized. Techniques for data with a single time-dependent variable will precede those for multiple time-dependent variables. At the subsequent levels, we order the techniques by their affiliation to the sub-categories *arrangement*, *primitives*, *mapping of time*, and *dimensionality*.

| | time | | | | data | | | | vis | | | | | |
| | primitives | | arrangement | | frame of reference | | number of variables | | mapping of time | | dimensionality | | | |
	points	intervals	linear	cyclic	abstract	spatial	single	multiple	static	dynamic	2D	3D	page
Point Plot	■		■		■		■		■		■		232
Line Plot	■		■		■		■		■		■		233
Bar Graph, Spike Graph	■		■		■		■		■		■		234
Sparklines	■		■		■		■		■		■		235
TrendDisplay	■		■		■		■		■		■		236
Decision Chart	■		■		■		■		■		■		237
TimeTree	■		■		■		■		■		■		238
Arc Diagrams	■		■		■		■		■		■		239
SparkClouds	■		■		■		■		■		■		240
Growth Matrix	■		■		■		■		■		■		241
Multi-Resolution Visualization of Time Series	■		■		■		■		■		■		242
Pinus View	■		■		■		■		■		■		243
Ripple Graph	■		■		■		■		■		■		244

continued on next page

Table 7.1: Overview and categorization of visualization techniques.

| | time | | | | data | | | | vis | | | | |
| | primitives | | arrangement | | frame of reference | | number of variables | | mapping of time | | dimensionality | | |
	points	intervals	linear	cyclic	abstract	spatial	single	multiple	static	dynamic	2D	3D	page
Small MultiPiles	■		■		■		■		■		■		245
Time Maps	■		■		■		■		■		■		246
TimeDensityPlots	■		■		■		■		■		■		247
Interactive Parallel Bar Charts	■		■		■		■		■			■	248
TimeHistogram 3D	■		■		■		■		■			■	249
Intrusion Monitoring	■		■		■		■			■	■		250
Anemone	■		■		■		■			■	■		251
Dynamic Word Clouds	■		■		■		■			■	■		252
Gantt Chart	■	■	■		■		■		■		■		253
Set Streams	■	■	■		■		■		■		■		254
TimeSets	■	■	■		■		■		■		■		255
Perspective Wall	■	■	■		■		■		■			■	256
DateLens		■	■		■		■		■		■		257
Timeline		■	■		■		■		■		■		258
Paint Strips		■	■		■		■		■		■		259
PlanningLines		■	■		■		■		■		■		260
Time Annotation Glyph		■	■		■		■		■		■		261
SOPO Diagram		■	■		■		■		■		■		262
TimeNets		■	■		■		■		■		■		263
Storyline Visualization		■	■		■		■		■		■		264
Temporal Mosaic		■	■		■		■		■		■		265
Train Delay Uncertainty		■	■		■		■		■		■		266
Triangular Model		■	■		■		■		■		■		267
Cycle Plot	■		■	■	■		■		■		■		268
Tile Maps	■		■	■	■		■		■		■		269
Multi Scale Temporal Behavior	■		■	■	■		■		■		■		270
GROOVE	■		■	■	■		■		■		■		271
SolarPlot	■		■	■	■		■		■		■		272
Cluster and Calendar-Based Visualization	■		■	■	■		■		■		■	■	273
Enhanced Interactive Spiral	■			■	■		■		■		■		274
ClockMap	■			■	■		■		■		■		275
SpiraClock		■		■	■		■			■	■		276
Horizon Graph	■		■		■		■	■	■		■		277
VizTree	■		■		■		■	■	■		■		278
Time Curves	■		■		■		■	■	■		■		279
TimeSlice	■	■	■		■		■	■	■		■		280
Silhouette Graph	■		■	■	■		■	■	■		■		281
Recursive Pattern	■		■	■	■		■	■	■		■		282

continued on next page

Table 7.1: Overview and categorization of visualization techniques.

| | time | | | | data | | | | vis | | | | |
| | primitives | | arrangement | | frame of reference | | number of variables | | mapping of time | | dimensionality | | |
	points	intervals	linear	cyclic	abstract	spatial	single	multiple	static	dynamic	2D	3D	page
Lin-spiration	■		■	■	■		■	■	■		■		283
Spiral Graph	■			■	■		■	■	■		■		284
Spiral Display	■	■		■	■		■	■	■		■	■	285
Stacked Graphs	■		■		■			■	■		■		286
TimeSearcher 3, River Plot	■		■		■			■	■		■		287
Timeline Trees	■		■		■			■	■		■		288
Layer Area Graph	■		■		■			■	■		■		289
TimeSearcher	■		■		■			■	■		■		290
BinX	■		■		■			■	■		■		291
MultiComb	■		■		■			■	■		■		292
ThemeRiver	■		■		■			■	■		■		293
history flow	■		■	■	■			■	■		■		294
LifeLines2	■		■		■			■	■		■		295
Similan	■		■		■			■	■		■		296
VIE-VISU	■		■		■			■	■		■		297
TimeWheel	■		■		■			■	■		■		298
LiveRAC	■		■		■			■	■		■		299
CareCruiser	■		■		■			■	■		■		300
Braided Graph	■		■		■			■	■		■		301
CiteSpace II	■		■		■			■	■		■		302
ChronoLenses	■		■		■			■	■		■		303
Connected Scatterplot	■		■		■			■	■		■		304
DimpVis	■		■		■			■	■		■		305
FluxFlow	■		■		■			■	■		■		306
Line Density Plot	■		■		■			■	■		■		307
Matrix-Based Comparison	■		■		■			■	■		■		308
MultiStream	■		■		■			■	■		■		309
netflower	■		■		■			■	■		■		310
Optimized Stream Graphs	■		■		■			■	■		■		311
RankExplorer	■		■		■			■	■		■		312
Sankey Diagram, Alluvial Diagram	■		■		■			■	■		■		313
TACO	■		■		■			■	■		■		314
WireVis	■		■		■			■	■		■		315
MOSAN	■		■		■			■	■		■	■	316
3D ThemeRiver	■		■		■			■	■			■	317
Data Tube Technique	■		■		■			■	■			■	318
Kiviat Tube	■		■		■			■	■			■	319
Temporal Star	■		■		■			■	■			■	320

continued on next page

Table 7.1: Overview and categorization of visualization techniques.

	time				data				vis				
	primitives		arrangement		frame of reference		number of variables		mapping of time		dimensionality		
	points	intervals	linear	cyclic	abstract	spatial	single	multiple	static	dynamic	2D	3D	page
Time-tunnel	■		■		■		■		■			■	321
Worm Plots	■		■		■		■		■			■	322
Software Evolution Analysis	■		■		■		■		■			■	323
3D TimeWheel	■		■		■		■		■			■	324
Vanishing-Point Plot	■		■		■		■		■			■	325
InfoBUG	■		■		■			■	■	■	■		326
Gravi++	■		■		■			■	■	■	■		327
CircleView	■		■		■			■	■	■	■		328
CloudLines	■		■		■			■	■	■	■		329
Animated Scatter Plot	■		■		■			■		■	■		330
Process Visualization	■		■		■			■		■	■		331
TimeRider	■		■		■			■		■	■		332
Flocking Boids	■		■		■			■		■		■	333
Time Line Browser	■	■	■		■			■	■		■		334
PatternFinder	■	■	■		■			■	■		■		335
FacetZoom	■	■	■		■			■	■		■		336
KNAVE II	■	■	■		■			■	■		■		337
Continuum	■	■	■		■			■	■		■		338
VisuExplore	■	■	■		■			■	■		■		339
Midgaard	■	■	■		■			■	■		■		340
LifeLines	■	■	■		■			■	■		■		341
EventRiver	■	■	■		■			■	■		■		342
Story Curves	■	■	■		■			■	■		■		343
TextFlow	■	■	■		■			■	■		■		344
Event-Flow Visualization	■	■	■		■			■	■		■		345
Co-Bridges		■	■		■			■	■		■		346
LiveGantt		■	■		■			■	■		■		347
PeopleGarden	■		■	■	■			■	■		■		348
PostHistory	■		■	■	■			■	■		■		349
Pixel-Oriented Network Visualization	■		■	■	■			■	■		■		350
Parallel Glyphs	■		■	■	■			■	■			■	351
Circos	■	■	■	■	■			■	■		■		352
Kaleidomaps	■			■	■			■	■		■		353
KAVAGait	■			■	■		■		■		■		354
SentiCompass	■			■	■			■	■		■		355
TreeRose	■			■	■			■	■		■		356
Intrusion Detection	■			■	■			■	■			■	357
KronoMiner	■	■		■	■			■	■		■		358

continued on next page

Table 7.1: Overview and categorization of visualization techniques.

	primitives		arrangement		frame of reference		number of variables		mapping of time		dimensionality		
	points	intervals	linear	cyclic	abstract	spatial	single	multiple	static	dynamic	2D	3D	page
Small Multiples	■		■	■	■	■	■	■	■		■	■	359
EventViewer	■	■	■	■	■	■	■	■	■		■		360
Ring Maps	■	■	■	■	■	■	■	■	■		■	■	361
ThermalPlot	■		■			■		■		■	■		362
Circular	■	■	■			■		■	■		■		363
ViDX	■	■	■	■		■		■	■	■	■	■	364
Value Flow Map	■		■			■	■		■		■		365
Flowstrates	■		■			■	■		■		■		366
Traffigram	■		■			■	■		■		■		367
Time-Varying Hierarchies on Maps	■		■			■	■		■			■	368
Great Wall of Space-Time	■		■			■	■		■			■	369
MoSculp	■		■			■	■		■			■	370
Flow Map	■	■	■			■	■		■		■		371
Visits	■	■	■			■	■		■		■		372
Time-Oriented Polygons on Maps	■		■	■		■	■		■		■		373
Growth Ring Maps	■		■	■		■	■		■		■		374
Trajectory Wall	■		■	■		■	■		■		■	■	375
Spatio-Temporal Event Visualization	■		■			■	■	■	■			■	376
Space-Time Cube	■	■	■			■	■	■	■			■	377
Space-Time Path	■	■	■			■	■	■	■			■	378
Icons on Maps	■	■	■	■		■	■	■	■		■	■	379
VIS-STAMP	■		■			■		■	■		■		380
Time-Ray Maps	■		■			■		■	■		■		381
Wakame	■		■			■		■	■			■	382
GeoTime	■		■			■		■	■	■		■	383
Multiple Temporal Axes Model	■	■	■			■		■	■		■		384
Data Vases	■	■	■			■		■	■		■	■	385
Pencil Icons	■	■	■			■		■	■			■	386
Temporal Focus+Context		■	■			■		■	■			■	387
Chro-Ring	■	■		■		■		■	■		■		388
Helix Icons	■	■		■		■		■	■			■	389
count	143	49	145	35	133	31	72	102	149	15	132	35	

Table 7.1: Overview and categorization of visualization techniques.

Table 7.1 provides an overview of the available techniques based on which we can already make some interesting observations about the state of the art. Next, we take a closer look at possible explanations for the observable balances and imbalances with regard to our categorization.

Time – primitives Points in time are the most commonly used time primitive in our corpus of techniques. This seems natural because data are often measured at a particular point in time. Intervals occur less often, for example, in planning scenarios, where it is important to know how long certain activities will take as in PlanningLines (↪ p. 260). Intervals also become relevant when data, while measured at a particular point in time, are valid in local scope, or when data need to be aggregated from individual points to intervals in order to deal with bigger and bigger datasets as in Midgaard (↪ p. 340).

Time – arrangement Most techniques support linear time; the approaches for cyclic time are significantly outnumbered. Reasons for this might be that users are usually interested in trends evolving from past, to present, to future, rather than in finding cycles in the data. The latter aspect, however, is important to fully understand the data, and therefore, expert data analysts need effective cyclic representations as offered, for example, by the spiral display (↪ p. 285) or Circos (↪ p. 352).

Data – number of variables The number of techniques for a single variable is roughly balanced with the number of techniques for multiple variables. While classic techniques often consider simpler univariate data, modern approaches take on the challenge of dealing with multiple variables. Our corpus also contains several techniques that cope with multiple variables simply by the repetition of a basic visualization design that only addresses a single variable. An example is the recursive pattern (↪ p. 282) technique.

Data – frame of reference In this book, we mainly focus on abstract data, which is also reflected in the list of techniques. Showing time-oriented data in a spatial frame of reference significantly increases the design efforts because more information has to be packed into the visual mapping. Particularly, the disciplines of cartography and geo-visualization, which are established, independent fields of research, have developed approaches to combining the visualization of temporal and spatial aspects of data (see Andrienko and Andrienko, 2006; Kraak and Ormeling, 2020).

Visual representation – mapping of time Apparently, the printed pages of a book are better suited for showing static techniques. In this sense, our corpus of techniques is biased in that it contains mostly static approaches. However, dynamic animation is equally important and often it is the first solution offered when time-oriented data have to be visualized. Animation can also be an option in combination with static methods to extend the capacity of a technique in terms of the data that can be handled. This strategy is for example followed by GeoTime (↪ p. 383).

Visual representation – dimensionality Two-dimensional visual representations are often preferred over three-dimensional ones, because they are more abstract and thus tend to be easier to understand. Especially techniques developed in the early days of computer graphics tend to stick with two dimensions simply due to the limited computing power available then. However, modern technologies have made it easier both for visualization designers to implement three-dimensional visualization and for visualization users to navigate and explore virtual 3D visualization spaces. This is particularly useful when data with spatial references have to be visualized (e.g., space-time cube (↪ p. 377) or space-time path (↪ p. 378) techniques).

There is another significant fact that can be derived from the corpus of existing visualization techniques: Most approaches address the model of an *ordered* time domain, while only a few of them explicitly consider the visualization of *branching* alternative strings of time, and none of them is capable of visualizing data that are based on the model of *multiple perspectives*. Although there is one technique, called story curves (↪ p. 343), which at least include two perspectives on time in the visualization. Still, branching time and time with multiple perspectives deserve more research attention in the future.

It is also visible that the majority of techniques are specific to a particular *what* and as a consequence represent tailored solutions in terms of *how* the data are visualized. An advantage of such specific solutions is that they are fine-tuned to be successful in supporting a specific category of time-oriented data. The downside, however, is that these solutions are hard to adapt and reuse for other visualization problems, even when a new problem is similar to the original one and differs only in one aspect of our categorization schema.

In summary, there are very many different techniques for visualizing time and time-oriented data. This can be both a blessing and a burden. While we can choose from a variety of options, the decision may not be easy.

7.4 Guided Search for Visualization Techniques

Next, we look at concrete example scenarios that illustrate how one can reduce the 158 visualization techniques included in our corpus to a much smaller set of promising candidates. The TimeViz Browser can guide the search for appropriate techniques based on the simplified categorization schema.

As already mentioned, the schema includes selected aspects of the time dimension and the data (what) as well as the visual representation (how). Aspects of analysis tasks (why) are not included. Therefore, in the second part of this section, we outline a theoretical framework for a multi-faceted selection process that covers all relevant aspects. We also indicate the challenges of applying this framework in practice.

7.4.1 Example Scenarios

Let us start with two concrete examples illustrating the process of selecting appropriate visualization techniques with respect to time and data (what) and also with respect to preferred visual representations (how).

Visual Exploration of Work Time Patterns of Employees

For our first example, we take the role of a manager of a manufacturing facility. As a manager, we would like to visually explore time patterns related to the work times of employees in order to better plan work schedules. The data originate from a time tracking system, which provides employee information along with time interval data for the start and end times of work sessions. The time primitives we need to analyze are intervals and as we do have a 24/7 production schedule with regular work schedules, we would like to mainly focus on cyclic time. Our data include multiple variables and have no spatial context.

With this scenario in mind, we now use the TimeViz Browser to find suitable visualization techniques. We start with the TimeViz Browser showing all 158 techniques. The search process is based on adjusting the filters according to the time and the data at hand.

What First, we filter for visualization techniques that support interval-based data. This already reduces the number of techniques by about 60%, but there are still very many. Second, we focus on techniques for cyclic time, which reduces the result set substantially. Third, as we need to show multiple aspects of employee data, we filter for techniques supporting multiple variables. However, there is only one techniques that does not fulfill this criterion, so we do not get much of a reduction with this filtering step. Fourth, as we do not have spatial data, we hide the techniques for data with a spatial frame of reference. As a result, the TimeViz Browser eventually lists three techniques suitable for our data, including Circos ($\hookrightarrow$ p. 352), KronoMiner ($\hookrightarrow$ p. 358), and Spiral Display ($\hookrightarrow$ p. 285). We can now further assess these techniques with respect to their visual representation (how), which is easily doable using further filters.

How As we want to use the visualization of our employees' work times as an argument in a managerial report, we must use a static mapping of the dimension of time. Moreover, printed visualizations in a report are preferably kept in 2D. We set the corresponding filters in the TimeViz Browser to match these requirements. However, no further reduction of the suitable techniques is possible.

Still, our interactive search was successful overall. We could reduce the problem of choosing from 158 possible techniques to the much easier task of making a choice among three techniques, for which we know they match our needs in terms of time, data, and visual representation.

Dynamic Visualization of Financial Time Series in an Immersive Display

In the second example, we take the role of a visualization designer. As a visualization designer, we have been commissioned to create an immersive experience of stock market data to be on display for an event of a bank. Our users will be interested customers and stakeholders, who will be provided with virtual reality (VR) headsets at an information booth in the lobby of the bank. The data to be visualized consist of daily stock prices of the national stock exchange over the last three years. Our client wants the dynamics of stock price changes to be made visible and for the bank's strategic actions to visually match the patterns in the data.

The time primitives to be displayed are points, as we get stock price readings for each trading day. The time series are given on a linear time scale and we have multiple stocks to represent. Furthermore, the data are not spatially anchored. Again, we adjust the filters in the TimeViz Browser accordingly.

What First, we select points as the time primitives and linear time as the time arrangement. However, since most techniques meet both of these criteria, the reduction is rather small at about 10%. With still too many techniques being listed, we next hide visualization techniques that can show only a single variable and those that deal with spatial data. Unfortunately, still several dozens of techniques are in the pool of potential candidates. We need to distill them further by filtering with respect to the visual representation.

How The dynamics in time-oriented data can be represented nicely with the use of a dynamic mapping. We set the filter in the TimeViz Browser accordingly. Now we get a substantial reduction in the displayed candidates to less than ten. Moreover, our visualization is to be used in an immersive VR setting. Therefore, we filter for techniques that use a 3D visual representation. Eventually, this results in only a single suitable technique: Flocking Boids (↪ p. 333).

Again, with a few clicks in the TimeViz Browser's filter interface, we were able to find suitable visualization techniques, exactly one in this case, for our problem at hand. These two successful examples, however, should not hide the fact that we may be less successful in other scenarios. Therefore, we need additional leverage points to further support the guided search. For this purpose, we next outline a multi-faceted selection process.

7.4.2 Towards a Multi-Faceted Selection Process

The selection of appropriate visualization techniques for a particular application purpose is a multi-faceted and iterative decision-making process that can be delicate and overwhelming. It is necessary to take into account a number of aspects concerning the data, the users, their tasks, the used devices, as well as the usage contexts at hand. To cope with this complexity, it makes sense to follow a structured, multi-faceted selection process that narrows down the number of visualization options step by step. Figure 7.3 provides an overview of the aspects to be considered. It extends the what-why-how structure introduced at the beginning of this book along the Five W's and How tool (see Hart, 1996):

1. What: time and data
2. Why: tasks
3. Who: users
4. Where: devices
5. When: contexts
6. How: visual representation

What is presented? – Time & data   As can be seen in Table 7.1, literally hundreds of different visualization techniques exist. In a first step, this set can be narrowed down by discarding techniques unable to visualize the data and time characteristics present in the data. A prerequisite for this is the identification of data and time characteristics present in the data to be visualized. To support this step semi-automatically, we can use the TimeViz Browser (`https://browser.timeviz.net`) with its interactive filter interface to narrow down a candidate set of visualization techniques. Usually, such an initial filtering based on the *what* aspect leads to a reduction to dozens of potentially suitable techniques.

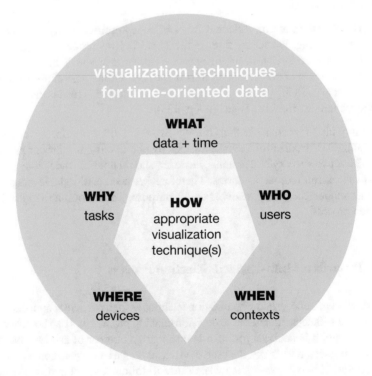

Fig. 7.3: Multi-faceted selection of appropriate visualization techniques. ©① *The authors.*

Why is it presented? – User tasks To narrow down the number of suitable solutions further, we must consider why a user needs or is provided with a visualization. A prerequisite for this is the process of *task abstraction*, that is, the identification of user tasks to be supported by the visualization. A useful framework to systematically consider user tasks particularly in the context of time-oriented data was discussed in Section 4.1.2. Depending on the set of tasks relevant for the problem under consideration, further visualization approaches can be discarded that do not provide the necessary identification, lookup, or comparison support. It might also be the case that a combination of different visualization techniques in a multiple-views approach is necessary to address all the data and task requirements. A difficulty though is to categorize existing techniques with respect to the tasks they support. More research is needed to come up with validated categorizations, which could then also be picked up by the TimeVizBrowser to further support users in their search.

Who is the presentation for? – Users In order to be able to cater to the needs of a specific user group, it is necessary to take into account who the users will be. This includes gathering as much information as possible to better understand the target group. For example, aspects about the application domain, age group, educational background, visualization literacy, domain knowledge, or already known visualization concepts can provide essential rationales for further prioritization of techniques. Again it is also necessary to categorize existing techniques according to the *who* aspect, which also represents a formidable task for future research.

Where is it presented? – Target devices In addition to the data and human-centered aspects of the previous questions, technical aspects of display and interaction devices to be targeted are crucial. For example, visualization techniques that are appropriate for certain *what* and *why* aspects might after all not be feasible for being used on small smartphone screens or with touch interaction. In the context of visualization, the most important aspects are the display capabilities for visual output (e.g., screen size, aspect ratio, resolution, and available colors) as well as the input modalities offered (e.g., pointing device, touch interaction, and physical interaction).

When is it presented? – Contexts Questions regarding the concrete usage environments, i.e., the specific contexts in which a visualization is going to be used, tie together all the other aspects considered so far. Consider, for example, data variables of a car (*what*, e.g., rpm, velocity, and fuel level) to be used by engineers (*who*) for monitoring the variables' changes over time (*why*) on a tablet device (*where*). Depending on whether the intended usage context is next to a test stand inside a factory building or inside a car while driving it at the same time, suitable visualization techniques will be quite different from each other for either case. Practically, the *where* and *when* aspects are often closely interrelated. For example, if a visualization for informing bike riders on the go should be provided, small screen mobile devices appear to be most suitable. What's more, if we have a visualization on a small screen mobile device, it lends itself to be used in contexts where users are moving.

How is it presented? – Visual representation Applying the considerations that result from the answers to the five W's leads to a candidate set of appropriate visualization techniques. Each of these techniques provides an answer to the final *how* question: how is it represented? As we've learned in Chapter 4 and from the TimeVizBrowser, the how aspect for the visualization of time-oriented data is mainly categorized by the differentiation of static and dynamic as well as 2D and 3D representations. These characteristics can also be used to further narrow down the set of appropriate visualization techniques (e.g., if a stereoscopic, head-mounted display is the target device, a 3D representation might be more suitable).

The multi-faceted selection process outlined above is informed by user research as well as data and task abstraction steps. Parts of the process can be semi-automated, such as selecting appropriate techniques using the TimeViz Browser. Other stages have to be individually curated by the visualization designers or users themselves who are responsible to decide which of the appropriate techniques match their needs.

It is important to understand that the selection (as much as the design) of appropriate visualization techniques is usually not a linear, clearly defined sequence of steps. The involved decisions are often not independent and may influence each other. All the mentioned W-questions contribute to a well-informed decision-making process of a visualization designer or user. Moreover, in different situations, different subsets of facets might be given as constraints while in other cases these are up for the user or designer to decide. As an example, the target devices to be used might be fixed (e.g., as service technicians are already equipped with small screen handheld devices of some kind) whereas in other cases the choice of target devices might be up to the designer who makes this selection based on the other given facets (e.g., based on a specific user group that needs to be supported in a specific context such as health workers on their ward). Answers to one of the questions might also be in conflict with answers to other questions and lead to reconsiderations, iterative refinements, or the need to revisit facets. In other cases, it might not be possible or necessary to consider all facets for the selection process, for example, if the set of suitable techniques is already down to a single possible one.

So far, we have provided an overview and checklist of aspects to possibly consider when selecting visualization techniques for time-oriented data. Although there is no strict order to be followed when considering the W-questions, starting with the *what*, the characteristics of data and time, as well as the *why*, the tasks to be supported, makes sense in many cases. For further narrowing down this set of available solutions, considering the aspects of users, devices, contexts, as well as the visual representations allows for distilling down to visualization techniques suitable for a given domain problem. Figure 7.4 provides an overview of this process for guiding the selection of suitable visualization techniques.

What's more, we primarily focused on the *what*, *why*, and *how* aspects so far, since these are to a large extent influenced and determined by the special characteristics of time, which this book is about. However, it is equally important to take into account the aspects of the users (who), target devices (where), and the usage contexts (when). Given our time-oriented focus, we do not go into the details of these more general facets, but refer the interested reader to the general literature on visualization design (see Munzner, 2014; Kirk, 2019).

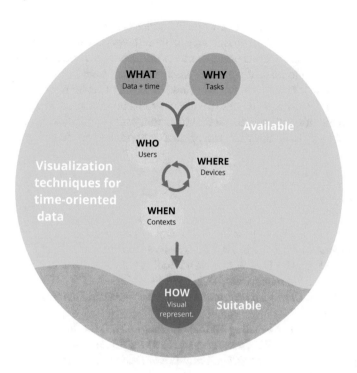

Fig. 7.4: Process of guiding the selection of suitable visualization techniques. ©① *The authors.*

7.5 Summary

This chapter was concerned with the practical question of selecting visualization techniques for time-oriented data for a given data analysis problem. We introduced a simplified categorization schema for existing techniques and employed this schema in the TimeViz Browser to guide the selection of suitable visualization techniques. To this end, the TimeViz Browser maps the categorization schema to an interactive filter interface allowing users to narrow down their search for visualization techniques.

The TimeViz Browser is based not only on the simplified categorization schema, but also on a large corpus of 158 categorized visualization techniques. While the individual techniques of this corpus are described in Appendix A, this chapter looked at the corpus as a whole and derived some interesting observations with respect to the distribution of techniques in the different categories. The detailed descriptions in the appendix will reveal that some general concepts reoccur in several instantiations, as for instance the general application of line plots as the most basic visualization of time-dependent data, the utilization of the third display dimension to encode time, or the mapping of time to spiral shapes in order to visualize cyclic aspects.

While the TimeVizBrowser is a great tool to support the selection of visualization techniques, it does not cover all aspects that might be relevant for such a selection.

Therefore, we also outlined a conceptual framework for a multi-faceted selection process that addresses the relevant aspects more broadly. We indicated the benefits of such an elaborate selection process and also mentioned the challenges to be tackled in future research to make the process practically applicable.

In the next and final chapter of this book, based on the insight gained so far about existing visualization techniques, we discuss remaining practical concerns, outline the tight integration of visualization, interaction, and computational methods under the umbrella of visual analytics, and present opportunities for future research.

References

Andrienko, N. and G. Andrienko (2006). *Exploratory Analysis of Spatial and Temporal Data*. Springer. DOI: 10.1007/3-540-31190-4.

Hart, G. (1996). "The Five W's: An Old Tool for the New Task of Task Analysis". In: *Technical Communication* 43.2, pp. 139–145.

Kirk, A. (2019). *Data Visualisation: A Handbook for Data Driven Design*. 2nd edition. SAGE.

Kraak, M.-J. and F. Ormeling (2020). *Cartography: Visualization of Geospatial Data*. 4th edition. CRC Press. DOI: 10.1201/9780429464195.

Munzner, T. (2014). *Visualization Analysis and Design*. A K Peters/CRC Press. DOI: 10.1201/b17511.

Norman, D. A. (1993). *Things That Make Us Smart: Defending Human Attributes in the Age of the Machine*. Addison-Wesley Longman Publishing.

Chapter 8
Conclusion

> All the pieces are there – huge amounts of information, a great need
> to clearly and accurately portray them, and the physical means for
> doing so. What has been lacking is a broad understanding of how
> best to do it.

Wainer (1997, p. 112)

This book has dealt with concepts and methods for visualizing time and time-oriented data. This chapter will briefly summarize the key aspects that have been discussed in the previous chapters and shed some light on practical concerns when applying the described solutions to real-world data analysis problems. We will also consider going one step further from visualization to visual analytics of time-oriented data, for which we outline a basic framework. We conclude with a list of research opportunities for future work.

8.1 Book Summary

Computational analysis and visualization often deal with data that are anchored in space and time. Depicting a spatial frame of reference and the data within it are topics of cartography and geo-visualization, which are independent disciplines with their own books and scientific communities. As there are no such independent disciplines for the temporal frame of reference, this book focused deliberately on the visualization of time-oriented data. In fact, visual depictions of time have a long and venerable history, which has been illustrated by means of several classic examples from the pre-computer era in Chapter 2.

In order to design appropriate visual representations for time-oriented data, one needs to consider the characteristics of time and of the data that are related to time. In Chapter 3, we introduced what these characteristics are and how they can be categorized. A discussion of the issue of data quality provided some insight into

© The Author(s) 2023
W. Aigner et al., *Visualization of Time-Oriented Data*, Human–Computer
Interaction Series, https://doi.org/10.1007/978-1-4471-7527-8_8

what problems one might have to deal with before any reasonable data visualization can take place.

General principles of how time and time-oriented data can be visualized were presented in Chapter 4. Yet, to apply the principles successfully, it is necessary to understand why a visualization is needed, that is, to understand the users' tasks. In terms of user tasks, we distinguished the goals to be achieved, the analytical questions involved, the targets being relevant, and the means to be applied to actually accomplish a task. In terms of general visualization principles for time-oriented data, two basic strategies were introduced: mapping time to space and mapping time to time, which result in static and dynamic visual representations, respectively. On top of these basic strategies, we explained various examples of concrete visualization designs addressing specific aspects of the data, the task, and the presentation itself.

Visual exploration and analysis of time-oriented data also require interaction methods allowing users to manipulate the visual representation in a variety of ways, including navigation of time and data, adjustment of the visual encoding and the spatial arrangement, selection of data of interest, filtering out irrelevant data, and many more. Chapter 5 provided a compact overview of such interaction concepts and techniques.

Moreover, analytical methods have to be provided for supporting the generation of expressive visual representations. Among other purposes, analytical methods are useful for computing data abstractions that may serve to cope with large volumes of data or to enable visual analysis at different levels of granularity. Chapter 6 was dedicated to the aspect of analytical support.

As diverse and varied as time and data characteristics and the choice of visualization design, interaction concepts, and analysis methods are, as diverse is the range of visualization techniques for time and time-oriented data. In Chapter 7, we categorized many state-of-the-art techniques according to six major criteria and proposed some ideas to guide the process of selecting appropriate visualization solutions via an easy-to-use interactive tool, the TimeViz Browser. For reference, brief descriptions and illustrations of all techniques listed in the TimeViz Browser are given in Appendix A.

In conclusion, this book suggests that time is indeed an important dimension that deserves special treatment in visualization with appropriate support for interaction and analytical computation. In the next sections, we will take a look at selected issues that are worth considering further but could not be discussed in this book.

8.2 Practical Concerns

A main concern from an application perspective is the gap between the development of powerful visualization methods on the one hand, and their integration into the real-life data analysis workflows in different application scenarios on the other hand. Bridging this gap requires addressing a variety of software-related and user-related aspects.

Software systems and research prototypes There are a variety of commercial and open source visualization systems, for instance, Tableau,[1] Spotfire,[2] Qlik Sense,[3] Redash,[4] or vtk.[5] Many available systems provide excellent support for visual exploration and analysis of multivariate data. However, the specifics of time are not always considered comprehensively. Quite contrary, support for the wide range of characteristics that are relevant when dealing with time (e.g., support for cyclic time or for different time primitives) is often lacking. Consequently, it can be difficult or actually not feasible for users to apply existing visualization systems. As a result, users may have to design and implement custom solutions that emphasize the dimension of time as necessary for the task at hand.

On the other hand, the visualization community has developed useful research prototypes that provide dedicated support for the time aspect. A prominent example in this regard is the TimeSearcher[6] project for visual exploration of time-series data (↪ p. 290). However, the integration of such prototypes into the infrastructure of the day-to-day business is usually problematic and requires additional effort. Furthermore, research prototypes are usually not designed to cover all aspects of time, but instead address only particular cases – mostly the visualization of linear and ordered time domains.

Data interfaces Another significant problem to be solved is caused by the diversity of existing data formats and interfaces. Processes that generate or collect data and tools that manipulate or analyze the data often use specific databases and data formats that meet the requirements of the particular application scenario. Software tools for visualizing the data and interacting with them often use different formats. This circumstance requires individual and possibly complex data transformations, which can represent a substantial obstacle. To overcome this obstacle more comprehensible and simplified data interfaces need to be developed. Moreover, appropriate tools for data wrangling (see Kandel et al., 2011; Bors, 2020) can be considered to assist users in preparing time-oriented data for visual analysis.

Visualization literacy In addition to improving the technical basis of software, it is also important to take the needs of the users into account. In this regard, an important point is to improve awareness of modern visualization and interaction methods. Most of the time, people rely on traditional visualization techniques such as line plots or bar graphs. These techniques are well-established and have proven to be useful. However, new innovative visualization methods such as the line density plot (↪ p. 307) or the DimpVis (↪ p. 305) approach go beyond what is possible with classic techniques. Modern approaches often can represent a larger number of variables and data values, provide comprehensive interaction functionality, and take

---

[1] https://www.tableau.com

[2] https://www.spotfire.com

[3] https://www.qlik.com/us/products/qlik-sense

[4] https://redash.io

[5] https://vtk.org

[6] https://www.cs.umd.edu/hcil/timesearcher

the specific aspects of time into account. These new possibilities can improve the data analysis and lead to new findings.

Another point to mention is that users who are experts in a specific application domain are not necessarily experts in visualization. However, to be successful in their data analysis work, the domain experts have to know which visual representation should be used for which task. If users were better supported in choosing expressive, effective, and efficient visualization techniques, the quality of information display and analysis results could greatly improve. Enabling people to browse and filter for suitable visualization techniques according to different criteria as suggested in Chapter 7 is only a first step. Visualization recommendation (see Kriglstein et al., 2014; Wongsuphasawat et al., 2016) and guidance approaches (see Ceneda et al., 2017; Ceneda et al., 2018) can offer additional support during the data analysis.

Workflow integration In many application domains, visual methods are primarily used to present previously generated analysis results. That is, the power of visualization is merely used to communicate results at the end of the analysis process. This can be a sufficient strategy, but only for those analytical problems for which a solution can be computed automatically without involving the user. However, many practically-relevant analysis problems are ill-defined and open-ended and as such require a human-in-the-loop interactive visual exploration process. In such cases, visual methods can support all stages of data analysis workflows, from data wrangling to hypothesis generation and falsification to collaborative discussion of findings, and of course, the final presentation of results.

Yet, such a comprehensive and tight workflow integration is often not achieved by existing solutions. Instead, data must be transformed and transferred manually between different analysis and visualization tools. Such extensive switching between different applications is a substantial obstacle to smooth data analysis workflows. Ideally, interactive visualization methods for time-oriented data would integrate seamlessly into existing application portals and systems. Approaching this ideal is the goal of current research on unified data analysis interfaces (see Nonnemann et al., 2021; Nonnemann et al., 2022).

To summarize, bridging the gap between research on interactive visualization methods and their application requires both imparting an awareness of the variety of possibilities and providing means to effectively use them within a given application infrastructure.

8.3 From Visualization to Visual Analytics

This book is entitled *visualization of time-oriented data.* And indeed, we focused on visualization. Interaction and computational analysis were considered as well, but merely to support the visualization. In order to optimally facilitate exploration and analysis of time-oriented data, we should strive for a tight interconnection of visual, interactive, and computational methods, effectively utilizing their strengths and compensating for their weak spots. The field of research that addresses such tight integration of visualization, interaction, and computational analysis is called *visual analytics*. At its core, visual analytics is defined as follows:

> Visual analytics is the science of analytical reasoning facilitated by interactive visual inter-faces. People use visual analytics tools and techniques to synthesize information and derive insight from massive, dynamic, ambiguous, and often conflicting data; detect the expected and discover the unexpected; provide timely, defensible, and understandable assessments; and communicate assessment effectively for action.
>
> Thomas and Cook (2005, p. 4)

Analytical reasoning for real-world problem-solving usually involves the analysis of huge amounts of heterogeneous, possibly incomplete, conflicting, inconsistent, and dynamic information (see Andrienko et al., 2020). For this, human judgment is required to deal with ill-defined problems, synthesize knowledge, and make decisions based on complex data. Thus, a major tenet of visual analytics is that analytical reasoning is not a routine activity that can be automated completely (see Wegner, 1997). Instead, it depends heavily on analysts' initiative and domain experience. Thus, visual analytics aims to facilitate the collaboration of humans and machines by combining:

> [..] automated analysis techniques with interactive visualisations for an effective understand-ing, reasoning and decision making on the basis of very large and complex datasets.
>
> Keim et al. (2010, p. 7)

Thus, the discipline puts its focus on the information discovery process and aims to enable the exploration and understanding of large and complex datasets by com-bining interactive visualization, automated data analysis, and human-computer inter-action. Visual analytics is an inherently multi-disciplinary field that aims to combine the findings of various research areas such as human-computer interaction (HCI), usability engineering, cognitive and perceptual science, information visualization, scientific visualization, databases, data mining, statistics, knowledge discovery, data management, and knowledge representation. Application domains benefiting from visual analytics are for example health care, biotechnology, security and disaster management, environmental science, or climate research.

The basic idea of visual analytics is the integration of the outstanding capabilities of humans in terms of visual information exploration and the enormous processing power of computers to form a powerful knowledge discovery environment. Both visual as well as automated methods are combined in an intertwined manner to fully support this process. Most importantly, the human users are not merely passive elements who interpret the outcome of visual and automated methods, but rather they are the core elements.

Visual analytics process and spaces Keim et al. (2010) propose a process-oriented view of visual analytics as illustrated in Figure 8.1. It focuses on the tight integration of visual data exploration and automated data analysis and describes the dynamic process of synthesizing knowledge from data, following the visual analytics mantra *"Analyze First – Show the Important – Zoom and Filter, and Analyze Further – Details on Demand"* as formulated by Keim et al. (2006a).

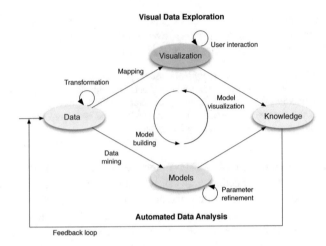

Fig. 8.1: The visual analytics process. ⓒⓘ *The authors. Adapted from Keim et al. (2010).*

Large and usually complex *data* are the basis of visual analytics systems. In many cases, these data come in forms that cannot be directly visualized or automatically analyzed. Therefore, *transformation* steps are applied to perform data cleansing, reformatting, preprocessing, and integration measures. After this initial data transformation, analysts can perform visual exploration and automated analysis. For automated data analysis, *data mining* can be applied to create *models*, which may need to be adjusted through *parameter refinement*. This involves *user interaction* with *visualizations* of the models. For visual data exploration, visual *mapping* is applied to the input data. Based on the *visualization*, *model building* can be performed via *user interactions* with the visual interface. In this sense, the interplay between automated data analysis and visual data exploration inform and support each other throughout the visual analytics process. By interacting with visualizations and models, analysts create new *knowledge* about the data and the underlying phenomena.

Visual analytics can further be conceptualized by considering the different spaces involved in the visual analytics process. Sedig et al. (2012) proposed a conceptual model that involves the five spaces shown in Figure 8.2:

- The **information space** is concerned with modeling, abstracting, and characterizing the sources of information to be studied.
- The **computing space** deals with encoding and storing internal representations of elements from the information space and includes computational operations carried out on such representations.
- The **representation space** makes the internal representations accessible to users using interactive visual representations (IVRs).
- In the **interaction space**, the dyad of action-reaction takes place and perception connects to the mental space.
- The **mental space** is concerned with internal mental events and operations of human analysts.

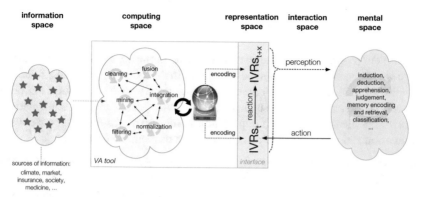

Fig. 8.2: Spaces involved in visual analytics. ©① *The authors. Adapted from Sedig et al. (2012).*

Design issues Through the combined human and computational effort in these five spaces, visual analytics aims to amplify cognition. But simply producing images is no guarantee that complex visual representations will be understood and are useful for gaining insights. Therefore, a human-centered approach is essential and should follow four main principles:

- **Early focus on users and tasks.** Understanding the users, the tasks they perform, and the environment in which users perform these tasks is vital early in the process (see Munzner, 2009; Kerren et al., 2007).
- **Design for human perception and cognition.** Artifacts (methods, techniques, tools, and systems) need to be designed based on established knowledge of human perception and cognition, including pre-attentive processing (see Ware, 2000), Gestalt principles (see Wertheimer, 1938), and sensemaking theory (see Pirolli and Card, 2005).
- **Continuous evaluation.** Visual analytics solutions should be evaluated continuously involving studies on effectiveness, efficiency, and usability to identify measurable benefits and understand limitations (see Lam et al., 2012).
- **Iterative design and refinement.** To improve visual analytics solutions, problems found by experts and users should be corrected iteratively throughout the design and development life cycle (see Shneiderman and Plaisant, 2004).

As already indicated, implementing human-centric visual analytics solutions is challenging, and first steps have been taken to tackle this challenge. Among them are new frameworks to better understand analysis tasks (see Schulz et al., 2013a; Brehmer and Munzner, 2013), concepts to help characterize data (see Schulz et al., 2017), guidance methods to assist users during the data analysis (see Ceneda et al., 2017; Ceneda et al., 2019; Collins et al., 2018), progressive and incremental methods to cope with large amounts of data (see Stolper et al., 2014; Schulz et al., 2016), modern ways of interacting with visual representations of data (see Lee et al., 2021), and onboarding techniques allowing users to get easy access to visual analytics solutions (see Stoiber et al., 2022).

8.4 Future Research Opportunities

Despite the progress that has already been made in the context of visual analytics, there are many open questions to be addressed. In the following, we take a brief look at topics for future research. Our list of topics is aligned with the contents of this book. We will be concerned with visualization, interaction, and analytical computations. Overall, the identified research opportunities aim to advance the interactive visual exploration and analysis specifically of time and time-oriented data.

Cover the specifics of time more broadly A large diversity of powerful visualization techniques for time-oriented data are known in the literature. In Chapter 7, we outlined a corpus of 158 techniques, each of which is also detailed in Appendix A. However, still, most of them support only certain parts of the introduced time and data categorization; in the particular case of visualizing multivariate data, usually linear, point-based, and ordered time domains. Further investigations are required, including the development of techniques for interval-based time, branching time, and multiple perspectives, for simultaneously displaying raw data and data abstractions, and for showing the time-oriented data in their spatial frame of reference.

Another aspect to be considered originates from the multi-scale nature of time. New visualization techniques are required to allow analysts to combine different levels of data and time and to switch between the levels. How this can be done with a basic line plot of linear time series was indicated in Section 5.4.3. But it remains unclear how multi-scale data exploration can be carried out with other visual representations for different categories of time-oriented data.

In light of a diversity of dedicated visual representations for time-oriented data, there is also the question of how a more comprehensive picture of the data can be drawn by combining multiple views. Classically, multi-view approaches work with side-by-side arrangement of views. A promising alternative is to consider smooth transitions between views as suggested by Tominski et al. (2021). Exploring the design space for transitioning between multiple non-trivial representations of time-oriented data while at the same time considering the human capabilities in perceiving and comprehending such transitions seems a formidable research challenge.

Consider more data aspects Throughout this book, we considered time-oriented data, either as abstract data or as data with a spatial frame of reference. However, while being important, space is not the only additional aspect that might be relevant in analyzing time-oriented data.

On top of temporal and spatial dependencies in the data, as a third data aspect, there can be semantic relationships between data items, typically modeled as edges between nodes in a graph. When these relationships change over time, we are dealing with dynamic graphs or temporal networks (see Holme and Saramäki, 2012). Beck et al. (2017) provide a comprehensive overview of visualization techniques for dynamic graphs. However, these techniques primarily represent changes in the graph topology (i.e., creation and removal of edges) rather than communicating the time-oriented data being associated with a graph's nodes and edges. Hence, Beck et al. (2017)

identified the consideration of multiple (dynamic) data dimensions as an open issue, which remains an open topic even today.

Parameter dependency and analytic provenance are two closely related and relevant data aspects that should also be considered. These aspects help users understand how obtained analysis results are influenced by the choice of parameters and how the involved choices were developed. General concepts for visually analyzing parameter spaces (see Sedlmair et al., 2014) and incorporating provenance information (see Xu et al., 2020a) have already been established. There are also techniques specifically developed in the context of time-oriented data (see Eichner et al., 2020). However, existing approaches still have difficulties in dealing with multi-scale dependencies of multiple data variables depending multiple parameters.

Finally, data quality and uncertainty are further important data aspects. We briefly discussed them in Section 3.4 and Section 6.3.2, respectively. Integrating these data aspects more tightly into the visual analysis process of time-oriented data is the goal of ongoing research. For example, Gschwandtner et al. (2016) study different visual representations for temporal uncertainty. How such representations can be employed to support the analysis of time-series segmentations was described by Bors et al. (2020). While these works illustrate the benefit of integrating data quality and related uncertainties into the visualization, we still do not know how to do this generally for different applications and different visualization techniques. This is where future work can improve the expressiveness of visual representations and also strengthen the user's confidence in the obtained findings by considering the aforementioned additional data aspects more broadly.

Communicate on more channels Apart from the numerous options for visualizing time-oriented data, other forms of communication are possible. The data could for example be communicated via sound or haptic sensations (e.g., with braille interfaces). Smell and flavor might also be candidates for alternative communication channels. Despite the fact that these mappings are in principle imaginable, their feasibility and usefulness have to be investigated.

Speeth (1961) already showed how seismographic data can be presented in an auditory display. Another example of attempts in this direction is a system for data sonification by Zhao et al. (2008a) to explore spatial data for users with visual impairments. It seems there are several similarities between the methods and design theories of visualization and sonification (e.g., perceptually encoding data attributes). However, comparatively little is known about combining auditory and visual representations for data analysis. First theoretical underpinnings were proposed by Enge et al. (2021), who conceptualize space as substrate for visualization marks and time as a substrate for auditory marks. Later, Enge et al. (2022) could instantiate their theoretical approach in an exploratory data analysis tool for multivariate data. However, further research is needed to explore these communication channels, particularly for time-oriented data.

Support a broader range of interactions Similar to using perceptual channels more broadly, it would also be interesting to study different ways of interacting with time-oriented data. Our Chapter 5 on interaction already hinted at new interaction modalities such as touch and tangible interaction. These are but two examples of modern approaches to interacting with data; there are more to be explored.

An interesting question is how to better support the interaction on the different scales of time-oriented data. New interaction techniques could address the combination of fine-grained and precise interaction on a local level with coarser, but also faster interaction on a global level. This would also include providing users the option to effortlessly switch between different scales of time or even work on multiple temporal scales simultaneously.

On the technical side, different interaction modalities (Lee et al., 2021) could be investigated for their usefulness in the context of analyzing time-oriented data. For example, to address the limited precision of touch interaction, it would make sense to consider more precise pen-based interaction to support navigation in time or to define temporal queries. Natural language interaction also seems a promising research direction. It allows users to formulate temporal queries via spoken commands, which would significantly reduce interaction costs.

Finally, interaction for collaborative data analysis is a hot research topic. Large high-resolution displays combined with small mobile displays appear to be particularly suited for collaboration (see Horak et al., 2018). Large displays naturally lend themselves to interactively exploring large time-oriented data. Small mobile displays can be used for detailed inspection of data subsets. However, interacting at large scale (e.g., via gaze or physical navigation) and small scale (e.g., via touch or wrist gestures) requires dedicated interaction designs. According to Brehmer et al. (2021), supporting a seamless interaction experience in such cross-device scenarios is a challenge for future work.

Better support for computational analysis Computational analysis plays an important role in understanding large time-oriented data. However, many analytical methods that are applied to time-oriented data treat time as a flat, ordered sequence of events. Thus, these methods are lacking information about the time intervals between events or the reoccurrence of particular temporal patterns. Only few existing analytical methods, like for example the seasonally adjusted autoregressive integrated moving average (SARIMA), model cyclic temporal behavior adequately. As a consequence, better support for dealing with the hierarchical and cyclical structures of time is needed.

Moreover, analytical methods usually behave like black boxes. They accept some input data and generate some analytic result as output, but it often remains unclear to users what happens inside the black box. Yet, as already mentioned, understanding the involved computations and how they are influenced by parameters is essential to appropriately configure analytical methods with regard to the given data and tasks. Therefore, Mühlbacher et al. (2014) demand that black boxes be opened to make computational methods more transparent and steerable for users. This typically involves parameter space analysis to facilitate understanding and progressive algorithms to make analytical computations steerable.

General challenges with respect to parameter space analysis were already identified by Sedlmair et al. (2014). But the particular problems associated with time-oriented data were not yet sufficiently addressed. In a recent survey, Piccolotto et al. (2023) investigate the visual parameter space exploration for spatial and temporal data in particular. They come to the conclusion that spatial and temporal parameters received comparably little attention in the literature and therefore recommend studying these types of parameters in particular in the future.

To make analytical computations steerable, long-running calculations must be split into smaller pieces. This can be achieved via progressive and incremental methods (see Stolper et al., 2014; Schulz et al., 2016). Given the typically large size of time-oriented data, such methods could greatly improve the analysis. However, it is still a task for future work to either develop new temporal analysis methods that are inherently progressive or to revise and adapt existing methods to give them progressive capabilities where possible.

Focus more on user needs While advancing individual visualization, interaction, and computational analysis methods and techniques, it is decisive to take the needs of users into account. Only if the users' data analysis workflows are sufficiently understood can appropriate solutions be developed in a user-centered manner.

In the field of software engineering, it is generally acknowledged that the first step in developing tools and user interfaces should be a sound requirements analysis of the given problem domain (see Hackos and Redish, 1998; Courage and Baxter, 2005). The same applies to designing visualization solutions, where a couple of design recommendations were introduced, general ones by Munzner (2014) and ones specifically for time-oriented data by Miksch and Aigner (2014). In the first place, it is necessary to appropriately characterize the visualization problem, which in the case of visually analyzing time-oriented data includes (1) the characteristics of time and time-oriented data, (2) the characteristics of users, and (3) the intentions and tasks of users. Federico et al. (2017) suggest that providing means to appropriately describe, store, and utilize this knowledge can help in automating the selection or the development of visualization solutions that suit the users, the data, and the tasks.

Approaches for automatic visualization design have been studied for decades already (see Mackinlay, 1986; Senay and Ignatius, 1994; Wills and Wilkinson, 2010; Moritz et al., 2019). In this book, we described the TimeViz Browser as a tool for selecting visualization techniques based on data characteristics. However, a selection based on analysis tasks is not yet possible, not to mention a selection based on user characteristics. These would require an easy-to-use way of specifying tasks and users, and also studies to determine which techniques are suitable for which specification. Both these aspects bear great potential for future research.

The overall aim is to support the users, rather than burden them with technical details. Thus, a significant shift could be realized from a technique-centered view to a user-centered view, where the user is in the focus, similar to what Shneiderman (2022) proposed for human-centered artificial intelligence.

Guide users better In addition to user-centered design, it is also of great relevance to support users during the use of visual data analysis solutions, especially in the context of visual analytics where several visualizations, interactions, and computations play in concert.

As we have seen, the special characteristics of time usually require more advanced visual representations than traditional business charts. However, Börner et al. (2016) and Börner et al. (2019) found that many people have difficulties interpreting novel visual representations and comprehending the underlying data. Limited visualization literacy skills can hamper access to valuable information and complicate problem-solving and decision-making. To mitigate this in the initial phase of the visual data analysis, visualization onboarding plays an increasingly important role. The goal of visualization onboarding is to make people familiar with unknown visual data representations and to empower them to extract the information they need. A few onboarding methods exist in the literature but further research is needed to identify effective designs of onboarding methods for time-oriented data visualization and to understand user behavior while using onboarding methods. This also involves creating flexible and adaptive approaches that take the prior knowledge of the users into account.

Even when users are familiar with a visualization solution, it can still be necessary to guide users while they are working on visual data analysis tasks. Ceneda et al. (2017) characterized guidance as a concept to support users in situations where they have difficulties making analytic progress on their own. In Section 5.4, we briefly illustrated how such guidance can look like for multi-scale exploration of large time-oriented data. But still many open questions need to be addressed to arrive at true mixed-initiative solutions wherein both the system and the user can contribute to the progress of the data analysis. The overall vision of Ceneda et al. (2020) is to make guidance effective, available, trustworthy, adaptive, controllable, and non-disruptive in the future. Yet, how to detect the point where users need assistance? How to determine an appropriate level of support and how to dynamically adjust it? How to generate guidance without actually knowing neither the elusive data analysis problem of the user nor its concrete answer, if it exists at all? While these questions are generic, the research to solve them must be tailored to the specifics of the problem domain, which in our case means time and time-oriented data.

Understand what works   Finally, to be able to guide users or to suggest appropriate techniques to users based on data and tasks, we need to know which techniques are *good*. This requires evaluation. Evaluation has to be conducted in terms of the three criteria expressiveness, effectiveness, and efficiency (see Chapter 1). Expressiveness and effectiveness are related to the data level and the task level, respectively. They require testing whether the characteristics of time and data are sufficiently communicated, and whether the visual representation, interaction techniques, and analysis methods match the tasks, expectations, and cognitive capabilities of users. With the efficiency criterion, a balance of required resources (technical and human) and gained benefits in an application domain comes into play.

The literature provides a wealth of methodologies to conduct evaluation studies in general (see Lazar et al., 2017). There are also specific methods for visualization, for example, to measure effectiveness (see Zhu, 2007). However, Plaisant (2004) points out that thorough evaluation is challenging as it requires the combined consideration of multiple criteria. Addressing this challenge, Munzner (2009) introduced a nested model for design and validation where each nested level (i.e., characterization of data and tasks, abstraction into operation and data types, design of encoding and interaction, and development of algorithms) is associated with a dedicated methodology for evaluation. Lam et al. (2012) outline seven typical scenarios for empirical studies in the context of visual data analysis. However, neither Munzner's model nor Lam et al.'s scenarios consider time and time-oriented data and tasks explicitly. More research is needed to close this gap in the literature.

Even more advanced evaluation strategies are needed for understanding complex visual analysis solutions with their interplay of visual, interactive, and computational components. Adding to standard solutions more sophisticated methods such as uncertainty visualization, cross-device interaction, or user guidance can make evaluation studies extremely challenging and expensive. Therefore, it seems reasonable to investigate new evaluation strategies for their applicability to selected aspects first. For example, one could look into evaluation heuristics like ICE-T (see Wall et al., 2019) with a particular focus on guidance-enhanced visual analysis solutions for time-oriented data. Based on the results of multiple such focused evaluations one could then develop a better understanding of how to evaluate complex intertwined visual analysis solutions.

—

With these ideas for future work, we close this book on *visualizing time-oriented data* hoping that the next decade of research brings us closer to *visual analytics for time-oriented data*, which then might be the title for a potential third edition – or a completely new book.

References

Andrienko, N., G. Andrienko, G. Fuchs, A. Slingsby, C. Turkay, and S. Wrobel (2020). *Visual Analytics for Data Scientists*. Springer. DOI: 10.1007/978-3-030-56146-8.

Beck, F., M. Burch, S. Diehl, and D. Weikopf (2017). "A Taxonomy and Survey of Dynamic Graph Visualization". In: *Computer Graphics Forum* 36.1, pp. 133–159. DOI: 10.1111/cgf.12791.

Börner, K., A. Bueckle, and M. Ginda (2019). "Data Visualization Literacy: Definitions, Conceptual Frameworks, Exercises, and Assessments". In: *Proceedings of the National Academy of Sciences* 116.6, pp. 1857–1864. DOI: 10.1073/pnas.1807180116.

Börner, K., A. Maltese, R. N. Balliet, and J. Heimlich (2016). "Investigating Aspects of Data Visualization Literacy Using 20 Information Visualizations and 273 Science Museum Visitors". In: *Information Visualization* 15.3, pp. 198–213. DOI: 10.1177/1473871615594652.

Bors, C. (2020). "Facilitating Data Quality Assessment Utilizing Visual Analytics: Tackling Time, Metrics, Uncertainty, and Provenance". PhD thesis. Institute of Visual Computing and Human-Centered Technology, TU Wien.

Bors, C., C. Eichner, S. Miksch, C. Tominski, H. Schumann, and T. Gschwandtner (2020). "Exploring Time Series Segmentations Using Uncertainty and Focus+Context Techniques". In: *Proceedings of the Eurographics / IEEE Conference on Visualization (EuroVis) - Short Papers*. Eurographics Association, pp. 7–11. DOI: 10.2312/evs.20201040.

Brehmer, M., B. Lee, J. Stasko, and C. Tominski (2021). "Interacting with Visualization on Mobile Devices". In: *Mobile Data Visualization*. Edited by Lee, B., Dachselt, R., Isenberg, P., and Choe, E. K. CRC Press, pp. 67–110. DOI: 10.1201/9781003090823-3.

Brehmer, M. and T. Munzner (2013). "A Multi-Level Typology of Abstract Visualization Tasks". In: *IEEE Transactions on Visualization and Computer Graphics* 19.12, pp. 2376–2385. DOI: 10.1109/TVCG.2013.124.

Ceneda, D., N. Andrienko, G. Andrienko, T. Gschwandtner, S. Miksch, N. Piccolotto, T. Schreck, M. Streit, J. Suschnigg, and C. Tominski (2020). "Guide Me in Analysis: A Framework for Guidance Designers". In: *Computer Graphics Forum* 39.6, pp. 269–288. DOI: 10.1111/cgf.14017.

Ceneda, D., T. Gschwandtner, T. May, S. Miksch, H.-J. Schulz, M. Streit, and C. Tominski (2017). "Characterizing Guidance in Visual Analytics". In: *IEEE Transactions on Visualization and Computer Graphics* 23.1, pp. 111–120. DOI: 10.1109/TVCG.2016.2598468.

Ceneda, D., T. Gschwandtner, and S. Miksch (2019). "A Review of Guidance Approaches in Visual Data Analysis: A Multifocal Perspective". In: *Computer Graphics Forum* 38.3, pp. 861–879. DOI: 10.1111/cgf.13730.

Ceneda, D., T. Gschwandtner, S. Miksch, and C. Tominski (2018). "Guided Visual Exploration of Cyclical Patterns in Time-series". In: *Proceedings of the IEEE Symposium on Visualization in Data Science (VDS)*. IEEE Computer Society.

Collins, C., N. Andrienko, T. Schreck, J. Yang, J. Choo, U. Engelke, A. Jena, and T. Dwyer (2018). "Guidance in the Human-Machine Analytics Process". In: *Visual Informatics* 2.3. DOI: 10.1016/j.visinf.2018.09.003.

Courage, C. and K. Baxter (2005). *Understanding Your Users*. Morgan Kaufmann. DOI: 10.1016/B978-1-55860-935-8.X5029-5.

Eichner, C., H. Schumann, and C. Tominski (2020). "Making Parameter Dependencies of Time-Series Segmentation Visually Understandable". In: *Computer Graphics Forum* 39.1, pp. 607–622. DOI: 10.1111/cgf.13894.

Enge, K., A. Rind, M. Iber, R. Höldrich, and W. Aigner (2021). "It's about Time: Adopting Theoretical Constructs from Visualization for Sonification". In: *Proceedings of the International Audio Mostly Conference (AMI)*. ACM Press, pp. 64–71. DOI: 10.1145/3478384.3478415.

Enge, K., A. Rind, M. Iber, R. Höldrich, and W. Aigner (2022). "Towards Multimodal Exploratory Data Analysis: SoniScope as a Prototypical Implementation". In: *Proceedings of the Eurographics / IEEE Conference on Visualization (EuroVis)*

- *Short Papers*. Eurographics Association, pp. 67–71. DOI: 10.2312/evs.20221095.

Federico, P., M. Wagner, A. Rind, A. Amor-Amoros, S. Miksch, and W. Aigner (2017). "The Role of Explicit Knowledge: A Conceptual Model of Knowledge-Assisted Visual Analytics". In: *Proceedings of the IEEE Conference on Visual Analytics Science and Technology (VAST)*. IEEE Computer Society, pp. 92–103. DOI: 10.1109/VAST.2017.8585498.

Gschwandtner, T., M. Bögl, P. Federico, and S. Miksch (2016). "Visual Encodings of Temporal Uncertainty: A Comparative User Study". In: *IEEE Transactions on Visualization and Computer Graphics* 22.1, pp. 539–548. DOI: 10.1109/TVCG.2015.2467752.

Hackos, J. T. and J. C. Redish (1998). *User and Task Analysis for Interface Design*. John Wiley & Sons, Inc.

Holme, P. and J. Saramäki (2012). "Temporal Networks". In: *Physics Reports* 519.3, pp. 97–125. DOI: 10.1016/j.physrep.2012.03.001.

Horak, T., S. K. Badam, N. Elmqvist, and R. Dachselt (2018). "When David Meets Goliath: Combining Smartwatches with a Large Vertical Display for Visual Data Exploration". In: *Proceedings of the SIGCHI Conference on Human Factors in Computing Systems (CHI)*. ACM Press. DOI: 10.1145/3173574.3173593.

Kandel, S., A. Paepcke, J. M. Hellerstein, and J. Heer (2011). "Wrangler: Interactive Visual Specification of Data Transformation Scripts". In: *Proceedings of the SIGCHI Conference on Human Factors in Computing Systems (CHI)*. ACM Press, pp. 3363–3372. DOI: 10.1145/1978942.1979444.

Keim, D., J. Kohlhammer, G. Ellis, and F. Mansmann, eds. (2010). *Mastering the Information Age – Solving Problems with Visual Analytics*. Eurographics Association. URL: https://diglib.eg.org/handle/10.2312/14803.

Keim, D. A., F. Mansmann, J. Schneidewind, and H. Ziegler (2006a). "Challenges in Visual Data Analysis". In: *Proceedings of the International Conference Information Visualisation (IV)*. IEEE Computer Society, pp. 9–16. DOI: 10.1109/IV.2006.31.

Kerren, A., A. Ebert, and J. Meyer, eds. (2007). *Human-Centered Visualization Environments*. Vol. 4417. Lecture Notes in Computer Science. Springer. DOI: 10.1007/978-3-540-71949-6.

Kriglstein, S., M. Pohl, and M. Smuc (2014). "Pep Up Your Time Machine: Recommendations for the Design of Information Visualizations of Time-Dependent Data". In: *Handbook of Human Centric Visualization*. Edited by Huang, W. Springer, pp. 203–225. DOI: 10.1007/978-1-4614-7485-2_8.

Lam, H., E. Bertini, P. Isenberg, C. Plaisant, and S. Carpendale (2012). "Empirical Studies in Information Visualization: Seven Scenarios". In: *IEEE Transactions on Visualization and Computer Graphics* 18.9, pp. 1520–1536. DOI: 10.1109/TVCG.2011.279.

Lazar, J., J. H. Feng, and H. Hochheiser (2017). *Research Methods in Human-Computer Interaction*. 2nd edition. John Wiley & Sons, Ltd.

Lee, B., A. Srinivasan, P. Isenberg, and J. T. Stasko (2021). "Post-WIMP Interaction for Information Visualization". In: *Foundations and Trends in Human-Computer Interaction* 14.1, pp. 1–95. DOI: 10.1561/1100000081.

Mackinlay, J. (1986). "Automating the Design of Graphical Presentations of Relational Information". In: *ACM Transactions on Graphics* 5.2, pp. 110–141. DOI: 10.1145/22949.22950.

Miksch, S. and W. Aigner (2014). "A Matter of Time: Applying a Data-Users-Tasks Design Triangle to Visual Analytics of Time-Oriented Data". In: *Computers & Graphics* 38, pp. 286–290. DOI: 10.1016/j.cag.2013.11.002.

Moritz, D., C. Wang, G. L. Nelson, H. Lin, A. M. Smith, B. Howe, and J. Heer (2019). "Formalizing Visualization Design Knowledge as Constraints: Actionable and Extensible Models in Draco". In: *IEEE Transactions on Visualization and Computer Graphics* 25.1, pp. 438–448. DOI: 10.1109/TVCG.2018.2865240.

Mühlbacher, T., H. Piringer, S. Gratzl, M. Sedlmair, and M. Streit (2014). "Opening the Black Box: Strategies for Increased User Involvement in Existing Algorithm Implementations". In: *IEEE Transactions on Visualization and Computer Graphics* 20.12, pp. 1643–1652. DOI: 10.1109/TVCG.2014.2346578.

Munzner, T. (2009). "A Nested Process Model for Visualization Design and Validation". In: *IEEE Transactions on Visualization and Computer Graphics* 15.6, pp. 921–928. DOI: 10.1109/TVCG.2009.111.

Munzner, T. (2014). *Visualization Analysis and Design*. A K Peters/CRC Press. DOI: 10.1201/b17511.

Nonnemann, L., M. Hogräfer, M. Röhlig, H. Schumann, B. Urban, and H.-J. Schulz (2022). "A Data-Driven Platform for the Coordination of Independent Visual Analytics Tools". In: *Computers & Graphics* 106, pp. 152–160. DOI: 10.1016/j.cag.2022.05.023.

Nonnemann, L., M. Hogräfer, H. Schumann, B. Urban, and H.-J. Schulz (2021). "Customizable Coordination of Independent Visual Analytics Tools". In: *Proceedings of the EuroVis Workshop on Visual Analytics (EuroVA)*. Eurographics Association. DOI: 10.2312/eurova.20211094.

Piccolotto, N., M. Bögl, and S. Miksch (2023). "Visual Parameter Space Exploration in Time and Space". In: *Computer Graphics Forum* 42.6. DOI: 10.1111/cgf.14785.

Pirolli, P. and S. Card (2005). "The Sensemaking Process and Leverage Points for Analyst Technology as Identified Through Cognitive Task Analysis". In: *Proceedings of the International Conference on Intelligence Analysis*. Vol. 5, pp. 2–4.

Plaisant, C. (2004). "The Challenge of Information Visualization Evaluation". In: *Proceedings of the Conference on Advanced Visual Interfaces (AVI)*. ACM Press, pp. 106–119. DOI: 10.1145/989863.989880.

Schulz, H.-J., M. Angelini, G. Santucci, and H. Schumann (2016). "An Enhanced Visualization Process Model for Incremental Visualization". In: *IEEE Transactions on Visualization and Computer Graphics* 22.7, pp. 1830–1842. DOI: 10.1109/TVCG.2015.2462356.

Schulz, H.-J., T. Nocke, M. Heitzler, and H. Schumann (2013a). "A Design Space of Visualization Tasks". In: *IEEE Transactions on Visualization and Computer Graphics* 19.12, pp. 2366–2375. DOI: 10.1109/TVCG.2013.120.

Schulz, H.-J., T. Nocke, M. Heitzler, and H. Schumann (2017). "A Systematic View on Data Descriptors for the Visual Analysis of Tabular Data". In: *Information Visualization* 16.3, pp. 232–256. DOI: 10.1177/1473871616667767.

Sedig, K., P. Parsons, and A. Babanski (2012). "Towards a Characterization of Interactivity in Visual Analytics". In: *Journal of Multimedia Processing and Technologies, Special issue on Theory and Application of Visual Analytics* 3 (1), pp. 12–28. URL: https://www.dline.info/jmpt/fulltext/v3n1/2.pdf.

Sedlmair, M., C. Heinzl, S. Bruckner, H. Piringer, and T. Möller (2014). "Visual Parameter Space Analysis: A Conceptual Framework". In: *IEEE Transactions on Visualization and Computer Graphics* 20.12, pp. 2161–2170. DOI: 10.1109/TVCG.2014.2346321.

Senay, H. and E. Ignatius (1994). "A Knowledge-Based System for Visualization Design". In: *IEEE Computer Graphics and Applications* 14.6, pp. 36–47. DOI: 10.1109/38.329093.

Shneiderman, B. (2022). *Human-Centered AI*. Oxford University Press.

Shneiderman, B. and C. Plaisant (2004). *Designing the User Interface: Strategies for Effective Human-Computer Interaction*. 4th edition. Pearson Addison Wesley.

Speeth, S. D. (1961). "Seismometer Sounds". In: *The Journal of the Acoustical Society of America* 33.7, pp. 909–916. DOI: 10.1121/1.1908843.

Stoiber, C., D. Ceneda, M. Wagner, V. Schetinger, T. Gschwandtner, M. Streit, S. Miksch, and W. Aigner (2022). "Perspectives of Visualization Onboarding and Guidance in VA". In: *Visual Informatics* 6 (1), pp. 68–83. DOI: 10.1016/j.visinf.2022.02.005.

Stolper, C. D., A. Perer, and D. Gotz (2014). "Progressive Visual Analytics: User-Driven Visual Exploration of In-Progress Analytics". In: *IEEE Transactions on Visualization and Computer Graphics* 20.12, pp. 1653–1662. DOI: 10.1109/TVCG.2014.2346574.

Thomas, J. J. and K. A. Cook (2005). *Illuminating the Path: The Research and Development Agenda for Visual Analytics*. IEEE Computer Society.

Tominski, C., G. Andrienko, N. Andrienko, S. Bleisch, S. I. Fabrikant, E. Mayr, S. Miksch, M. Pohl, and A. Skupin (2021). "Toward Flexible Visual Analytics Augmented through Smooth Display Transitions". In: *Visual Informatics* 5.3, pp. 28–38. DOI: 10.1016/j.visinf.2021.06.004.

Wainer, H. (1997). *Visual Revelations: Graphical Tales of Fate and Deception from Napoleon Bonaparte to Ross Perot*. Copernicus.

Wall, E., M. Agnihotri, L. Matzen, K. Divis, M. Haass, A. Endert, and J. Stasko (2019). "A Heuristic Approach to Value-Driven Evaluation of Visualizations". In: *IEEE Transactions on Visualization and Computer Graphics* 25.1, pp. 491–500. DOI: 10.1109/TVCG.2018.2865146.

Ware, C. (2000). *Information Visualization: Perception for Design*. Morgan Kaufmann.

Wegner, P. (1997). "Why Interaction Is More Powerful Than Algorithms". In: *Communications of the ACM* 40.5, pp. 80–91. DOI: 10.1145/253769.253801.

Wertheimer, M. (1938). "Laws of Organization in Perceptual Forms". In: *A Sourcebook of Gestalt Psychology*. Edited by Ellis, W. D. London, UK: Routledge and Kegan Paul, pp. 71–88.

Wills, G. and L. Wilkinson (2010). "AutoVis: Automatic Visualization". In: *Information Visualization* 9.1, pp. 47–69. DOI: 10.1057/ivs.2008.27.

Wongsuphasawat, K., D. Moritz, A. Anand, J. D. Mackinlay, B. Howe, and J. Heer (2016). "Voyager: Exploratory Analysis via Faceted Browsing of Visualization Recommendations". In: *IEEE Transactions on Visualization and Computer Graphics* 22.1, pp. 649–658. DOI: 10.1109/TVCG.2015.2467191.

Xu, K., A. Ottley, C. Walchshofer, M. Streit, R. Chang, and J. Wenskovitch (2020a). "Survey on the Analysis of User Interactions and Visualization Provenance". In: *Computer Graphics Forum* 39.3, pp. 757–783. DOI: 10.1111/cgf.14035.

Zhao, H., C. Plaisant, B. Shneiderman, and J. Lazar (2008a). "Data Sonification for Users with Visual Impairment: A Case Study with Georeferenced Data". In: *ACM Transactions on Computer-Human Interaction* 15.1, 4:1–4:28. DOI: 10.1145/1352782.1352786.

Zhu, Y. (2007). "Measuring Effective Data Visualization". In: *Proceedings of the International Symposium on Visual Computing (ISVC)*. Springer, pp. 652–661. DOI: 10.1007/978-3-540-76856-2_64.

Appendix A
Survey of Visualization Techniques

This appendix is dedicated to surveying existing visualization techniques for time and time-oriented data. As there are different types of time and time-oriented data as well as many different questions one may ask about the data, there are numerous visualization techniques taking the peculiarities of data and tasks into account. Some of them are specific to a particular application, others are more general with potential applicability in various domains.

The following pages list the techniques on a per-page basis. Each page briefly describes a particular technique's background, explains its main idea and concepts, and indicates its application. Each page also includes relevant references to publications that originally proposed or substantially extended a technique. Illustrating figures demonstrate the techniques in use or their conceptual construction. Additionally, each page has a side-bar for categorizing the described visualization technique. Here we use the simplified categorization schema introduced in Chapter 7. Accordingly, the side-bar information follows this pattern:

- **Time**
    - *Primitives* – points vs. intervals
    - *Arrangement* – linear vs. cyclic

- **Data**
    - *Number of variables* – single vs. multiple
    - *Frame of reference* – abstract vs. spatial

- **Visual representation**
    - *Mapping of time* – static vs. dynamic
    - *Dimensionality* – 2D vs. 3D

Where possible, a distinct classification will be given. However, this is not always achievable, particularly for more general and flexible visualization approaches. In such cases, we will indicate that a technique exhibits multiple characteristics per category.

While the list of techniques included in our survey is extensive, it is clear that we cannot claim completeness. This is due to the fact that visualization of time-oriented data is a hot research area constantly yielding new techniques. Moreover, visualization solutions can be highly application-dependent, and it is virtually impossible to dig out every tiny variation of existing visualization approaches that might be hidden in the vast body of scientific literature across application domains. Therefore, we took care to include a wide spectrum of key techniques, both classic ones with proven usefulness and contemporary ones with potential impact.

In what follows, you will find the one-page descriptions for all the techniques listed in Table 7.1 from page 201 of Chapter 7. Below all techniques are listed alphabetically with their corresponding page numbers for easy reference. Visit `https://browser.timeviz.net` for an online version of the survey.

A.1 List of Techniques

primitives: **points**
arrangement: **linear**

A.2 Techniques for Abstract Time-Oriented Data

Point Plot

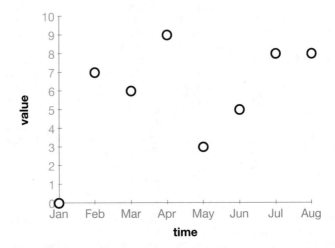

number of variables: **single**
frame of reference: **abstract**

Fig. A.1: Data are displayed as points in a Cartesian coordinate system where time and data are mapped to the horizontal axis and the vertical axis, respectively. ⓒⒾ *The authors.*

One of the most straightforward ways of depicting time-series data is using a Cartesian coordinate system with time on the horizontal axis and the corresponding value on the vertical axis. A point is plotted for every measured time-value pair. This kind of representation is called point plot, point graph, or scatter plot, respectively. Harris (1999) describes it as a 2-dimensional representation where quantitative data aspects are visualized by distance from the main axis. Many extensions of this basic form such as 3D techniques (layer graph) or techniques that use different symbols instead of points are known. This technique is particularly suited for emphasizing individual values. Moreover, depicting data using position along a common scale can be perceived most precisely by the human perceptual system.

mapping of time: **static**
dimensionality: **2D**

Relevant references: Harris (1999)

Line Plot

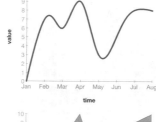
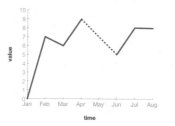
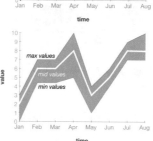

Fig. A.2: Successive data points are connected with lines to visualize the overall change over time. Top-left: straight lines; top-right: Bézier curves; bottom-left: missing data; bottom-right: band graph. ⓒⓘ *The authors.*

The most common form of representing time series are line plots. They extend point plots (↪ p. 232) by linking the data points with lines, which emphasizes their temporal relation. Consequently, line plots focus on the overall shape of data over time. This is in contrast to point plots where individual data points are emphasized. As illustrated in the figure, different styles of connections between the data points such as straight lines, step lines (instant value changes), or Bézier curves can be used depending on the phenomenon under consideration. However, what has to be kept in mind is that one can not be sure in all cases about the data values in the time interval between two data points and that any kind of connection between data points reflects only an approximation. A further point of caution is missing data. Simply connecting subsequent data points might lead to false conclusions regarding the data. Therefore, this should be made visible to the viewer, for instance by using dotted lines (bottom left). Harris (1999) lists many extensions and subtypes of line plots, including fever graphs, band graphs (bottom right), layer line graphs, surface graphs, index graphs, and control graphs.

Relevant references: Harris (1999)

time

Bar Graph, Spike Graph

primitives: **points**
arrangement: **linear**

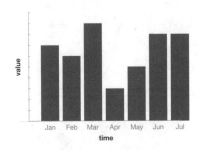 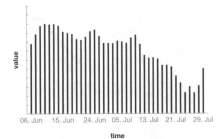

Fig. A.3: Time-dependent data values are represented as bars of different lengths. Left: regular bars; right: spike graph where bars are narrowed to spikes. ⓒ① *The authors.*

data

number of variables: **single**
frame of reference: **abstract**

Bar graphs are a well-known and widely used type of representation where bars are used to depict data values. This makes comparisons easier than with point plots. As bar length is used to depict data values, only variables with a ratio scale (having a natural zero) can be represented. Consequently, the value scale also has to start with zero to allow for a fair visual comparison. In contrast to line plots, bar graphs emphasize individual values as do point plots. A variant of bar graphs often used for graphing larger time series (e.g., stock market data such as price or volume) are spike graphs. As illustrated in right-hand figure, the vertical bars are reduced so as to appear as spikes, where spike height is again used to encode data values. This way, a good visual balance is achieved between focusing on individual values and showing the overall development of a larger number of data values.

Relevant references: Harris (1999)

vis

mapping of time: **static**
dimensionality: **2D**

Sparklines

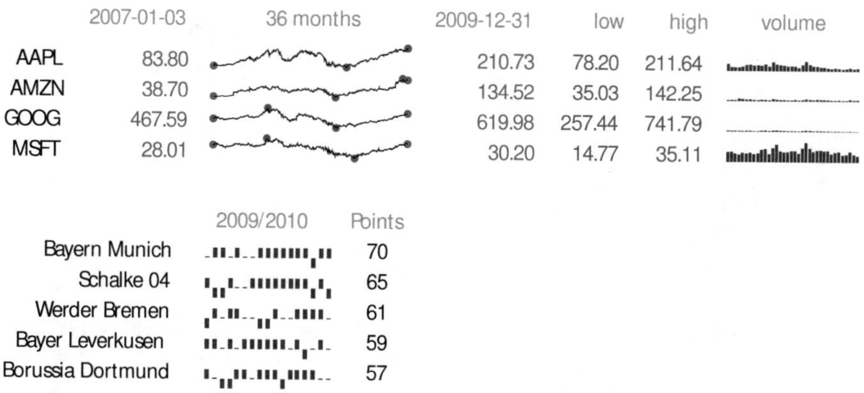

time

	2007-01-03	36 months	2009-12-31	low	high	volume
AAPL	83.80		210.73	78.20	211.64	
AMZN	38.70		134.52	35.03	142.25	
GOOG	467.59		619.98	257.44	741.79	
MSFT	28.01		30.20	14.77	35.11	

	2009/2010	Points
Bayern Munich		70
Schalke 04		65
Werder Bremen		61
Bayer Leverkusen		59
Borussia Dortmund		57

Fig. A.4: Sparklines are miniature charts to be integrated into text. Top: Sparklines visualizing stock market data; bottom: Soccer results visualized using ticks (up = win, down = loss, base = draw). ©① *The authors. Generated with the sparklines package for LaTeX.*

Tufte (2006) describes sparklines as simple, word-like graphics intended to be integrated into text. This adds richer information about the development of a variable over time that words themselves could hardly convey. The visualization method focuses mainly on giving an overview of the development of values for time-oriented data rather than on specific values or dates due to their small size and the omission of axes and labels. Sparklines can be integrated seamlessly into paragraphs of text, can be laid out as tables, or can be used for dashboards. They are increasingly adopted to present information on web pages (such as usage statistics) in newspapers (e.g., for sports statistics), or in finance (e.g., for stock market data). Usually, miniaturized versions of line plots (↪ p. 233) and bar graphs (↪ p. 234) are employed to represent data. For line plots, the first and last values can be emphasized by showing red dots and printing the associated values to the left and right of the sparkline (top figure). Additionally, the minimum and maximum values can be marked by blue dots. Besides this, colored bands in the background of the plot can be used to show normal value ranges 4.8 8.3. For the special case of binary or three-valued data, special bar graphs can be applied that use ticks extending up and down a horizontal baseline. One use for this kind of representation is the wins and losses of sports teams where the history of a whole season can be presented using very little space (bottom figure).

Relevant references: Tufte (2006)

TrendDisplay

primitives: **points**
arrangement: **linear**

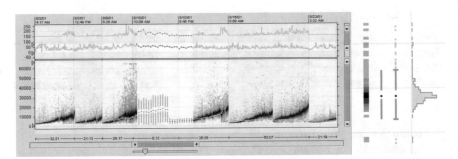

Fig. A.5: TrendDisplay shows data (center) and derived statistics (top) at four different levels of visual abstraction. Depending on the available screen space, the visual representation is dynamically chosen to be either color-coded density distributions, thin box plots, box plots plus outliers, or bar histograms (right). © *2003 IEEE. Reprinted, with permission, from Brodbeck and Girardin (2003).*

number of variables: **single**
frame of reference: **abstract**

The TrendDisplay technique by Brodbeck and Girardin (2003) allows the analysis of trends in larger time series. The technique is used for the drug discovery process and in quality control. Basically, the TrendDisplay window is composed of two panels. The main panel in the center shows the measured (raw) data and the top panel depicts derived statistical values. Four different levels of detail (right) are used in order to cope with large numbers of time points: color-coded density distributions, thin box plots, box plots plus outliers, and bar histograms (from low to high level of detail). In the temporal dimension, bifocal focus+context functionality is used for enlarging areas of interest without losing context information about neighboring data. The different levels of detail are chosen automatically depending on the available screen space. Moreover, brushing & linking as well as smooth transitions complete the highly interactive interface.

Relevant references: Brodbeck and Girardin (2003)

mapping of time: **static**
dimensionality: **2D**

Decision Chart

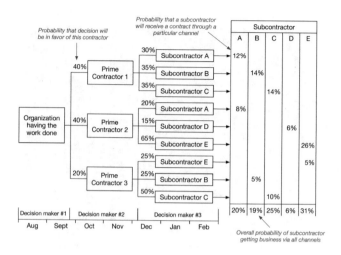

Fig. A.6: Future decisions and corresponding alternative outcomes along with their probabilities are depicted over time in a decision chart. ⓒⓕ *The authors. Adapted from Harris (1999).*

Harris (1999) describes decision charts as a graphical representation for depicting future decisions and potential alternative outcomes along with their probabilities over time. Decision charts are one of the very few techniques for time-oriented data that use the *branching time* model. Decision charts use a horizontal time axis along which information elements (decisions and probabilities) are aligned. Multiple decisions for a particular time interval are stacked on top of each other, indicating that they are possible alternatives for that interval. However, the temporal context itself is not of prime interest and is just indicated by a simple time scale at the bottom of the chart. The main advantage of the decision chart is that it allows planners to investigate possible outcomes and implications before decisions are made.

Relevant references: Harris (1999)

time

primitives: **points**
arrangement: **linear**

data

number of variables: **single**
frame of reference: **abstract**

vis

mapping of time: **static**
dimensionality: **2D**

TimeTree

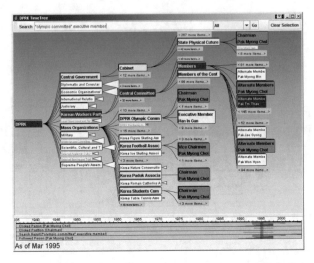

Fig. A.7: A TimeTree showing changes in the organizational structure of officials in the DPR Korea with focused and important nodes highlighted. The time slider (currently set to March 1995) shows selected personal events and serves for interactive navigation. © *2006 IEEE. Reprinted, with permission, from Card et al. (2006).*

TimeTree by Card et al. (2006) is a visualization technique to enable the exploration of changing hierarchical organizational structures and of individuals within such structures. The visualization consists of three parts: a time slider (bottom), a tree view(center), and a search interface (top). The time slider's main purpose is to allow users to navigate to any point in time. Additionally, it shows information strips with events for a selected set of individuals, and thus, provides insight into where in time interesting things have happened. The tree view shows the snapshot of the organizational structure corresponding to the selected time point. The tree visualization uses a degree-of-interest (DoI) approach to highlight important information. To this end, a specific color scheme is used to indicate certain data characteristics, as for instance, important nodes, nodes matching with search queries, or recently clicked nodes. Low-interest nodes, with regard to the user's current focus and search query, are ghosted, and entire sub-trees may be represented as triangular abstractions in order to de-clutter the display and maintain the readability of important nodes. The search interface supports a textual search for individuals in the represented organization.

Relevant references: Card et al. (2006)

Arc Diagrams

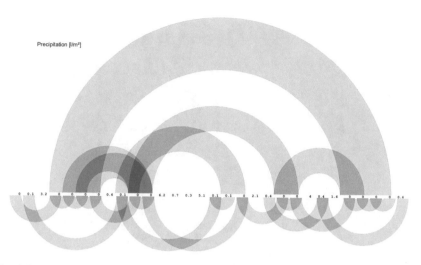

Precipitation [l/m²]

Fig. A.8: A sequence of data values is shown along the horizontal axis. Matching subsequences are connected by arcs, where arc thickness and height encode subsequence size and occurrence distance, respectively. © *Courtesy of Michael Zornow.*

Patterns in sequences of data values can be visualized using arc diagrams. They were introduced as an interactive visualization technique by Wattenberg (2002). Given a sequence of values, the goal is to extract significant subsequences that occur multiple times in the original sequence. The visualization displays the sequence of data values in textual form along the horizontal (time) axis. Occurrences of significant subsequences are visually connected by spanning arcs. The arcs' thickness represents the size of the subsequence, that is, the number of data values in the subsequence. The height of an arc indicates the distance between two successive occurrences of the subsequence. To express data-specific aspects, one can separately use the space above or below the data sequence. This also helps to reduce the overlap of arcs. Additionally, transparency is used to allow users to see through overlapping arcs. The visualization can be controlled interactively via several parameters (e.g., minimum size of subsequences or tolerance threshold for fuzzy pattern extraction) to keep the number of arcs at an interpretable level.

Relevant references: Wattenberg (2002)

SparkClouds

primitives: **points**
arrangement: **linear**

data

number of variables: **single**
frame of reference: **abstract**

Fig. A.9: Display of important keywords in time-varying data. Font size encodes the keyword importance. Keywords being less relevant for the studied point in time are attenuated by using dimmed color and smaller font size. © *2010 IEEE. Reprinted, with permission, from Lee et al. (2010).*

vis

mapping of time: **static**
dimensionality: **2D**

Tag clouds visualize a set of keywords weighted by their importance. To this end, a layout of the keywords is computed. By varying font size, color, or other visual variables important keywords are emphasized over less-important keywords. Classic tag clouds, however, are incapable of representing the evolution of keywords. Lee et al. (2010) integrate sparklines (↪ p. 235) into tag clouds in order to visualize temporal trends in the development of keywords. The idea is to visually combine a keyword (or tag) and its temporal evolution. The keyword's importance is encoded with the font size used to render the text, where the size can correspond either to the overall importance of the keyword for the entire time series or to the importance at a particular point in time. Attached to the keyword is a sparkline that represents the keyword's trend. A color gradient is shown in the background of each keyword-sparkline pair to make this design perceivable as a visual unit. Lee et al. (2010) conducted user studies with sparkclouds and could confirm that sparkclouds are useful and have advantages over alternative standard methods for visualizing text and temporal information.

Relevant references: Lee et al. (2010)

Growth Matrix

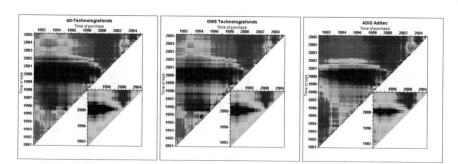

Fig. A.10: Growth Matrices are used for the analysis of financial data. For each triangle, all possible subintervals between time of purchase (x-axis) and time of sale (y-axis) are shown by a single pixel. Each pixel's color indicates financial gain (green) or loss (red) for each subinterval. The example shows the performance of three different funds (large triangles) compared to the market performance of similar assets (small triangles). ⓒⓘⓢⓔ *Growth Matrix by Keim et al., also see Keim et al.* *(2006b).*

Analysis of the performance of assets is a core task in financial analytics. To support this, Keim et al. (2006b) developed Growth Matrix, a heatmap-based visualization technique for analyzing asset return rates over all possible subintervals in a given time frame. It is based on the Return Triangle visualization technique in financial analytics and extends it to a dense, pixel-based visualization approach. In the triangle, the horizontal and vertical dimensions represent time from left to right and bottom to top, respectively. Color-coding is applied to show growth rate quantities. Growth rates are normalized to achieve a normalized color-mapping scheme using red hues for losses and green hues for gains. Smaller Growth matrices might be added to allow for comparisons, e.g., against the overall market performance of similar assets (see figure for an example). In subsequent work, Ziegler et al. (2007) extended the approach by adding Performance Matrices that directly depict performance values for different holding periods in a rectangular matrix by mapping time of sale on the x-axis and holding periods on the y-axis. Moreover, improvements of color mappings are introduced by extending the original approach to a 2-dimensional color map that is able to depict both, inter-asset and intra-asset performance.

Relevant references: Keim et al. (2006b) • Ziegler et al. (2007)

primitives: **points**
arrangement: **linear**

data

number of variables: **single**
frame of reference: **abstract**

vis

mapping of time: **static**
dimensionality: **2D**

Multi-Resolution Visualization of Time Series

time

primitives: **points**
arrangement: **linear**

Raw data

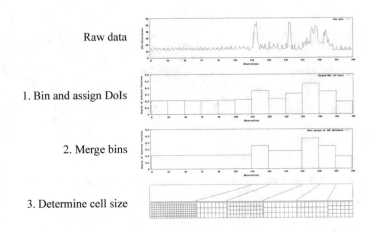

1. Bin and assign DoIs

2. Merge bins

3. Determine cell size

data

number of variables: **single**
frame of reference: **abstract**

Fig. A.11: Multi-resolution visualization of time-series data is based on the idea of assigning more display space to relevant portions of the data and less space to less relevant parts. © *Courtesy of Tobias Schreck.*

Hao et al. (2007) address the problem of visualizing large time series with many time points. Extending earlier work on importance-driven layouts of time series (see Hao et al. (2005)), the authors propose a degree-of-interest (DoI) approach to generate a multi-resolution visualization. First, the raw data are binned and a DoI score is assigned to each bin. Optionally, neighboring bins with similar scores can be merged. The DoI scores are then used to determine the cell size for the different parts of the visualization. Data with low DoI will be represented at lower resolution (smaller cells), whereas interesting data with high DoI will be shown at higher resolution (larger cells). In this way, less relevant data take up less display space and relevant data will be assigned more display space. This makes it easier for the user to analyze the relevant parts in more detail. While the authors propose a matrix-like color-based visualization, the multi-resolution approach is generally applicable to other kinds of visual representations of time-oriented data as well.

vis

mapping of time: **static**
dimensionality: **2D**

Relevant references: Hao et al. (2007) • Hao et al. (2005)

Pinus View

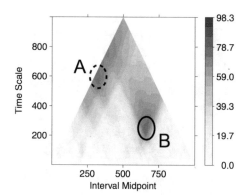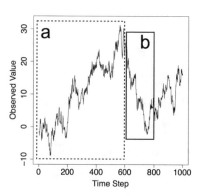

time

primitives: **points, intervals**
arrangement: **linear**

Fig. A.12: The matrix-based Pinus view. Users can identify regions A and B with variance values in the Pinus view and their corresponding temporal patterns a and b in the time series. Color indicates variance values of subsequences. © *2012 IEEE. Reprinted, with permission, from Sips et al. (2012).*

data

number of variables: **single**
frame of reference: **abstract**

Detecting interesting patterns in numerical time series according to various time scales and starting points an analyst is looking at is an important tasks, in particular when exploring environmental time series. Sips et al. (2012) propose a visual analytics approach to tackle this research problem, by providing (1) an algorithm to compute statistical values for all possible time scales and starting positions of intervals, (2) visual identification of potentially interesting patterns in a matrix visualization, and (3) interactive exploration of detected patterns. The core matrix-based visualization is *Pinus* view, which presents the variation of a user-chosen statistical measure across all possible time scales and starting positions of intervals. As such, the pinus view is similar to the triangular model (↪ p. 267). The utility of the pinus view was demonstrated with two use cases (regime changes of the earth's climate system and ocean modeling).

Relevant references: Sips et al. (2012)

vis

mapping of time: **static**
dimensionality: **2D**

Ripple Graph

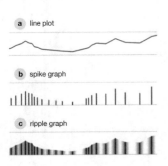

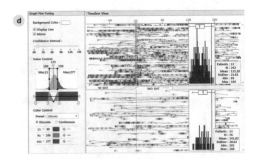

Fig. A.13: Ripple graphs visualize measurement frequency as well as uncertainties of values between measurements. The left part of the figure shows how ripple graphs are constructed: starting from a line plot (a), vertical lines are drawn at the time of measurements resulting in a spike graph (b), finally, color gradients are added to represent uncertainties between measurements (c). On the right, ripple graphs are shown in the context of the Stroscope visualization system. © *2014 IEEE. Reprinted, with permission, from Cho et al. (2014).*

Although most modeling, visualization, and automated methods expect time-series data at regular time intervals, measurements at irregular time intervals are frequently present in many domains. Irregular measurements are particularly challenging due to varying (un)certainty of what can be said about the values between measurements. With Ripple graphs, Cho et al. (2014) developed a visualization technique that makes the aspects of measurement frequency as well as uncertainty between points of measurement explicit. Ripple graphs are based on spike graphs (↪ p. 234) which are spaced according to the time of measurements. To represent the level of uncertainty, color gradients are added with decreasing opacity for increasing distances from the points of measurement. Ripple graphs are applied within the Stroscope system, a multi-scale visualization approach for irregularly measured time-series data. In addition to the ripple graphs, color-coding is applied to highlight values above or below user-defined regions of interest (d). On the left side of the Stroscope interface (d), a control panel allows for the configuration of the ripple graphs. Here, the user can adjust the temporal confidence intervals, i.e., the extent of the color gradients displayed around the spikes, and define regions of interest for the data values along with a user-defined color-coding.

Relevant references: Cho et al. (2014)

Small MultiPiles

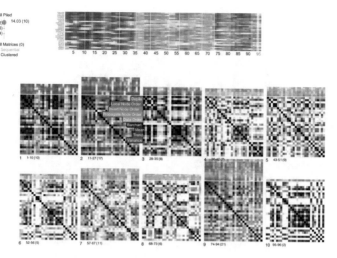

Fig. A.14: The Small MultiPiles approach shows dense networks as piles of matrices. A slider can be used to adjust the number of piles. An overview at the top represents node connectivity over time. ©① *The authors. Generated with the Small MultiPiles software by Benjamin Bach.*

A common approach to visualize dense networks that vary over time is to show them as matrices in a small multiples arrangement (↪ p. 359). Each matrix shows the network for a particular time point and each matrix cell color-codes the weight of an edge between two network nodes. However, when the number of time points is large, the matrices consume much display space and scrolling is required to get an overview of the data. The Small MultiPiles (note the exact spelling) by Bach et al. (2015) are a technique for generating compact overviews by creating piles of matrices. The piles are formed by means of hierarchical clustering, where the number of visible clusters (i.e., piles) can be adjusted via a slider in the interface (top left in the figure). If necessary, the piles can be edited manually using simple drag and drop gestures. With the Small MultiPiles approach, each pile groups similar matrices, and because there are fewer piles than original time steps, the arrangement of piles is compact and grants an overview of the data. Hovering the matrix preview that is attached to the top of each pile grants access to the details of each individual matrix. Moreover, for each pile, the cover matrix (i.e., the topmost matrix of a pile) can show different aggregated information about the matrices in a pile, including summary, trend, difference, and variability. An overview at the top represents node connectivity over time.

Relevant references: Bach et al. (2015)

Time Maps

primitives: **points**
arrangement: **linear**

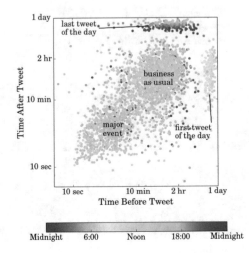

number of variables: **single**
frame of reference: **abstract**

Fig. A.15: Time map visualization of tweets. The x-position of a point corresponds to a tweet's temporal distance to its preceding tweet. The y-position shows the distance to a tweet's subsequent tweet. Color visualizes the time of the day a tween was posted. © *Courtesy of Max C. Watson.*

Watson (2015) proposes *Time Maps* to visualize discrete event data. A time map is basically a point plot ($\hookrightarrow$ p. 232), where each point corresponds to an event in time. Time maps are special in that an event's x-coordinate is the time between the event itself and the preceding event. The y-coordinate is the time between the event itself and the subsequent event. Plotted this way, time maps allow the viewer to identify critical features, no matter if they occur on a timescale of milliseconds or months. Each point in a time map can also be colored to encode additional information, such as the time of day as in the figure.

mapping of time: **static**
dimensionality: **2D**

Relevant references: Watson (2015)

TimeDensityPlots

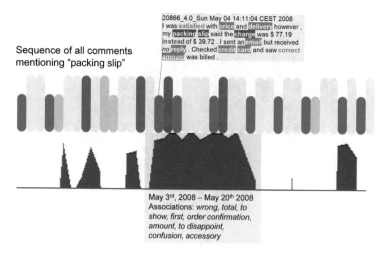

Sequence of all comments
mentioning "packing slip"

20866_4.0_Sun May 04 14:11:04 CEST 2008
I was satisfied with price and delivery, however ,
my packing slip said the charge was $ 77.19
instead of $ 39.72 . I sent an email but received
no reply . Checked credit card and saw correct
amount was billed .

May 3rd, 2008 – May 20th 2008
Associations: *wrong, total, to
show, first, order confirmation,
amount, to disappoint,
confusion, accessory*

Fig. A.16: TimeDensityPlots combine the visualization of a qualitative sequence of events (colored bars) with a density plot (dark silhouette curve) providing some quantitative information of the temporal distribution of data items. © *Courtesy of Christian Rohrdantz.*

In the context of text stream visualization, Rohrdantz et al. (2012) developed a technique called TimeDensityPlots. It is an point-based 2D visualization. Data items (here documents) are shown as a sequence of colored bars (red, gray, green in the figure). The order of the bars corresponds to their qualitative position in time, meaning we know that a data item is before another one, but we do not know the distance between the two. In order to maintain some quantitative information about the temporal distribution of data items, a density plot is aligned with the bar panel. The density plot visualizes the density of data items with respect to the time scale in the bars panel as a silhouette graph ($\hookrightarrow$ p. 281). The higher the curve below the bars of two items, the closer they are in time. In the figure, the highlighted interval represents a relatively dense part, which means that the associated bars represent data items being close together in time.

Relevant references: Rohrdantz et al. (2012)

data

number of variables: **single**
frame of reference: **abstract**

vis

mapping of time: **static**
dimensionality: **2D**

Interactive Parallel Bar Charts

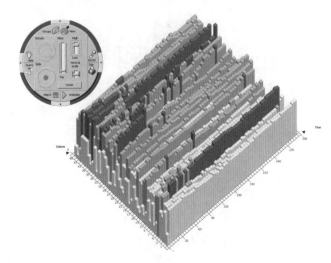

Fig. A.17: Visualization of clinical time-dependent data where one axis represents different hemodialysis sessions and the other axis represents the series of time steps. One time-dependent variable (e.g., blood pressure) is encoded to the height of the bars. © *2003 Elsevier. Reprinted, with permission, from Chittaro et al. (2003).*

Chittaro et al. (2003) present a technique for visualizing time-dependent hemodialysis data. To keep the visualization of multiple hemodialysis sessions simple and easy to use for physicians, the design is based on common 3D bar charts, where the height of bars encodes individual data values as in regular 2D bar graphs (↪ p. 234). Multiple bar charts (one per hemodialysis session) are arranged on a regular grid in a parallel fashion. This visual display is easy to interpret despite the 3D projection. Visual exploration and analysis are facilitated through various interaction tools. Dynamic filtering combined with a color-coding mechanism supports visual classification. To manage occlusions, interactive features are provided, such as flattening individual bars or groups of bars, or leaving only colored squares in the grid. The water level interaction is particularly helpful for comparison tasks: virtual water engulfs all bars with a height below a user-defined threshold. This eases the assessment of similarities and differences with respect to dialysis sessions and time. Additional visual cues support physicians in detecting anomalies or special events in the data and enable them to take necessary actions quickly. For multivariate analysis, an integration with parallel coordinate plots is supported.

Relevant references: Chittaro et al. (2003)

TimeHistogram 3D

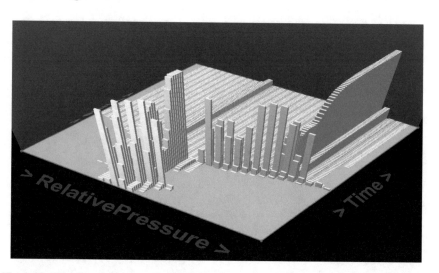

Fig. A.18: Time is encoded along the x-axis, while a time-dependent variable (RelativePressure) is represented along the y-axis. For each time-value pair in the resulting grid in the x-y plane, the height of a cuboid represents the frequency of data items per grid cell. © *2004 IEEE. Reprinted, with permission, from Kosara et al. (2004).*

TimeHistogram 3D is an interactive extension of well-known histograms. The Time-Histogram 3D is specially designed for time-oriented data. It has been developed to give an overview of complex data in the application context of computational fluid dynamics (CFD). A design goal of this technique was to show temporal information in static images while maintaining the easy readability of standard histograms. In the figure, the x-axis encodes time and the y-axis encodes a time-dependent variable (RelativePressure), effectively creating a grid in the x-y plane, where each grid cell corresponds to a unique time-value pair. In order to visualize the number of data items (i.e., their frequency) per time-value pair, cuboids are shown for each cell, where the height of a cuboid encodes the frequency. This way, the user can see where in time and in which value range data items accumulate. Several interactive features such as brushing, scaling, and a 2D context display, which is shown in the background of the histogram, are part of this technique.

Relevant references: Kosara et al. (2004)

time

primitives: **points**
arrangement: **linear**

data

number of variables: **single**
frame of reference: **abstract**

vis

mapping of time: **static**
dimensionality: **3D**

time

primitives: **points**
arrangement: **linear**

Intrusion Monitoring

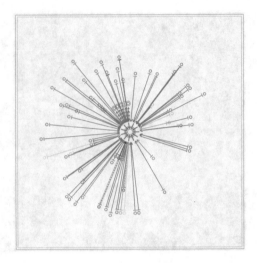

Fig. A.19: The glyph in the center represents system being monitored. Connections to remote hosts are depicted as radially arranged lines. Critical and suspicious connections are visualized by red and yellow color, respectively. © *Courtesy of Robert F. Erbacher.*

data

number of variables: **single**
frame of reference: **abstract**

Erbacher et al. (2002) describe a system that visualizes time-stamped network-related log messages that are dynamically generated by a monitored system. These messages correspond to events in a linear continuous time domain. The visualization shows the monitored server as a central glyph encoding the number of users and the server's load. Events are shown as radially arranged lines at whose end the remote host is shown as a smaller glyph. Regular network activities are drawn with a shade of gray. Unexpected or suspicious activities result in a change of color: Hosts that try to open privileged connections are colored in red, hosts that fail to respond turn yellow, lines representing timed-out connections or connections that failed the authentication procedure are shown in red, and connections that have been identified as intrusions are represented with even brighter red. To preserve a history of connections that have been terminated, the corresponding lines are faded out gradually. This kind of visual representation helps administrators in observing network communication. Presenting colored (red or yellow) lines among gray lines attracts the attention of administrators to suspicious activity and actions can be taken quickly to counter network attacks from remote hosts.

vis

mapping of time: **dynamic**
dimensionality: **2D**

Relevant references: Erbacher et al. (2002)

Anemone

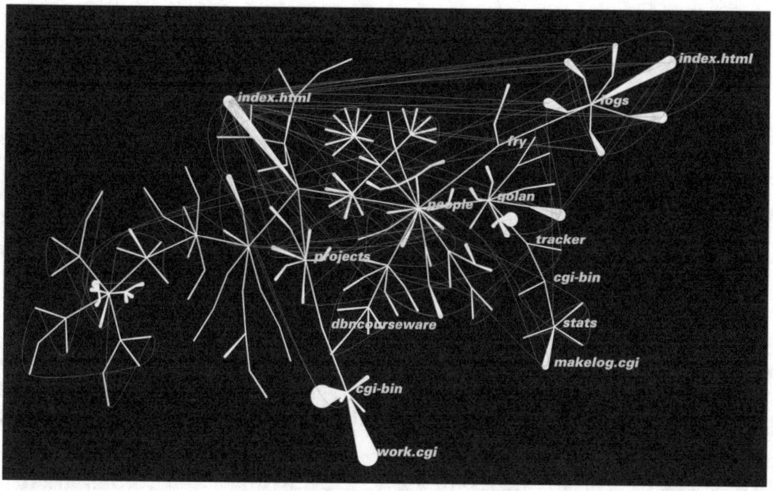

time

primitives: **points**
arrangement: **linear**

Fig. A.20: Snapshot of Anemone showing traffic patterns of people visiting the web site of the Aesthetics & Computation Group at the M.I.T. Media Lab. The structure of the web site is shown as a node-link representation. Nodes vary in size, depending on how frequently a page is visited by users. Rarely-visited parts fade out slowly. © *Courtesy of Ben Fry, MIT Media Laboratory, Aesthetics + Computation Group.*

data

number of variables: **single**
frame of reference: **abstract**

Anemone by Fry (2000) is a technique related to the visualization of structured information. It is a dynamic, organic representation designed to reveal not only the static structure of a website, which is based on its organization into folders and files, but also to reveal dynamic usage patterns. To this end, a classic node-link representation is visually enriched with dynamically updated usage statistics to form a living representation that truly reflects the restless nature of a website. The static structure is shown as nodes that are connected via straight branches. At the tip of a branch resides the actual web page. Additional labels can be used to identify nodes by their corresponding page's name. Nodes dynamically change size depending on how often they are visited by users. When a user follows a link from one page to another, a thin curved line is drawn connecting both pages. If parts of a site have not been visited for a long time, they shrink in size and slowly fade out. To allow users to concentrate on particular items of interest, it is possible to select nodes and lock them to a dedicated position. This is quite useful because the dynamic character of the technique implies that visual representation constantly changes its appearance.

vis

mapping of time: **dynamic**
dimensionality: **2D**

Relevant references: Fry (2000)

Dynamic Word Clouds

time

primitives: **points**
arrangement: **linear**

Fig. A.21: Dynamic word clouds are primarily concerned with keeping the layout of words coherent between subsequent time steps of a dynamic visualization. Addition visual cues (e.g., sparklines and color) can be used to show additional temporal information. ⓒⓘ *Courtesy of Weiwei Cui.*

data

number of variables: **single**
frame of reference: **abstract**

Word clouds are visual summaries of text where important words are shown in larger font sizes, while context words are smaller. Word clouds are popular for static texts. However, standard algorithms are not suitable for generating word clouds for texts that change over time. The reason is that the word clouds are generated independently for each time step, which may cause word positions to change drastically from one time point to the next. This makes it difficult to follow words when studying a sequence of word clouds. Dynamic word clouds tackle this problem in different ways. A key concern is to lay out the words coherently between subsequent time steps. To this end, one can use the adapted force-directed algorithm by Cui et al. (2010) or the spiral-based layout strategy in combination with collision detection by Seyfert and Viola (2017). In addition to coherent word positions, further pieces of information can be encoded visually. The figure shows sparklines (↪ p. 235) attached to the words, which is similar to SparkClouds (↪ p. 240). Moreover, colors indicate if words are appearing, disappearing or unique. It is also possible to tilt words to indicate words with increasing or decreasing importance with respect to previous time steps.

vis

mapping of time: **dynamic**
dimensionality: **2D**

Relevant references: Cui et al. (2010) • Seyfert and Viola (2017)

Gantt Chart

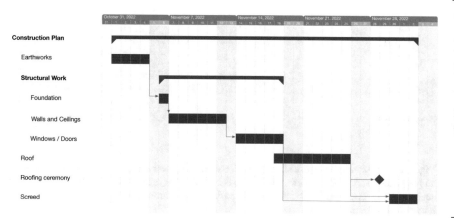

Fig. A.22: The Gantt chart shows a project plan for construction works. To the left, the chart provides an indented list of tasks. In the main panel, timelines show position and duration of tasks in time, where black and blue bars stand for groups of tasks and individual tasks, respectively. Additionally, diamonds indicate milestones. ©① *The authors.*

Planning activities, people, and resources is a task that is particularly important in the field of project management. One of the common visualization techniques used for such tasks is Gantt charts. This kind of representation was originally invented by Gantt (1910) who studied the order of steps in work processes. Mainly work tasks with their temporal location and duration as well as milestones are depicted. The tasks are displayed as a textual list in the left part of the diagram and might be augmented by additional textual information such as resources, for example. Related tasks can be grouped to form a hierarchy, which is reflected by indentation in the task list. For displaying the position and duration of tasks in time, timelines (↪ p. 258) are drawn at the corresponding vertical position of the task list. This leads to an easily comprehensible representation of information from the past, present, and future. Hierarchically grouped tasks can be expanded and collapsed interactively. Summary lines are used to maintain an overview of larger plans. Sequence relationships are represented by arrows that connect tasks (e.g., an arrow from the end of task A to the beginning of task B shows that task B may start only after task A is finished). Milestones indicating important time points for synchronization within a project plan are visually represented by diamonds. The fact that tasks are mostly ordered chronologically, typically leads to a diagonal layout from the upper left to the lower-right corner of the display.

Relevant references: Gantt (1910)

time

primitives: **points, intervals**
arrangement: **linear**

data

number of variables: **single**
frame of reference: **abstract**

vis

mapping of time: **static**
dimensionality: **2D**

Set Streams

time

primitives: **points, intervals**
arrangement: **linear**

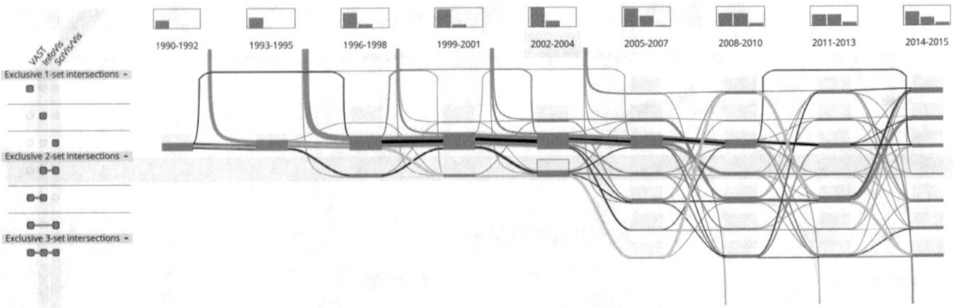

Fig. A.23: Set Streams visualize changes in set membership over time. Here, elements are authors, and sets are the three big visualization venues SciVis, InfoVis, and VAST. The time axis has been segmented into three-year intervals. ©① *The authors. Generated with the Set Streams software by Shivam Agarwal.*

data

number of variables: **single**
frame of reference: **abstract**

Set Streams by Agarwal and Beck (2020) is a technique for visualizing how set memberships of elements change over time. In the case of Set Streams, elements can be members of multiple sets, that is, the sets may overlap. Therefore, set intersections and their affiliated members are of particular interest. The visualization is based on a grid. Each row corresponds to a so-called exclusive set intersection, where intersections involve one, two, or three sets. The grid columns correspond to the time axis, where individual columns may represent time points or segmented time intervals, as shown in the figure. The fill level of the grid cells indicates the number of elements per corresponding exclusive set intersection and time primitive. Changes in set membership are visualized by curved streams between adjacent columns similar to Sankey diagrams (↪ p. 313). The streams may branch and merge depending on the underlying data changes. These streams make it possible for users to trace the path of elements through time and the various set intersections. On top of that, interaction techniques can be used to investigate the data in detail. For example, the grid rows can be sorted according to different criteria and a selected element can be highlighted across the entire grid. Moreover, two groups of elements can be defined using a dynamic query mechanism. The two groups are then distinctly colored in the visualization to facilitate their detailed comparison.

vis

mapping of time: **static**
dimensionality: **2D**

Relevant references: Agarwal and Beck (2020)

TimeSets

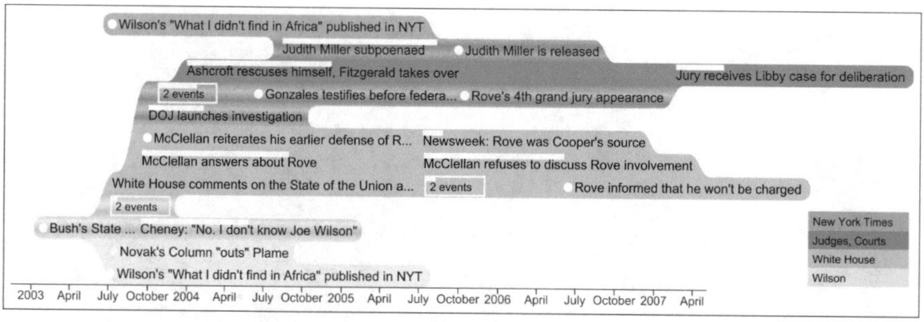

Fig. A.24: TimeSets visualization using the CIA leak case. The timeline contains events based on time points or intervals from 2002 to 2007. Each event has a label and topics. Events are positioned along the horizontal time axis based on time points and are vertically grouped by topics. A time-point event is shown with a white circle to its left and an interval-based event with a horizontal bar on top showing its duration. Each topic has a unique color, and events shared by two topics have gradient backgrounds. © *2015 SAGE. Reprinted, with permission, from Nguyen et al. (2016).*

TimeSets by Nguyen et al. (2016) visualize set relationships among events using a timeline design ($\hookrightarrow$ p. 258). Following two Gestalt principles of grouping, namely proximity and uniform connectedness, the TimeSets technique groups temporal events vertically with colored backgrounds according to their set memberships. Events shared by two sets are visualized using layers with a color gradient background transitioning between the colors of the two topics. TimeSets also dynamically adjust the event labels between three levels of detail to scale with the number of events. The number of displayed event labels can be reduced to easily follow events chronologically, which is controlled by a traceability layout algorithm. Xu et al. (2020b) incorporated additional analysis capacities in the TimeSets visualization to support open-source intelligence analysis with Twitter data, particularly the challenge of finding the right questions to ask, and to facilitate uncertainty analysis involving fake news. In a controlled experiment, TimeSets were significantly more accurate, and the participants preferred TimeSets for aesthetics and readability.

Relevant references: Nguyen et al. (2016) • Xu et al. (2020b)

time

primitives: **points, intervals**
arrangement: **linear**

data

number of variables: **single**
frame of reference: **abstract**

vis

mapping of time: **static**
dimensionality: **2D**

Perspective Wall

primitives: **points, intervals**
arrangement: **linear**

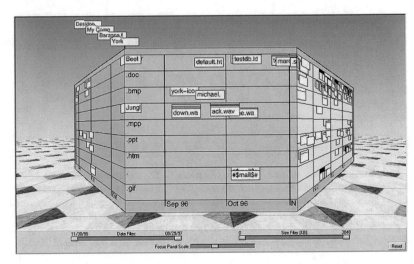

number of variables: **single**
frame of reference: **abstract**

Fig. A.25: A perspective wall representing time-related information of a file system for a period of several months. The focus (currently set to September/October 1996) shows detailed text labels for files, whereas the context regions only indicate files as yellow boxes. © *Inxight Federal Systems. Used with permission.*

mapping of time: **static**
dimensionality: **3D**

Time-oriented data that are linked to a longer time axis (i.e., wide span in time or many time primitives) are usually difficult to represent visually because the image becomes very wide and exhibits an aspect ratio that is not suited for common displays. The perspective wall by Mackinlay et al. (1991) is a technique that addresses this problem by means of a focus+context approach. The key idea is to map time-oriented data to a 3D wall. For a user-selected focus, full detail is provided in the center of the display. Two context representations show the data in the past (to the left) and in the future (to the right) with regard to the current focus. The context is bent perspectively to reduce the display space occupied by these regions, effectively allowing for better space utilization in the central focus. Interaction methods are provided to enable users to navigate in time in order to bring different time spans into focus. The actual data representation on the wall may vary across applications; the only requirement is that time is mapped linearly from left to right. For example, one can use bars as in the figure or more advanced visual representations such as the ThemeRiver (↪ p. 293).

Relevant references: Mackinlay et al. (1991)

DateLens

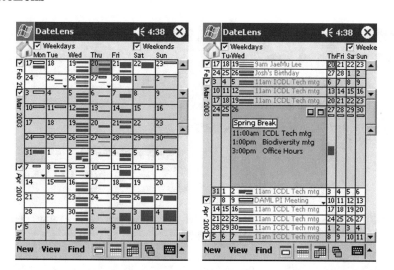

Fig. A.26: A calendar grid shows the items of one's personal schedule as colored bars (left). Fisheye distortion is applied to show detailed textual information at the point the user is focusing on, and to maintain the context with less graphical detail (right). © *Courtesy of Ben Bederson.*

Most people use calendars to plan their daily life, for instance, to maintain a list of appointments or bookmark future events. Bederson et al. (2004) developed the DateLens to make it easier to work with a personal schedule on small displays. According to Langner et al. (2021), the DateLens is a classic technique for mobile data visualization. Because display space is usually limited on mobile devices (compared to common desktop displays), focus+context mechanisms are applied to present temporal information at different levels of detail. Based on a common tabular representation of a calendar (left), the DateLens magnifies selected table cells (right) so as to provide more display space for important information that is currently in the user's focus. The fisheye distortion magnifies the focus and reduces graphical detail in the context of the display. If sufficient display space is available, calendar entries are shown in textual form. Otherwise, temporal intervals of calendar entries are indicated by bars that visualize the starting point and temporal extent of appointments and events stored in the calendar. Various interaction mechanisms allow users to view the calendar at different temporal granularities and to navigate forward and backward in time.

Relevant references: Bederson et al. (2004) • Langner et al. (2021)

Timeline

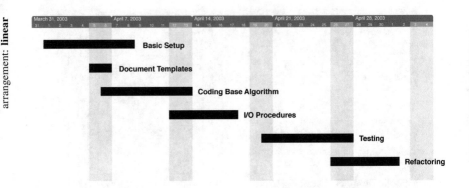

Fig. A.27: Bars are arranged relative to a time axis to visualize both the location and duration of intervals. One can also see how intervals are related to each other. ⓒ① *The authors.*

When the time primitives of interest are not points but intervals, the visualization has to communicate not only where in time a primitive is located, but also how long it is. A simple and intuitive way of depicting incidents with a duration is by marking them visually along a time axis. This form of visualization is called a timeline. Most commonly, a visual element such as a line or a bar represents an interval's starting point and duration (and consequently its end). The figure shows an example with bars. If multiple intervals share a common time axis, as in this example, it is even possible to discern how the various intervals are related to each other. Timelines are a very powerful visualization technique that, according to Tufte (1983), had been used long before computers even appeared. Many different variants of timelines exist in diverse visualization tools, where contemporary timelines come with a number of design options that go beyond the original approach (see Brehmer et al., 2017). Additional interaction techniques often allow users not only to view time intervals, but also to create and edit them. Prominent examples are LifeLines (↪ p. 341) and Gantt charts (↪ p. 253).

Relevant references: Tufte (1983) • Brehmer et al. (2017)

Paint Strips

time

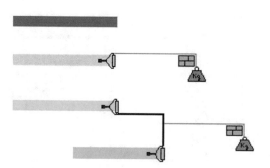

primitives: **intervals**
arrangement: **linear**

Fig. A.28: Paint strips indicate the location and duration of time intervals, effectively allowing users to assess relationships of intervals. Temporal indeterminacy of intervals is indicated by paint rollers that can move flexibly within certain constraints, which are represented by wall elements. © *Courtesy of Luca Chittaro.*

data

Chittaro and Combi (2003) designed paint strips to represent relations between time intervals for visualizing queries on medical databases. The technique is strongly related to timelines (↪ p. 258), but here paint strips are used as equivalents of bars to indicate time intervals, and optionally, the indeterminacy of intervals is communicated by placing paint rollers at either end of the paint strips. A paint roller with a weight attached to it means this interval can possibly extend in time. Graphical depictions of wall elements represent constraints on the extension. This way, the maximum duration and earliest start or latest end of intervals are defined, depending on which end of the painting strip the paint rollers are attached to. It is also possible to link strips, which means if one strip moves, the other one moves to the same extent as well. This relationship is indicated graphically by connecting the involved paint rollers before attaching them to the rope that holds the weight. Paint strips were specially developed for medical applications but can be used anywhere where indeterminate time intervals have to be visualized. Thanks to the simplicity of the paint strip metaphor, there is room for application-dependent enhancements, such as textual annotations for start and end points as well as for durations of intervals.

number of variables: **single**
frame of reference: **abstract**

Relevant references: Chittaro and Combi (2003)

vis

mapping of time: **static**
dimensionality: **2D**

time

primitives: **intervals**
arrangement: **linear**

data

number of variables: **single**
frame of reference: **abstract**

vis

mapping of time: **static**
dimensionality: **2D**

PlanningLines

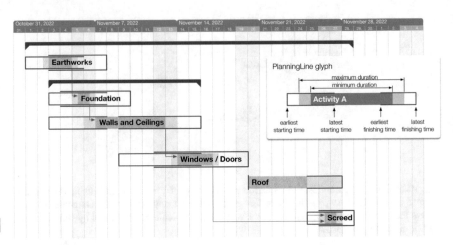

Fig. A.29: Project plan of construction works that represents temporal uncertainties via PlanningLines. A PlanningLine glyph consists of two encapsulated bars that represent minimum and maximum duration. The bars are bounded by two caps encoding the start and end intervals. ⓒⓘ *The authors. Adapted from Aigner et al. (2005).*

Since the future is always inherently connected with possible uncertainties, delays, and the unforeseen, these issues need to be dealt with in many domains like project management or medical treatment planning. PlanningLines by Aigner et al. (2005) allow the representation of temporal uncertainties, thus supporting project managers in their difficult planning and controlling tasks. PlanningLines have been designed to be easily integrated into well-known timeline-based visualization techniques such as Gantt charts (↪ p. 253). As indicated in the figure, a single glyph provides a visual representation of the temporal indeterminacy of a single activity, facilitates the identification of (un)defined attributes, supports in maintaining logical constraints (e.g., bars may not extend caps), and gives a visual impression of the individual and overall uncertainties. Uncertainties might be introduced by explicit specifications, usually connected with future planning (e.g., "The meeting will start between 12 p.m. and 2 p.m." – which might be any point in time between noon and 2 p.m.) or is implicitly present in cases where data are given with respect to multiple temporal granularities (e.g., data given on a granularity of days and shown on an hourly scale).

The original concept of PlanningLines has been extended by Höhn et al. (2022) to allow for visualizing both, variable and fixed activities with possibly open start and end. Moreover, the probability distribution of the activities can be defined and input methods for mouse and keyboard as well as stylus interaction are offered.

Relevant references: Aigner et al. (2005) • Höhn et al. (2022)

Time Annotation Glyph

time

Definition:
[[ESS, LSS], [EFS, LFS], [MinDu, MaxDu], Reference]

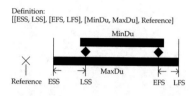

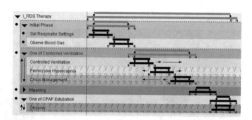

primitives: **intervals**
arrangement: **linear**

Fig. A.30: The time annotation glyph was designed to represent temporal constraints. It uses the metaphor of bars that lie on pillars. Left: single glyph and associated parameters; right: glyphs being applied in a tool for representing medical treatment plans. © *Courtesy of Robert Kosara.*

The time annotation glyph by Kosara and Miksch (2001) uses the simple metaphor of bars that lie on pillars to represent a complex set of time attributes. Four vertical lines on the base specify the earliest and the latest starting and ending times. Supported by these pillars lies a bar that is as long as the maximum duration. On top of the maximum duration bar, a bar that represents the minimum duration lies upon two diamonds indicating the latest start and the earliest end. Furthermore, undefined parts and different granularities are indicated visually. Because of this metaphor, a few simple time parameter constraints can be understood intuitively. For example, the minimum duration cannot be shorter than the interval between the latest start and earliest end – if it was, the minimum duration bar would fall down between its supports. All of the parameters might be defined relative to a reference point that is also represented graphically. In summary, the following parameters are shown: earliest starting shift (ESS), latest starting shift (LSS), earliest finishing shift (EFS), latest finishing shift (LFS), minimum duration (MinDu), and maximum duration (MaxDu). The technique is used to represent the time annotations of medical treatment plans within the AsbruView application.

Relevant references: Kosara and Miksch (2001)

data

number of variables: **single**
frame of reference: **abstract**

vis

mapping of time: **static**
dimensionality: **2D**

SOPO Diagram

time

primitives: **intervals**
arrangement: **linear**

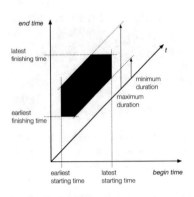
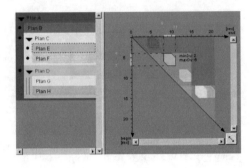

Fig. A.31: A SOPO diagram (left) shows the possible configurations of the begin and end times of an event via a constrained polygonal shape. SOPOView (right) is an interactive visualization tool for working with SOPOs applied for medical treatment plans. © *Courtesy of Robert Kosara.*

data

number of variables: **single**
frame of reference: **abstract**

For planning and scheduling, the temporal extents of events can be characterized by sets of possible occurrences (SOPOs), i.e., a set of possible begin and end times during which an event may happen. Rit (1986) defined a theoretical model for the definition and propagation of temporal constraints for scheduling problems. A graphical representation of SOPOs was introduced as a visual aid for understanding and solving such problems. In this representation, the extent of temporal uncertainty is expressed via a polygonal shape. The axes of a SOPO diagram represent begin time (x-axis) and end time (y-axis). Similar to the triangular model ($\hookrightarrow$ p. 267), points in this diagram do not represent points in time, but complete intervals specified by their begin (x-coordinate) and end time (y-coordinate). Hence, the extent of an interval is represented by its position, not its visual extent. The area an item covers reflects all intervals that fit the specification given by means of earliest start, latest start, earliest end, latest end, minimum, and maximum duration. Moreover, the exact occurrence of an event may be constrained by other related events which further modify the sets of possible occurrences. The propagation of such constraints is aided graphically, e.g., via overlaps of individual SOPOs. Later, this idea was interactively enhanced and further developed to be applied for visualizing medical treatment plans in the tool SOPOView by Kosara and Miksch (2002).

vis

mapping of time: **static**
dimensionality: **2D**

Relevant references: Rit (1986) • Kosara and Miksch (2002)

TimeNets

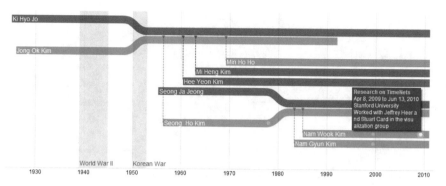

Fig. A.32: TimeNets visualize temporal and structural aspects of genealogical data. Bands that extend along the horizontal time axis visualize individuals. Marriage and divorce are indicated by converging and diverging bands, respectively. Children are connected to their parents via drop lines. Labels are shown for the persons' names as well as for historical and personal events. © *Courtesy of Jeffrey Heer and Nam Wook Kim.*

Genealogical data are an interesting source of time-oriented information. In such data, not only family structures are of interest, but also temporal relationships. Kim et al. (2010) propose the TimeNets approach, which aims to visualize both of these aspects. TimeNets are similar to storylines (↪ p. 264) and represent persons as individual bands that extend horizontally along a time axis from left to right. Each band shows a label of the person's name and different colors are used to encode sex: red is reserved for females, and males are shown in blue. Marriage of persons is visualized by converging the corresponding bands, while divorce is indicated by diverging bands. When a child is born, a new band is added to the display. A so-called drop line connects the band of the child to the parents' bands to convey the parent-child relationship. In order to allow users to focus on relevant parts of the data, a degree-of-interest (DoI) algorithm is applied. Bands below the DoI threshold are filtered out or smoothly fade in where they are linked to bands of relevant persons. Users can select multiple persons to focus on. On each change of the focus, the visualization shows a smooth transition of the display to keep users oriented.

Relevant references: Kim et al. (2010)

time

primitives: **intervals**
arrangement: **linear**

data

number of variables: **single**
frame of reference: **abstract**

vis

mapping of time: **static**
dimensionality: **2D**

Storyline Visualization

primitives: **intervals**
arrangement: **linear**

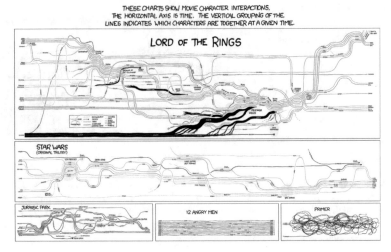

Fig. A.33: Storyline visualization of character co-occurrence in movies. Time is shown from left to right and each band represents a movie character. The proximity of bands indicates character co-occurrence. ⊚①⑤ *"Movie Narrative Charts" from xkcd comics by Randall Munroe.*

number of variables: **single**
frame of reference: **abstract**

Storylines have become a popular means to visualize the relationships of objects over time (see Tang et al., 2019b). They are similar to timelines (↪ p. 258) in that each object is represented as a band that extends from left to right along the horizontal time axis. Yet, for storylines, individual bands can bend to change their vertical position in the visualization. This way, the bands can be bundled to converge to denser clusters, or they can be thinned out to create distance from other bands. This basic encoding is useful in a variety of scenarios. The figure shows storyline visualizations of character co-occurrence in movies. Here, bands running closely and in parallel represent times when the corresponding characters appear together. Storyline visualizations can also be useful for comparing the similarity of data objects over time. In this case, bands representing similar objects run in clusters. Changes in cluster affiliation, that is, individual bands turning away from their original cluster to a new one, may indicate significant events in the development of data objects over time. The TimeNets (↪ p. 263) technique employs the idea of storylines to visualize genealogical data. Despite their simplicity, storyline visualizations are not trivial to generate. The key difficulty is to find a layout for the bands with minimal crossings, bend angles, and display space. Finding a globally optimal band layout is a computationally complex problem. Tanahashi and Ma (2012), Liu et al. (2013), Kostitsyna et al. (2015), and Tanahashi et al. (2015) investigate optimization and computational aspects of storyline visualizations in detail.

mapping of time: **static**
dimensionality: **2D**

Relevant references: Tanahashi and Ma (2012) ● Liu et al. (2013) ● Kostitsyna et al. (2015) ● Tanahashi et al. (2015) ● Tang et al. (2019b)

Temporal Mosaic

time

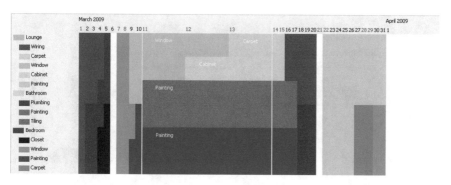

Fig. A.34: Temporal mosaic visualizing concurrent activities of a house renovation plan. Time is mapped from left to right, concurrent activities are stacked along the vertical axes and colors indicate the type of activity. © *Courtesy of Saturnino Luz.*

Temporal mosaic is a technique for the visualization of parallel time-based streams. It provides a compact way of representing concurrent events by allocating a fixed drawing area to time intervals and partitioning that area according to the number of events that co-occur in that time interval. The figure shows a temporal mosaic representation of a house renovation schedule with hierarchical event dependencies. Time is mapped along the horizontal axis. Differently colored regions in the mosaic indicate at which time a particular renovation is scheduled. One can see that some renovations can be executed in parallel, while others depend on each other and need to be carried out one after the other. The temporal mosaic technique has also been evaluated in the context of meeting browsing tasks.

Relevant references: Luz and Masoodian (2004) • Luz and Masoodian (2007) • Luz and Masoodian (2010)

primitives: **intervals**
arrangement: **linear**

data

number of variables: **single**
frame of reference: **abstract**

vis

mapping of time: **static**
dimensionality: **2D**

Train Delay Uncertainty

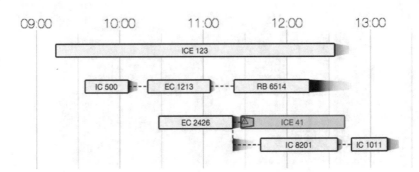

Fig. A.35: Train itinerary planning under consideration of uncertain train delays. Trains are shown by rectangles at whose right end a color gradient indicates potential delays. ⓒⓘ *Courtesy of Tatiana von Landesberger.*

While train delays are minimal in some countries, they can be a critical issue in other countries, especially when a train connection is missed due to a delayed train. While itinerary planning already takes some buffer into account for travelers to reach connecting trains, the consequences of delayed trains are usually not considered. Wunderlich et al. (2017) describe a visual solution to help travelers in planning and understanding itineraries under the influence of uncertain train delays. As illustrated in the figure, the solution uses a horizontal time axis and labeled rectangles representing train trips. The trains are positioned according to their scheduled departure and arrival times. Train delay, more specifically, the expected cumulative train delay is indicated via a color gradient at the arrival (right side of the rectangles). The cumulative delay represents the probability that a train is delayed no longer than a certain amount of minutes. Situations, where a train's expected cumulative delay could lead to missing a train connection, are marked with a small exclamation sign and the potentially missed train is shown in gray. If possible, the system automatically suggests alternative connections that would avoid missed transfers.

Relevant references: Wunderlich et al. (2017)

Triangular Model

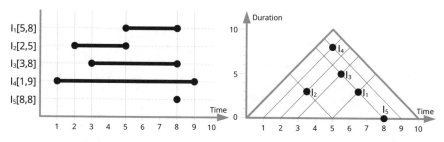

Fig. A.36: In contrast to common timelines (left), which represent intervals as lines or bars, intervals become points in the triangular model representation (right), which encodes interval start and end as well as interval duration. ⓒⓘ *The authors. Adapted from Qiang et al. (2012).*

Qiang et al. (2012) popularize the original idea by Kulpa (1997) and Kulpa (2006) of visualizing time intervals using a triangular model. In this model, an interval is represented as a dot with two attached arms. The dot is placed so that the arms connect the horizontal time axis exactly at the start and the end of the represented interval. The point's height corresponds to the interval's duration, which is mapped along the vertical axis. This representation is useful in many scenarios, in particular, when it comes to reasoning about properties and the relationships of multiple intervals. In that case, the triangular model generates easily distinguishable visual patterns for all possible interval relations. The figure compares a standard representation (intervals shown as bars) with a corresponding representation using the triangular model (intervals shown as dots).

Relevant references: Qiang et al. (2012) • Qiang et al. (2014) • Kulpa (1997) • Kulpa (2006)

time

primitives: **intervals**
arrangement: **linear**

data

number of variables: **single**
frame of reference: **abstract**

vis

mapping of time: **static**
dimensionality: **2D**

Cycle Plot

time

primitives: **points**
arrangement: **linear, cyclic**

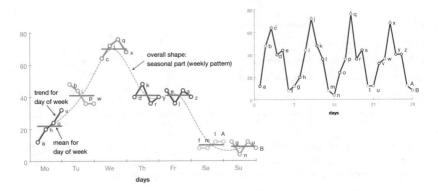

data

number of variables: **single**
frame of reference: **abstract**

Fig. A.37: Cycle plots make it easier to discern seasonal components and general trends from time-series data (left), which is considerably more difficult when using line plots (right). ⓒⓘ *The authors. Adapted from Cleveland (1993).*

vis

mapping of time: **static**
dimensionality: **2D**

Time-series data may contain seasonal components as well as general trends, which is also reflected in many statistical models. Cleveland (1993) describes cycle plots as a technique to make seasonal components and trends visually discernable. This is achieved by showing individual trends as line plots embedded within a plot that shows the seasonal pattern. For constructing a cycle plot, one has to define the time primitives to be considered for the seasonal component. The horizontal axis of the cycle plot is then subdivided accordingly. The left part of the figure demonstrates this using weekdays. We are interested in the trend for each weekday and the general weekly pattern. The data for a particular weekday are visualized as a separate line plot (e.g., data of the 1st, 2nd, 3rd, and 4th Monday). This allows the identification of individual trends for each day of the week. In the figure, we see an increasing trend for Mondays, but a decreasing trend for Tuesdays. Additionally, the cycle plot shows the mean value for each weekday (depicted as gray lines). Connecting the mean values as a line plot (dashed line in the figure) reveals the seasonal pattern, which in this case is a weekly pattern that clearly shows a peak on Wednesday.

Relevant references: Cleveland (1993)

Tile Maps

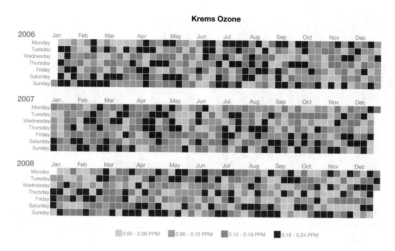

Fig. A.38: The matrices show ozone measurements made in the course of three years, where data values for individual days are encoded to the brightness of matrix cells. ⓒ① *The authors. Adapted from Mintz et al. (1997).*

Tile maps as described by Mintz et al. (1997) represent a series of data values along a calendar division. The idea behind this heatmap-based technique is to arrange data values according to different temporal granularities. For example, data values measured on a daily basis are displayed in a matrix where each cell (or tile) corresponds to a distinct day, a column represents a week, and a row represents all data values for a particular weekday. One additional level of granularity can be integrated by stacking multiple matrices as shown in the figure. Data values are visualized by varying the lightness of individual tiles. A visual representation constructed this way can be interpreted quite easily, because it corresponds to our experience of looking at calendars. The arrangement as a two-dimensional matrix allows users to identify long-term trends by considering the matrix as a whole, to discern individual trends for Mondays, Tuesdays, and so forth by looking at the matrix rows, and to derive weekly patterns by investigating matrix columns. For example, the U.S. Environmental Protection Agency provides a web tool that generates tile maps automatically for specified air pollutants and locations.

Relevant references: Mintz et al. (1997)

time

primitives: **points**
arrangement: **linear, cyclic**

data

number of variables: **single**
frame of reference: **abstract**

vis

mapping of time: **static**
dimensionality: **2D**

Multi Scale Temporal Behavior

time

primitives: **points**
arrangement: **linear, cyclic**

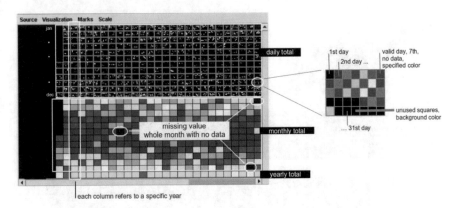

data

number of variables: **single**
frame of reference: **abstract**

Fig. A.39: Different levels of granularity are shown simultaneously in one display to allow users to explore patterns in precipitation data from Brazil at different temporal levels. © *Courtesy of Milton Hirokazu Shimabukuro.*

The Multi Scale Temporal Behavior technique by Shimabukuro et al. (2004) comprises different levels of granularity and aggregation to explore patterns at different temporal levels. The basis for the visualization is a matrix that is divided vertically into three regions, one for each of the three scale levels: daily data, monthly data, and yearly data (top to bottom in the figure). Each column of the matrix represents a year worth of data. The cells in the topmost region represent months. They show full detail by color-coding individual pixels within a cell according to daily total values. The middle region shows aggregated data. Here cells are no longer subdivided into pixels, but are colored uniformly, where the color represents the aggregated monthly total value. The same principle is applied to the bottom region (in fact, the bottom row). Twelve monthly values are aggregated into a single value for the year, which can again be represented by color. A significant and non-trivial problem in dealing with real-world datasets are missing data values. This issue is tackled here by preprocessing the data and marking missing values as such in the display.

vis

mapping of time: **static**
dimensionality: **2D**

Relevant references: Shimabukuro et al. (2004)

GROOVE

time

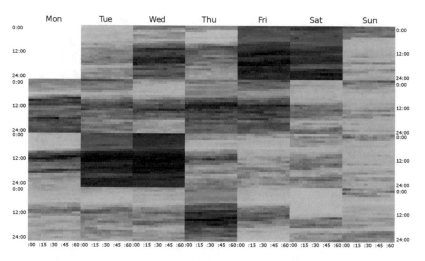

primitives: **points**
arrangement: **linear, cyclic**

Fig. A.40: GROOVE visualizations combine detail and overview readings and use regular layouts based on time granularities. Here, police assignments are shown over the course of one month. For each 5-minute interval, the number of units assigned to a duty is given. Each block shows a day, rows of blocks represent weeks, columns of blocks show the day of the week, and within each block, a row of pixels displays one hour. Color components are used to show detail values for pixels (lightness – bright: low, dark: high) and average values per block (hue – green: low, blue: high). ⓒⓘ *The authors.*

data

number of variables: **single**
frame of reference: **abstract**

GROOVE (Granularity Overview OVErlay) visualizations as presented by Lammarsch et al. (2009) utilize a user-configurable set of four time granularities to partition a dataset in a regular manner. A recursive layout shows columns and rows of larger blocks and a pixel arrangement within blocks for the details of the structure, where different arrangements might be chosen (e.g., row-by-row or back-and-forth) similar to the recursive patterns approach (↪ p. 282). GROOVE visualizations combine overview (aggregated values) and details in one place using one of three kinds of overlays. This allows micro and macro readings and avoids eye movements between the overview and detail representations. First, color components can be employed with color-based overlay as shown in the figure. Second, opacity overlay applies interactive crossfading between the overview and the detail display. Third, spatial overlay can be used for viewing the data selectively at different levels of aggregation by expanding and collapsing areas.

vis

mapping of time: **static**
dimensionality: **2D**

Relevant references: Lammarsch et al. (2009)

SolarPlot

time

primitives: **points**
arrangement: **linear, cyclic**

data

number of variables: **single**
frame of reference: **abstract**

vis

mapping of time: **static**
dimensionality: **2D**

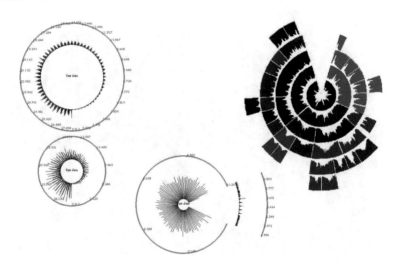

Fig. A.41: Daily values of ticket sales data over a period of 30 years are plotted around the circumference of a resizable circle (left). For a selected interval, a more detailed arc can be shown (center). The technique can be enhanced so as to combine the representation of data and a hierarchical structure, in this case, email traffic and company organization (right). © *1998 IEEE. Reprinted, with permission, from Chuah (1998).*

With the SolarPlot technique introduced by Chuah (1998), values are plotted around the circumference of a circle as shown left in the figure. Much like in a circular histogram, the first step is to partition the data series into a number of bins. Each bin is represented by a sunbeam whose length encodes the frequency of data items in the corresponding bin. The SolarPlot determines the number of bins dynamically depending on the size of the circle. Users are allowed to expand or contract the circle in order to get more or fewer bins, or in other words, to get a more or less detailed representation of the data. This way it is possible to explore the data at different levels of abstractions and to discern patterns globally across aggregation levels. The SolarPlot also supports switching locally to a more detailed plot for a user-selected focus interval as shown in the center of the figure. Chuah (1998) further suggests a variation of the SolarPlot called SolarPlot + Aggregate TreeMap, where the display of data is combined with a visual representation of a hierarchical structure as shown to the right.

Relevant references: Chuah (1998)

Cluster and Calendar-Based Visualization

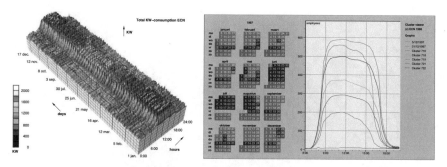

Fig. A.42: The 3D visualization (left) represents daily power consumption patterns for several weeks of a year. The calendar representation (center) color-codes cluster affiliation with respect to daily patterns of employees' office hours as shown in the line plots (right). © *1999 IEEE. Reprinted, with permission, from van Wijk and van Selow (1999).*

Temporal patterns can indicate at which time of the day certain resources are highly stressed. Relevant applications can be found in computing centers, traffic networks, or power supply networks. The cluster and calendar-based approach by van Wijk and van Selow (1999) allows analysts to find such temporal patterns at different temporal granularities. The starting point of the approach is to consider the course of a day as a line plot (↪ p. 233) covering the 24 hours of a day. Multiple daily courses of this kind are visualized as a three-dimensional height field (left part of the figure), where the hours of the day are encoded along one axis, individual days are encoded along the second axis, and data values are encoded as height (along the third axis). This allows users to detect short-term daily patterns and long-term trends of the data at higher temporal granularity. To assist in the analysis of the data, the approach further groups similar daily courses into clusters. The data belonging to a particular cluster are aggregated to define a representative for that cluster, i.e., the average line plot for all days belonging to a cluster. Cluster affiliation of individual calendar dates is then color-coded into a calendar (center) and individual or clustered daily patterns are shown as line plots (right). The user can adjust the number of clusters to be shown so as to find the level of abstraction that suits the data and the task at hand. The combination of analytical and visual methods as applied here is useful for identifying days of common and exceptional daily behavior.

Relevant references: van Wijk and van Selow (1999)

time

primitives: **points**
arrangement: **linear, cyclic**

data

number of variables: **single**
frame of reference: **abstract**

vis

mapping of time: **static**
dimensionality: **2D, 3D**

Enhanced Interactive Spiral

time

primitives: **points**
arrangement: **cyclic**

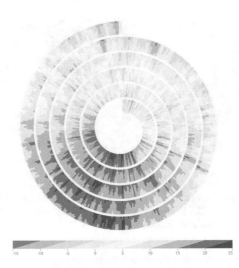

data

number of variables: **single**
frame of reference: **abstract**

Fig. A.43: The spiral shows the daily temperature measured in the city of Rostock from 2006 to 2010. The blue and green colors in the upper part of the spiral represent the colder winters of 2009 and 2010. ⓒ① *The authors. Generated with EnhancedSpiral.js by Christian Tominski.*

Tominski and Schumann (2008) apply the enhanced two-tone color-coding by Saito et al. (2005) to visualize time-dependent data along a spiral. Each time primitive is mapped to a unique segment of the spiral. Each segment is subdivided into two parts that are colored according to the two-tone coloring method. The advantage of using the two-tone approach is that it realizes the overview+detail concept by design. The two colors used per spiral segment allow users to quickly recognize the value range of that segment (quick overview). If a perceived value range is of interest to the user, the proportion of the two colors indicates the particular data value more precisely (detailed inspection). The enhanced spiral can be adjusted interactively in various ways. Setting a suitable number of segments per spiral cycle is crucial for being able to see cyclic patterns. The search for such a suitable parametrization can be supported by guidance methods (see Ceneda et al., 2018). Further parameters can be tuned to adjust the visualization, including the number of time primitives, the number of spiral cycles, the direction of the mapping of time (inward or outward), the color scale, and the number of color classes. Navigation in time is possible via direct manipulation of a time slider.

vis

mapping of time: **static**
dimensionality: **2D**

Relevant references: Tominski and Schumann (2008) • Ceneda et al. (2018) • Saito et al. (2005)

ClockMap

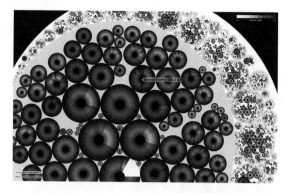
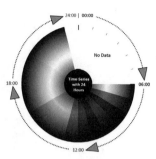

primitives: **points**
arrangement: **cyclic**

Fig. A.44: ClockMap is a visualization technique that supports the user in exploring and finding patterns in hierarchical time-series data. Left: Example shows 24 hours of network traffic of a large number of subnets in its hierarchical decomposition. Right: Clockeye glyph. Each glyph circle represents a 24-hour time series where color is mapped to the amount of network traffic in bytes. ©① *Fabian Fischer.* `https://ff.cx/clockmap/`

Hierarchical structures are commonplace for many datasets in a variety of application domains, such as IT networks, budget data, file systems, or regional decompositions. A well-known visualization technique for hierarchical data with quantitative attributes is the treemap. A treemap is a space-filling visualization technique based on nesting, i.e., an iterative subdivision of rectangular areas. However, in their basic form, treemaps focus on time-invariant data and are not suitable to represent changes over time. ClockMaps aim to fill this gap by combining circular glyphs (clockeyes) using a circular treemap layout. Individual glyphs (clockeyes) are based on the metaphor of an analog clock and represent a time series in a circle, whereas color is used to represent a quantity for each radial line. In doing so, the temporal changes of each node as well as the hierarchical structure of the data are combined in a single visualization technique. The size of a clockye glyph is determined by the amount of items that are nested within and does not represent a data attribute. Moreover, interactive features support the user in exploring and finding patterns in hierarchical time-series data. Users are able to drill-down, and apply semantic zoom, as well as details-on-demand.

Relevant references: Fischer et al. (2012)

data

number of variables: **single**
frame of reference: **abstract**

vis

mapping of time: **static**
dimensionality: **2D**

SpiraClock

Fig. A.45: The SpiraClock combines the classic analog clock with a spiral display of future events. The minute hand currently points to a meeting that has just started. Future events are aligned along a spiral within the clock face. ⓒⓘ *The authors. Generated with SpiraClock.js by Christian Tominski.*

The SpiraClock invented by Dragicevic and Huot (2002) visualizes time by using the clock metaphor. The visual representation consists of a clock face and clock hands indicating the current time. The interior of the clock shows a spiral that extends from the clock's circumference inward toward its center. Each cycle of the spiral represents one hour, with the current hour being shown at the outermost cycle and future hours displayed in the center (about five hours overall in the figure). Time intervals (e.g., meetings or appointments) are represented as spiral segments embedded into the spiral cycles exactly where the intervals start and end. This kind of representation makes it easy to see if the intervals are overlap-free and if the intervals are uniformly distributed in time (e.g., to confirm that there are sufficient breaks and no conflicting appointments). As time advances, the spiral is constantly updated and future intervals gradually move outward until they are current. Past intervals gradually fade out. Overall, the SpiraClock enhances classic clocks with a preview of the near future and a brief view of the past. The SpiraClock allows users to drag the clock handles to visit different points in time. Moreover, intervals of interest can be highlighted and corresponding textual annotations can be displayed.

Relevant references: Dragicevic and Huot (2002)

Horizon Graph

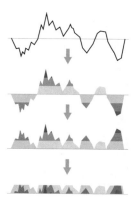
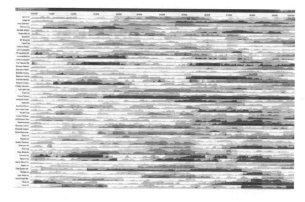

Fig. A.46: Horizon graphs require only little screen space and are very useful for comparing multiple time-dependent variables. Left: construction of a horizon graph from a line chart; right: stock market data visualized as stacked horizon graphs. *Left:* ⓒⓘ *The authors. Adapted from Reijner (2008). Right:* ⓒ *Courtesy of Hannes Reijner.*

Reijner (2008) describes horizon graphs as a visualization technique for comparing a large number of time-dependent variables. Horizon graphs are based on the two-tone pseudo coloring technique by Saito et al. (2005). The left part of the figure demonstrates the construction of horizon graphs (from top to bottom). Starting from a common line plot, the value range is divided into equally sized bands that are discriminated by increasing color intensity toward the maximum and minimum values while using different hues for positive and negative values. Then, negative values are mirrored horizontally at the zero line. Finally, the bands are layered on top of each other. This way, less vertical space is used, which means data density is increased while the resolution is preserved. A study by Heer et al. (2009) has shown that mirroring does not have negative effects and that layered bands are more effective than the standard line plot (↪ p. 233) for charts of small size. To improve the basic concept of horizon graphs, several extensions have been proposed since their introduction. Perin et al. (2013) extend horizon graphs with dedicated interaction techniques for adjusting the zero line and the number of colored bands. With their work on qualizon graphs, Federico et al. (2014) combine quantitative data with qualitative abstractions. Moreover, Dahnert et al. (2019) present further variations called collapsed horizon graphs, where the idea of horizon graphs is extended to two dimensions. That is, apart from the vertical quantitative variable, also the horizontal temporal dimension is being collapsed using bivariate color maps.

Relevant references: Reijner (2008) • Saito et al. (2005) • Heer et al. (2009) • Perin et al. (2013) • Federico et al. (2014) • Dahnert et al. (2019)

VizTree

time

primitives: **points**
arrangement: **linear**

data

number of variables: **single, multiple**
frame of reference: **abstract**

vis

mapping of time: **static**
dimensionality: **2D**

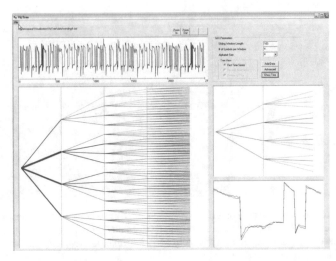

Fig. A.47: Time series of arbitrary lengths are represented as fixed-length subsequence trees. The figure shows a winding dataset that records the angular speed of a reel. Top left: input time series; top right: parameter setting area for discretization and subsequence length; center left: subsequence tree for the time series; center right: detail view of the tree shown in the left panel; bottom right: subsequences matching a particular string representation (e.g., subsequences starting with 'ab') whereas positions of matched subsequences are highlighted in the top-left panel. © *Courtesy of Eamonn Keogh.*

VizTree by Lin et al. (2005) is a time-series pattern discovery and visualization system for massive time-series datasets. It uses the time-series abstraction method SAX (Symbolic Aggregate approXimation) to discretize numeric time series into sequences of abstract symbols like 'abacacc' (see Lin et al., 2003; Lin et al., 2007). Subsequences (patterns) are generated by moving a sliding window along the sequence. These subsequences are combined and represented by a horizontal tree visualization where the frequency of a pattern is encoded by the thickness of a branch (or light gray if the frequency is zero). The VizTree interface consists of multiple coordinated views that show the input time series along with the subsequence tree as well as control and detail-on-demand panels. VizTree can be used to accomplish different pattern discovery tasks interactively: finding frequently occurring patterns (i.e., motif discovery) and surprising patterns (i.e., anomaly detection), query by content, and the comparison of two time series by calculating a difference tree.

Relevant references: Lin et al. (2005) ● Lin et al. (2003) ● Lin et al. (2007)

Time Curves

primitives: **points**
arrangement: **linear**

Fig. A.48: Time Curve showing global cloud circulation data. Time points are positioned based on their similarity, that is, similar points are close to each other, and distant points are dissimilar. Subsequent time points are connected to form a contiguous curve spanning the entire timeline. ⓒⓘ *The authors. Generated with the Time Curves software by Benjamin Bach.*

number of variables: **single, multiple**
frame of reference: **abstract**

Bach et al. (2016) describe a technique called Time Curves. The metaphor behind this technique is to fold a line plot visualization into itself so as to bring similar time points close to each other. This metaphor can be applied to any dataset where a similarity metric between time points can be defined. Technically, each time point is assigned a two-dimensional position based on the similarity metric. This can, for example, be implemented by dimensionality reduction via multi-dimensional scaling (MDS). Once positioned, successive time points are connected by a curve so as to visualize their temporal order. Time curves are a general approach to visualize patterns of evolution in temporal data, such as progression and stagnation, sudden changes, regularity and irregularity, reversals to previous states, temporal states and transitions, reversals to previous states, and others. It is important to mention that the interpretability of a Time Curves visualization heavily depends on the employed similarity metric and inherits all advantages and disadvantages of the implemented dimensionality reduction method.

mapping of time: **static**
dimensionality: **2D**

Relevant references: Bach et al. (2016)

TimeSlice

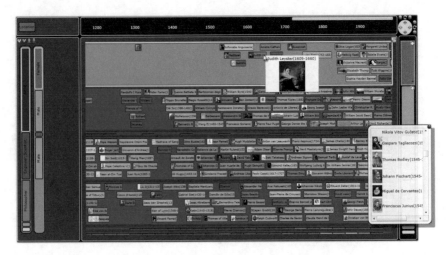

Fig. A.49: TimeSlice interface (left) for faceted browsing of a timeline visualization (center) of the lives of famous people. © *Courtesy of Jian Zhao.*

TimeSlice is an interactive faceted browsing tool (↪ p. 336) for time-oriented data. It provides a flexible approach for constructing, comparing, and manipulating multiple queries over faceted timelines (↪ p. 258). The main view shows time intervals (the life span of famous people in the figure) as vertical colored bars. Queries are organized in a dynamic filtering tree structure to the left of the interface. It displays the currently focused queries and also their contexts (such as queries that share the same attribute on one facet but that differ on another). Tree nodes (representing attributes) and tree levels (representing facets) can be manipulated directly, which offers efficient navigation across different perspectives of the data.

Relevant references: Zhao et al. (2012)

Silhouette Graph

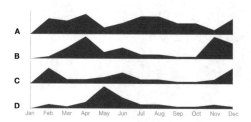

Fig. A.50: Silhouette graphs are filled line plots that can be used for enhancing the comparison of multiple time series that are shown side by side. ⓒⓘ *The authors. Adapted from Harris (1999).*

Silhouette graphs emphasize the visual impression of time series by filling the area below the plotted lines (see Harris, 1999). This leads to distinct silhouettes that enhance perception at wide aspect ratios of long time series compared to line plots (↪ p. 233) and allow an easier comparison of multiple time series. In the left figure, time is mapped to the horizontal axes and multiple time series are stacked upon each other. Other layouts of the axes might be used for reflecting different time characteristics. One example is circular silhouette graphs as shown in the right figure, where silhouette graphs are drawn on concentric circles in order to emphasize periodicity in time. The ideas of stacking graphs and showing them along concentric circles are not restricted to silhouette graphs, but can be found in several other techniques, for example, in horizon graphs (↪ p. 277) and in enhanced interactive spirals (↪ p. 274).

Relevant references: Harris (1999)

Recursive Pattern

primitives: **points**
arrangement: **linear, cyclic**

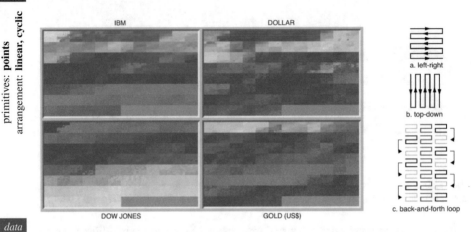

number of variables: **single, multiple**
frame of reference: **abstract**

Fig. A.51: Four time series are shown as recursive patterns, where the color of a pixel represents the stock price. Pixels are arranged to semantically match the hierarchical structure of the data: A row represents a year with its 12 months, and each month is further subdivided into weeks, which in turn consist of five workdays, each of which represents nine data values for a single day. Examples of possible pixel arrangements are shown to the right. © *1995 IEEE. Reprinted, with permission, from Keim et al. (1995).*

The most space-efficient way of visualizing data is to represent them on a per-pixel basis (see Lévy et al., 2007). Keim et al. (1995) propose a variety of pixel-based visualization approaches of which the recursive pattern technique is particularly suited to display large time series. The key idea behind the recursive pattern technique is to construct an arrangement of pixels that corresponds to the inherently hierarchical structure of time-oriented data given at multiple granularities. The figure shows financial data as a pixel-based visualization. The initial step is to map nine data values collected per day to a 3×3 pixel group. This group is then used to form a larger group for a week of workdays containing 5×1 day groups. Groups for months, years, and decades can be created by recursively arranging groups of the next lower granularity in a semantically meaningful way (e.g., 12 months are grouped into a year). In the resulting pattern, each pixel is color-coded with regard to a single data value in the time series. Multiple dense pixel displays of this kind can be combined to generate overviews of large multivariate datasets.

mapping of time: **static**
dimensionality: **2D**

Relevant references: Keim et al. (1995) • Lévy et al. (2007)

Lin-spiration

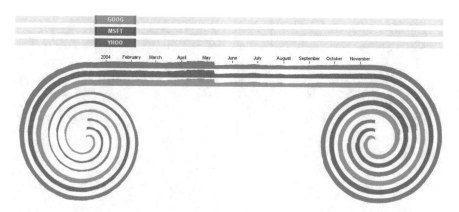

Fig. A.52: Lin-spiration combines linear (center) and cyclic (left and right) representations of multivariate time series. Time sliders (top) allow users to navigate the time axis of each variable independently. © *2012 ACM. Reprinted, with permission, from Graells and Jaimes (2012).*

Graells and Jaimes (2012) combine linear silhouette graphs (↪ p. 281) and spiral graphs (↪ p. 285) to a hybrid interactive visualization technique they call *Lin-spiration*. The basic idea is to show a focused subset of the data as multiple stacked linear silhouette graphs at full temporal resolution in the center of the display. To each side of the linear silhouette graphs, spiral graphs show in radial form the data preceding and succeeding the selected subset. The spirals' purpose is to provide context information about how the data looked like in the past and how it will look like in the future (with respect to the currently selected subset). To accommodate very large time-oriented data, the spirals show the time axis in compressed form. The resulting figure metaphorically resembles film rolls in classic photo cameras as already suggested by Tominski et al. (2003) and also bears resemblance with the Perspective Wall (↪ p. 256). The Lin-spiration approach is combined with linear time sliders (top) that allow users to navigate the time axis of each visualized variable independently and to compare the data not only for the same time period, but for different periods of interest.

Relevant references: Graells and Jaimes (2012) • Tominski et al. (2003)

primitives: **points**
arrangement: **linear, cyclic**

data

number of variables: **single, multiple**
frame of reference: **abstract**

vis

mapping of time: **static**
dimensionality: **2D**

Spiral Graph

time

primitives: **points**
arrangement: **cyclic**

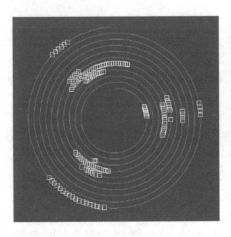 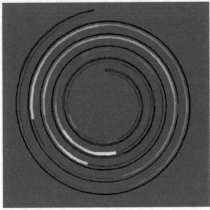

data

number of variables: **single, multiple**
frame of reference: **abstract**

Fig. A.53: A spiral graph encodes time-series data along a spiral, thus emphasizing the cyclic character of the time domain. Actual data values are visualized using symbols for nominal data (left) as well as color and line thickness for quantitative data (right). © *2001 IEEE. Reprinted, with permission, from Weber et al. (2001).*

vis

mapping of time: **static**
dimensionality: **2D**

The spiral graph developed by Weber et al. (2001) is a visualization technique that focuses on cyclic characteristics of time-oriented data. To this end, the time axis is represented by a spiral. Time-oriented data are then mapped along the spiral path. While nominal data are represented by simple icons, quantitative data can be visualized by color, line thickness, or texture. One can also visualize multivariate time series by intertwining several spirals as shown for the example of two variables in the spiral on the right. In this case, a distinct hue is used per spiral so that individual variables can be discerned. Weber et al. (2001) further envision extending the spiral to a three-dimensional helix, in order to cope with larger time series. The main purpose of the spiral graph is the detection of previously unknown periodic behavior of the data. The user can interactively adjust the spiral's cycle length (i.e., the number of data values mapped per spiral cycle) to explore the data for cyclic patterns. As an alternative, it is also possible to smoothly animate through possible cycle lengths. In this case, the periodic behavior of the data becomes immediately apparent by the emergence of a pattern. When such a pattern is spotted, the user stops the animation and an interesting cycle length has been found.

Relevant references: Weber et al. (2001)

Spiral Display

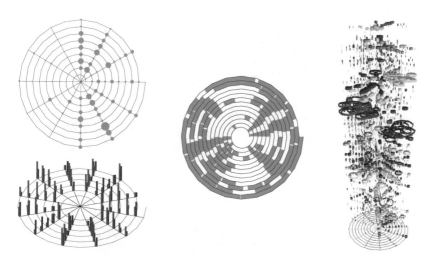

Fig. A.54: Data can be visualized along a spiral in different ways: by the area of circular elements (top left), by the sizes of multiple spikes (bottom left), as bars marking start and end of intervals (center), or by the volume of hollow cylinders aligned at different layers along the vertical axis (right). © *1998 ACM. Reprinted, with permission, from Carlis and Konstan (1998).*

The interactive spiral display by Carlis and Konstan (1998) uses Archimedean spirals to represent the time domain. Data values at particular time points are visualized as filled circular elements whose area is proportional to the data value (top left). In the case of interval-based data, filled bars are aligned with the spiral shape to indicate the start and end of intervals (center). If multivariate data are given at time points, the spiral is tilted and data values are visualized as differently colored spikes, where spike color indicates variable affiliation and spike height encodes the corresponding data value (bottom left). Alternatively, one can use the vertical z-axis to separate the display of multiple variables (right). In this case, each time-dependent variable has its own layer along the z-axis and is represented with a unique color. Within a layer, data values are encoded to the volume of cylinders, which are hollow to prevent occlusion. The system implemented by Carlis and Konstan (1998) allows users to display multiple linked spirals to perform comparison tasks. The cycle lengths of spirals can be adjusted interactively and can also be animated automatically for discovering periodic patterns.

Relevant references: Carlis and Konstan (1998)

time

primitives: **points, intervals**
arrangement: **cyclic**

data

number of variables: **single, multiple**
frame of reference: **abstract**

vis

mapping of time: **static**
dimensionality: **2D, 3D**

Stacked Graphs

primitives: **points**
arrangement: **linear**

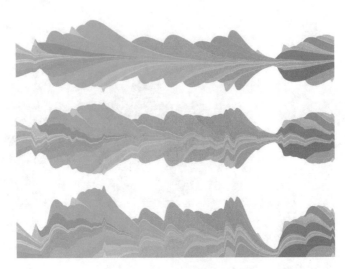

number of variables: **multiple**
frame of reference: **abstract**

Fig. A.55: Multivariate time series visualized as stacked graphs with different designs: stream graph design (top), ThemeRiver layout (center), and traditional stacking (bottom). ⓒ① *The authors. Generated with the streamgraph_generator code base by Lee Byron.*

Stacking multiple graphs on top of each other is a suitable approach to visualizing multiple time-dependent variables (see Harris (1999)). Elaborate variants of stacked graphs have been investigated in detail by Byron and Wattenberg (2008). To visualize the evolution of an individual variable, data values are encoded to the height of a so-called layer that extends along the horizontal time axis. A special color map is applied to visualize additional data variables and to make individual layers distinguishable. Several layers are then stacked on top of each other, effectively creating an overall graph that represents the visual sum of the entire dataset. Layout and sorting of layers can be done in various ways, resulting in quite different designs such as the so-called stream graph design, the ThemeRiver layout (↪ p. 293), or traditional layer area graphs (↪ p. 289), illustrated from top to bottom in the figure. The stream graph design (top) is notable because it received quite positive feedback when it appeared on the New York Times website as a visual representation of box office revenues. In that version, individual layers were also outfitted with text labels.

mapping of time: **static**
dimensionality: **2D**

Relevant references: Harris (1999) • Byron and Wattenberg (2008)

TimeSearcher 3, River Plot

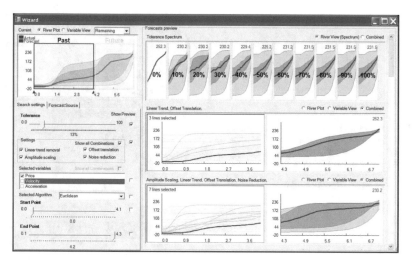

Fig. A.56: Forecasting of online auction data. Top left: selection of time interval for similarity search; bottom left: selection of variables to consider and parameters to vary in the previews; right: preview area that assists users in understanding the impact of parameters – varying tolerance levels as river plots and different combinations of applied transformations as line plots and river plots. © *Courtesy of Paolo Buono.*

TimeSearcher 3 is a tool to support similarity-based forecasting of multivariate time series. Similarity-based forecasting is a data-driven method using the similarity to a set of historical data for predicting future behavior. The outcome of the algorithm is affected by a number of options and parameters, for instance, the transformations applied or the tolerance threshold used for matching. As a result, the median of the matched subsets becomes the forecast and descriptive statistics measures reflect the uncertainty associated with the forecast. This is displayed graphically as a simplified, continuous box plot, called a river plot. It uses superimposed, colored regions, for which light gray indicates the range between the minimum and maximum and dark gray the range between the 25% and 75% percentiles, and a line in the center, where red indicates the forecast, brown shows the median during the matching period, and black is the median before this period. TimeSearcher 3 builds upon TimeSearcher (↪ p. 290) and adds a preview interface to allow users to interactively explore the effects of adjusting algorithm parameters and to see multiple forecasts simultaneously.

Relevant references: Buono et al. (2007)

time

primitives: **points**
arrangement: **linear**

data

number of variables: **multiple**
frame of reference: **abstract**

vis

mapping of time: **static**
dimensionality: **2D**

Timeline Trees

time

primitives: **points**
arrangement: **linear**

Day	Market basket and money spent
Mon:	milk $1, bananas $3
Tue:	cheese $1, apples $3
Wed:	milk $1, bananas $1, grapes $2
Thu:	milk $1
Fri:	milk $1, cheese $3

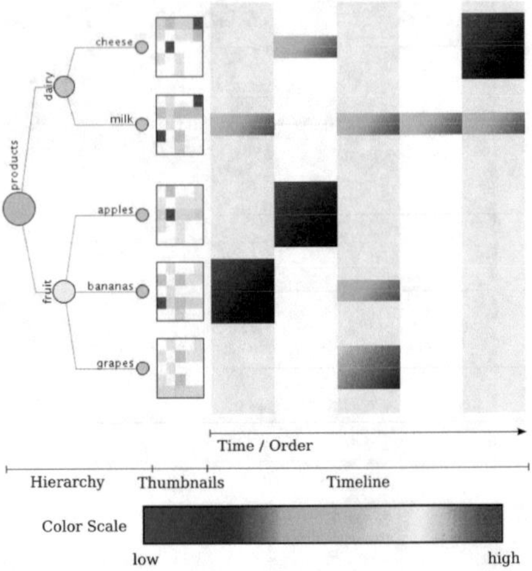

data

number of variables: **multiple**
frame of reference: **abstract**

Fig. A.57: A smaller set of products in a market basket is visualized using timeline trees. One can see that milk is bought regularly (green boxes for all but one day), and that cheese, apples, and bananas are more expensive (higher red-colored boxes). © *Courtesy of Michael Burch.*

vis

mapping of time: **static**
dimensionality: **2D**

Data that describe items being related to each other are quite common. An example of such data is transactions in online shopping systems where products being bought together are considered to be related. Burch et al. (2008) visualize temporal sequences of transactions by means of the so-called timeline trees. The visual representation consists of three parts: a display of an information hierarchy, a timeline representation of temporal sequences, and thumbnail pictures. The information hierarchy is a static hierarchical categorization of data items (e.g., a system of product groups), where groups can be expanded or collapsed interactively to view the data at different levels of detail. The timeline view shows multiple sequences of boxes for the current level of detail, where color and box size are used to encode data values (e.g., product price) of an item (or group) at a particular point in time. Thumbnails for each leaf of the information hierarchy show an overview of transactions masked by the corresponding leaf node. Enhanced with several interaction facilities, timeline trees help users to understand trends in the data and to find relations between different levels of abstractions (e.g., product groups and specific products).

Relevant references: Burch et al. (2008)

Layer Area Graph

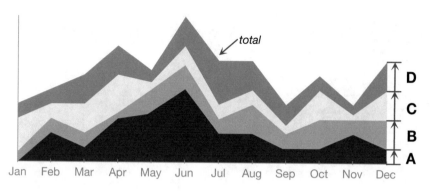

Fig. A.58: Layered area graphs show individual variables as bands stacked on top of each other emphasizing the total sum of all variables. ⓒⓘ *The authors. Adapted from Harris (1999).*

Layer area graphs might be used when comparing time series that share the same unit and can be summed up (see Harris, 1999). A layer area graph is a stacked visualization where time-series plots are drawn upon each other as layered bands. Caution needs to be exercised for this kind of representation because it is sensitive to the order of the layers (see Mathiesen and Schulz, 2021). Different orders influence the visual appearance of the individual layers because only the bottommost layer has a straight baseline. All subsequent layers are drawn relative to the layers below. An advantage of layer area graphs is the fact that they emphasize the total sum of values while providing information about the parts that constitute it. More advanced visualization techniques such as the ThemeRiver (↪ p. 293) or stacked graphs (↪ p. 286) build upon the basic principle of layer area graphs.

Relevant references: Harris (1999) • Mathiesen and Schulz (2021)

time

primitives: **points**
arrangement: **linear**

data

number of variables: **multiple**
frame of reference: **abstract**

vis

mapping of time: **static**
dimensionality: **2D**

TimeSearcher

time

primitives: **points**
arrangement: **linear**

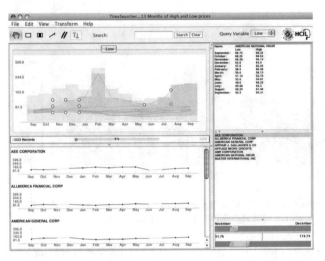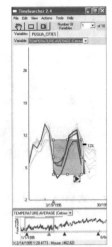

data

number of variables: **multiple**
frame of reference: **abstract**

Fig. A.59: Left: TimeSearcher showing stock price data (top left: multiple line plots and query boxes; bottom left: line plots for individual stocks matching the query). Right: TimeSearcher 2 with query-by-example using a searchbox (light red background) and matching items in red. ⓒⓘ *The authors. Generated with the TimeSearcher software from the University of Maryland Human-Computer Interaction Lab.*

vis

mapping of time: **static**
dimensionality: **2D**

Hochheiser and Shneiderman (2004) implemented TimeSearcher as a visual exploration tool for multiple time series. While employing a straightforward visual representation using line plots (↪ p. 233), its main objective is to enable users to identify and find patterns in the investigated data. To this end, the so-called *timebox query model* has been developed. It allows the specification of a rectangular query region that defines both a time interval and a value range of interest. Those time series that comply with a query (i.e., overlap with the timebox) are displayed, whereas all others are filtered out. Users can combine multiple timeboxes to refine the query further and other query functionalities such as leaders and laggers, angular queries, and variable timeboxes are also part of TimeSearcher. To provide contextual information, the data envelope and the query envelope can be displayed. Buono et al. (2005) extended these features in TimeSearcher 2 by allowing the representation of heterogeneous datasets and providing a *searchbox query model* that effectively implements a query-by-example functionality. Here, occurrences of a brushed portion of the time series are searched, whereas the similarity threshold of matches can be adjusted.

Relevant references: Hochheiser and Shneiderman (2004) • Buono et al. (2005)

BinX

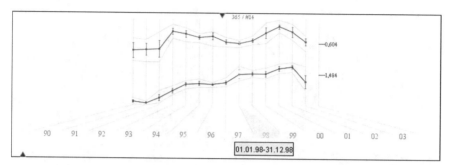

Fig. A.60: Exchange rates for two currencies are compared using the BinX tool. Each bin aggregates the daily rates for a whole year. A selected bin is highlighted and its position on the global time scale is marked accordingly. ⓒⓘ *The authors. Generated with the BinX tool by Lior Berry.*

Large time series require the application of abstraction methods in order to reduce the number of time points to be displayed, thus keeping visualization costs at a manageable level. Finding a suitable degree of abstraction, however, is not an easy task. The BinX tool developed by Berry and Munzner (2004) is interesting in that it supports the exploration of different aggregations of a time series. The aggregation is based on constructing bins, each of which holds a user-defined number of time points. Easy-to-use interaction is offered to quickly try out different bin sizes. BinX visualizes one or two quantitative time-dependent variables using common chart elements. An overview of the time axis is preserved at all times at the bottom of the BinX representation. The central chart view displays the two time series in an aggregated fashion according to the currently chosen bin size. In order to faithfully represent aggregated information, line plots ($\hookrightarrow$ p. 233), box plots, and a min-max band are used in combination. The correspondence between a point in the chart and a time span (bin) on the time axis is represented upon user request. BinX supports the clustering of bins as an additional mechanism for analytic abstraction. In this case, the cluster affiliation of bins is encoded via color.

Relevant references: Berry and Munzner (2004)

time
primitives: **points**
arrangement: **linear**

data
number of variables: **multiple**
frame of reference: **abstract**

vis
mapping of time: **static**
dimensionality: **2D**

MultiComb

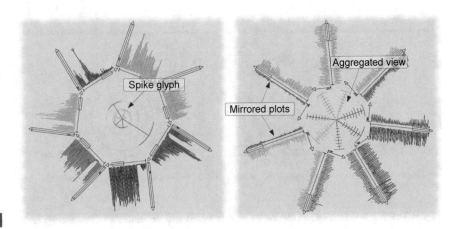

Fig. A.61: Two MultiComb representations visualize seven time-dependent variables. In the left MultiComb, line plots are arranged around the display center, in the right one, they extend outwards. The interior of the MultiCombs can be used to display additional information via a spike glyph or an aggregated view. ⓒ④ *The authors. Generated with the VisAxes software by Christian Tominski.*

Line plots (↪ p. 233) are expressive visual representations for univariate data. The rationale behind the MultiComb visualization is to utilize this expressiveness for representing multiple time-dependent variables. Tominski et al. (2004) describe the MultiComb as a visual representation that consists of multiple radially arranged line plots. Two alternative designs exist: time axes are arranged around the display center (left figure) or time axes extend outwards from the MultiComb's center (right figure). In the latter case, optional mirror plots duplicate the plots of neighbor variables to ease visual comparison. To maintain a certain aspect ratio for the separate plots, the axes do not start in the very center of the MultiComb. The screen space in the interior can thus be used to provide additional views: a spike glyph can be shown to allow a detailed comparison of data values for a selected time point, or an aggregated view might display the history of a temporal data stream in an aggregated fashion. Various possibilities for interaction allow users to browse in time, to zoom into details of the time axes, as well as to add, remove, and reorder plots, and to rotate the MultiComb.

Relevant references: Tominski et al. (2004)

ThemeRiver

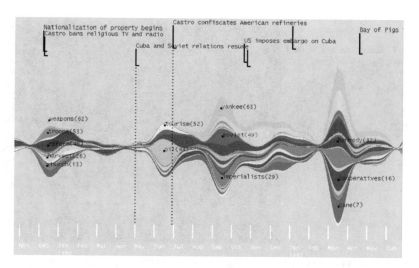

Fig. A.62: The ThemeRiver representation uses the metaphor of a river that flows through time. Colored currents within the river reflect thematic changes in a document collection, where the width of a current represents the relevance of its associated theme. © *2002 IEEE. Reprinted, with permission, from Havre et al. (2002).*

The ThemeRiver technique developed by Havre et al. (2000) represents changes in news topics in the media. Each topic is displayed as a colored current whose width varies continuously as it flows through time. The overall image resembles a river composed of multiple currents. The ThemeRiver provides an overview of the topics that were important at certain points in time. Hence, the main focus is directed toward establishing a picture of an easy to follow evolution over time using interpolation and approximation. Moreover, ThemeRiver representations can be annotated, e.g., with related major historical events, and raw data points with exact values can be shown. Even though the ThemeRiver was originally invented to visualize thematic changes in document collections, it is also suited to represent other multivariate, quantitative data. Because perception of data differs depending on where in the river individual variables are shown, it is important to provide interaction techniques to allow users to rearrange the horizontal position of variables. Generalizations of the ThemeRiver technique became known as stacked graphs or stream graphs (↪ p. 286). The ThemeRiver can also be used in combination with other views. Hashimoto and Matsushita (2012) combined the ThemeRiver technique with a heat map view, which illustrates the quantities at a given time point. Jiang et al. (2016) combines the ThemeRiver technique with spirals and scatter plots, to a so-called spiral theme plot.

Relevant references: Havre et al. (2000) • Havre et al. (2002) • Hashimoto and Matsushita (2012) • Jiang et al. (2016)

time

primitives: **points**
arrangement: **linear**

data

number of variables: **multiple**
frame of reference: **abstract**

vis

mapping of time: **static**
dimensionality: **2D**

time

history flow

primitives: **points**
arrangement: **linear**

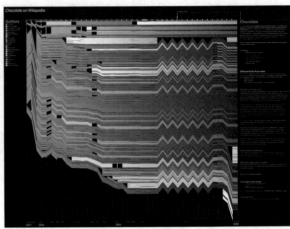

data

number of variables: **multiple**
frame of reference: **abstract**

Fig. A.63: history flow shows vertical revision lines, one for each revision, where colored sections reflect the different authors of a document as schematically depicted on the left. This method is applied to the visualization of the Wikipedia entry on chocolate as shown on the right. © *Courtesy of Fernanda B. Viégas.*

vis

mapping of time: **static**
dimensionality: **2D**

Viégas et al. (2004b) designed history flow as an exploratory wiki article analysis tool for finding author collaboration patterns, showing relations between document versions, revealing patterns of cooperation and conflict, as well as making broad trends immediately visible. The basis for the representation is the so-called revision lines. These top-aligned, vertical lines are displayed for every version of a document. The length of revision lines is proportional to the document length. Individual sections of a revision line are colored differently to visualize which authors worked on which parts of a document. The sections associated with a particular author are visually connected from one revision to the next. One can discern stable sections and splits of sections. Gaps in connections clearly indicate deletions and insertions. Two different layouts can be used for spacing revision lines: uniform spacing (space by occurrence/event-based) or spacing according to time (space by date/time-based). The first layout shows each document change equally spaced without showing time intervals between versions proportionally. The second alternative additionally gives information about the exact timing.

Relevant references: Viégas et al. (2004b)

LifeLines2

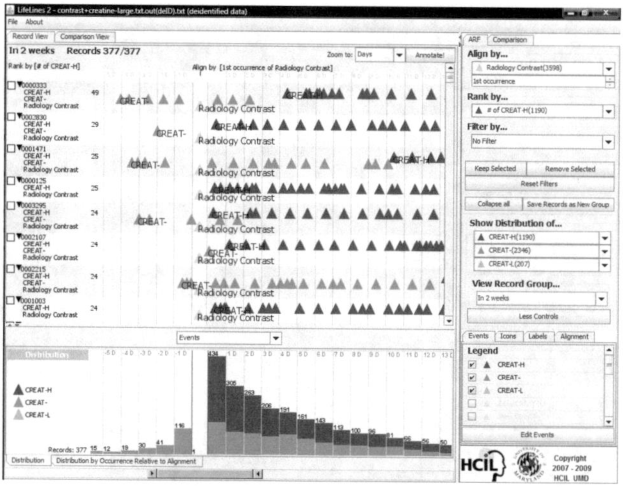

Fig. A.64: LifeLines2 show patient records in stacked rows, where triangles indicate health-related events. A histogram view visualizes the number of events over time. Interaction operators (i.e., align, rank, filter) support visual exploration. © *Courtesy of Taowei David Wang.*

LifeLines2 is an interactive visual exploration interface for instantaneous events based on categorical, health-related data (e.g., high, normal, or low body temperature). Events are displayed as triangles along a horizontal time axis, where color indicates event categories and data of different patient records are stacked vertically. An aggregation of events is represented as a histogram showing the number of occurrences over time. LifeLines2 introduces three powerful operators for interactive exploration: align, rank, and filter. The align operator can be used to arrange all records along a specific event type in temporal order, for example, to align a group of patients with regard to their first heart attack. Additionally, the time axis switches from an absolute time representation to relative time originating from the specified event (e.g., one week before, or two weeks after the first heart attack). The rank operator is useful for ordering records according to the number of occurrences of a specified event type. The filter operator allows users to search for particular sequences of events including both the presence of events and the absence of events (e.g., patients having had a heart attack but no stroke following it).

Relevant references: Wang et al. (2009)

time

primitives: **points**
arrangement: **linear**

data

number of variables: **multiple**
frame of reference: **abstract**

vis

mapping of time: **static**
dimensionality: **2D**

Similan

primitives: **points**
arrangement: **linear**

number of variables: **multiple**
frame of reference: **abstract**

mapping of time: **static**
dimensionality: **2D**

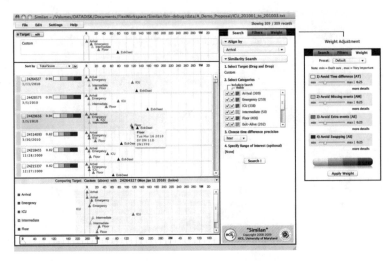

Fig. A.65: Similan ranks patient records (center) according to their similarity to a target record (top). Individual records can be compared directly with the target (bottom). Various interaction operators, including adjustment of similarity weights (right), can be used to refine the visualization. © *Courtesy of Krist Wongsuphasawat.*

Wongsuphasawat and Shneiderman (2009) describe Similan as a system for exploring patient records. Patient records are stacked upon each other and show health-related events as triangles, where color indicates event categories (e.g., arrival, emergency, ICU). Similan uses the same visual representation as LifeLines2 (↪ p. 295) but provides a different approach to data exploration. Instead of interactive filtering, records are ranked according to their similarity to a given event sequence (query-by-example). In the figure, the topmost record is the one that is most similar to the user-specified event sequence. This can be used to search for groups of patients who share similar temporal patterns. A dedicated view is provided to allow a direct comparison of the target query record with any particular record in the data. Another scenario is to search for an event sequence that the user is not certain whether it exists in the data. In this way, the tool can give the most similar results if the exact event sequence does not exist. For determining the similarity of event sequences a similarity measure (M&M measure) has been developed. The weights of factors that determine the similarity measure can be adjusted interactively by the user.

Relevant references: Wongsuphasawat and Shneiderman (2009)

VIE-VISU

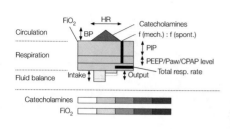

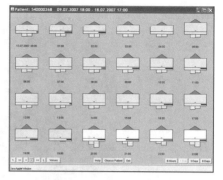

Fig. A.66: VIE-VISU encodes fifteen health-related patient parameters to different visual attributes of a glyph (left). The example on the right shows neonatal patient information on an hourly basis for the course of the day. *Left:* ⊚⊕ *The authors. Adapted from Horn et al. (2001). Right:* © *Courtesy of Werner Horn.*

Paper-based analysis of patient records is hard to conduct because many parameters are involved and an overall assessment of the patient's situation is difficult. Therefore, Horn et al. (2001) developed VIE-VISU, an interactive glyph-based visualization technique for time-oriented patient records. The glyph consists of three parts that represent circulation, respiration, and fluid balance parameters. All in all, 15 parameters are visualized using different visual attributes (i.e., length, width, color) as illustrated in the left part of the figure. For example, the circulation parameter heart rate (HR) is encoded to the width of the triangle on top of the glyph and the triangle's color encodes catecholamines (color legend is given at the bottom). Each glyph represents a one hour period and 24 glyphs are combined in a small multiples display (↪ p. 359) as shown in the right part of the figure. Interaction controls support navigation in time and switching to different periods for the small multiples view. VIE-VISU helps users to combine different measurements, maintain their relationships, show their development over time, and make specific, possibly life-threatening situations easy to spot.

Relevant references: Horn et al. (2001)

TimeWheel

primitives: **points**
arrangement: **linear**

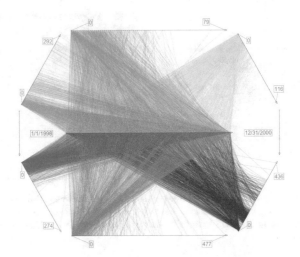

number of variables: **multiple**
frame of reference: **abstract**

Fig. A.67: The TimeWheel's central axis represents time. The axes in the periphery represent time-dependent variables. Here we see the number of cases for eight diagnoses. ⓒⓘ *The authors. Generated with the VisAxes software by Christian Tominski.*

Tominski et al. (2004) describe the TimeWheel as a technique for visualizing multiple time-dependent variables. Similar to the Parallel Coordinates technique for general multivariate data, the TimeWheel consists of several axes that are connected via lines. There is one time axis and multiple data axes for the data variables. The time axis is placed in the center of the display to emphasize the temporal character of the data. The data axes are associated with individual colors and are arranged circularly around the time axis. In order to visualize data, lines emanate from the time axis to each of the data axes to establish a visual connection between points in time and associated data values. These lines form visual patterns that allow users to identify positive or negative correlations with the time axis, trends, and outliers. Such patterns can be best discerned for those data axes that are parallel to the time axis. To bring data axes of interest into this focus, users can rotate the TimeWheel. Focused data axes are further emphasized by stretching them, effectively providing them with more drawing space. Data axes that are perpendicular to the time axis are more difficult to interpret and are, therefore, attenuated using color fading and shrinking. Interactive exploration, including navigation in time, is supported through different types of interactive axes. The idea of using linked interactive axes has been generalized by Claessen and van Wijk (2011). They propose a general model for linked axes that can be used to define a variety of axes-based visualization techniques of which the TimeWheel is only one example.

mapping of time: **static**
dimensionality: **2D**

Relevant references: Tominski et al. (2004) • Claessen and van Wijk (2011)

LiveRAC

time

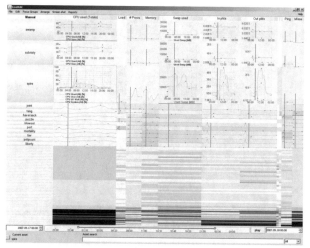

primitives: **points**
arrangement: **linear**

Fig. A.68: A full day of system management time-series data showing more than 4000 devices in rows and 11 columns representing groups of monitored parameters. Representations within cells adapt to the available screen space by using different representations with more or less detail. © *2008 ACM. Reprinted, with permission, from McLachlan et al. (2008).*

data

LiveRAC is a system for analyzing system management time-series data. It scales to dozens of parameters collected from thousands of network devices. Familiar representations such as line plots (↪ p. 233), bar graphs (↪ p. 234), and sparklines (↪ p. 235) appear as the cells of a spreadsheet-like matrix. Rows and columns of the matrix are associated with monitored network devices and monitored parameters, respectively. Each cell contains an area-aware chart showing time on the horizontal axis and parameters on the vertical axis. To ensure that all cells remain visible at all times (i.e., to avoid scrolling), LiveRAC uses a so-called *stretch and squish* layout, which dynamically compresses and expands cells according to user interaction. Moreover, the individual charts adapt to the available screen space. This semantic zoom functionality ranges from charts with detailed labels, to smaller charts with fewer curves and less labeling, and ultimately to colored blocks for the smallest view. The cell background color represents changeable thresholds of minimum, maximum, or average values of the displayed parameters. Aggregation is applied if cells would overlap due to space restrictions, which is reflected in color intensity.

number of variables: **multiple**
frame of reference: **abstract**

vis

Relevant references: McLachlan et al. (2008)

mapping of time: **static**
dimensionality: **2D**

CareCruiser

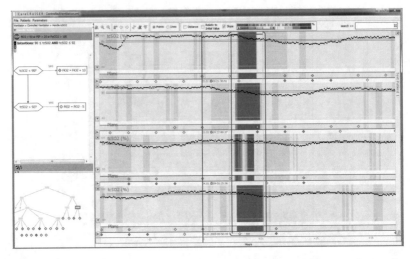

Fig. A.69: A patient's parameters are displayed together with the applied clinical actions. In the selected area on the right, a delayed drop of the patient's $tcSO_2$ values after applying a specific clinical action is revealed. Contextual views are shown on the left – top left: flow-chart like representation of the treatment plan logic; bottom left: hierarchical decomposition of the treatment plan. ©① *The authors. Generated with the CareCruiser software.*

CareCruiser by Gschwandtner et al. (2011) is a visualization system for exploring the effects of clinical actions on a patient's condition. It supports exploration via aligning, color highlighting, filtering, and providing focus and context information. Aligning clinical treatment plans vertically supports the comparison of the effects of different treatments or the comparison of different effects of one treatment plan applied to different patients. Three different color schemes are provided to highlight interesting portions of the development of a parameter: highlighting the distance of the actual values to the intended value helps to identify critical values; highlighting the progress of the actual values relative to the initial values shows to what extent the applied treatment plan has the intended effect; and highlighting the slope of a value helps to explore the immediate effects of applied clinical actions. A range slider is provided (at the top of the interface) to filter the color highlighting for selected events, and a focus window that grays out the color information outside its borders is used to support a focused investigation of a region of specific interest.

Relevant references: Gschwandtner et al. (2011)

Braided Graph

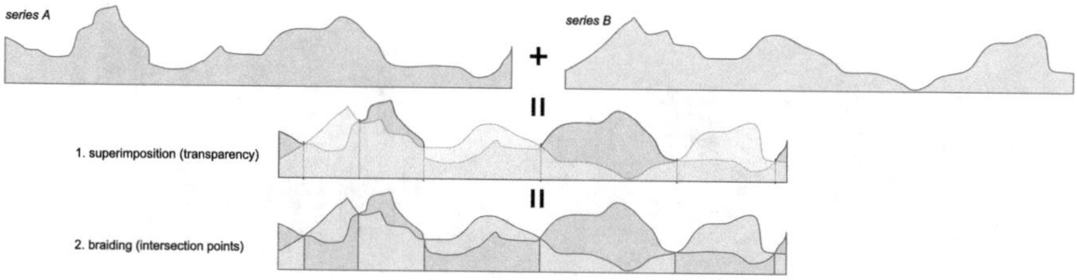

Fig. A.70: Construction scheme of a braided graph for two variables. Top: individual variables as silhouette graphs; center: superimposed silhouette graphs using transparency (dashed lines show intersection points); bottom: braided graph (segments between intersection points are sorted to ensure visibility of all fragments). © *2010 IEEE. Reprinted, with permission, from Javed et al. (2010).*

Braided graphs are a technique for superimposing silhouette graphs to show multivariate data. They take advantage of the expressiveness of silhouette graphs (↪ p. 281) and at the same time avoid the disadvantage of varying baselines of layered graphs (↪ p. 289). Simply drawing silhouette graphs on top of each other would lead to occlusion problems where a silhouette for larger data values occludes silhouettes for smaller values. The solution to this problem is to identify the points at which silhouettes intersect and to adapt the drawing order in between two intersections individually so that smaller silhouettes are always in front of larger ones. This ensures that all segments of all variables remain visible for the complete time series. In a user study, Javed et al. (2010) compared line plots (↪ p. 233), silhouette graphs (↪ p. 281), horizon graphs (↪ p. 277), and braided graphs along the three tasks of determining local maxima, comparing global slopes, as well as locating and comparing values at specific time points. Besides the visualization type, the number of displayed time series and the height of the representation were varied. Interestingly, the type of visualization was not found to have a significant effect on task correctness in all conditions. However, subjects using line plots and braided graphs were significantly faster when searching for local maxima. For value and slope comparison tasks this was not the case. In general, higher numbers of time series caused decreased correctness and increased completion time. Decreasing display space had a negative effect on correctness but little impact on completion time.

Relevant references: Javed et al. (2010)

primitives: **points**
arrangement: **linear**

number of variables: **multiple**
frame of reference: **abstract**

mapping of time: **static**
dimensionality: **2D**

CiteSpace II

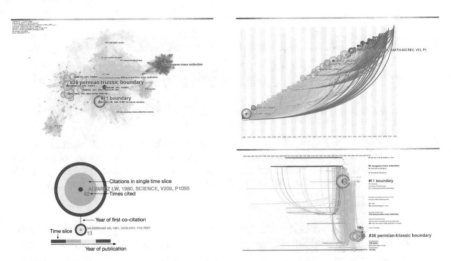

Fig. A.71: Visualization of a network of 750 most cited articles on mass extinction (1980–2010). Top left: cluster view; bottom left: legend – node size reflects overall amount of citations and colored rings show citations per time slice; top right: timezone view; bottom right: timeline view. © *Courtesy of Chaomei Chen.*

CiteSpace II by Chen (2006) is a system that supports the visual exploration of bibliographic databases. It combines rich analytic capabilities to analyze emerging trends in a knowledge domain with interactive visualization of co-citation networks. Three complementary views are provided for the visual representation: a cluster view, a time-zone view, and a timeline view. The cluster view represents a network as a node-link diagram using a force-directed layout. Node size shows how often an article or cluster was cited overall and citation tree rings of a node display the citation history from the center outward. The color of a ring represents a time period and its thickness is proportional to the number of citations in this period. The colors of links represent the time slice of the first co-citation. The time-zone view displays a network by arranging its nodes along vertical strips representing time zones using a modified spring-embedder layout that controls only the vertical positions of nodes freely. In the timeline view, time is mapped to the horizontal position and clusters are arranged along horizontal lines. Users can adjust a complex set of parameters to control the analysis process as well as interact and manipulate the visualization of a knowledge domain. CiteSpace II also provides clustering and labeling functions to help the user interpret various structural and temporal patterns.

Relevant references: Chen (2006)

ChronoLenses

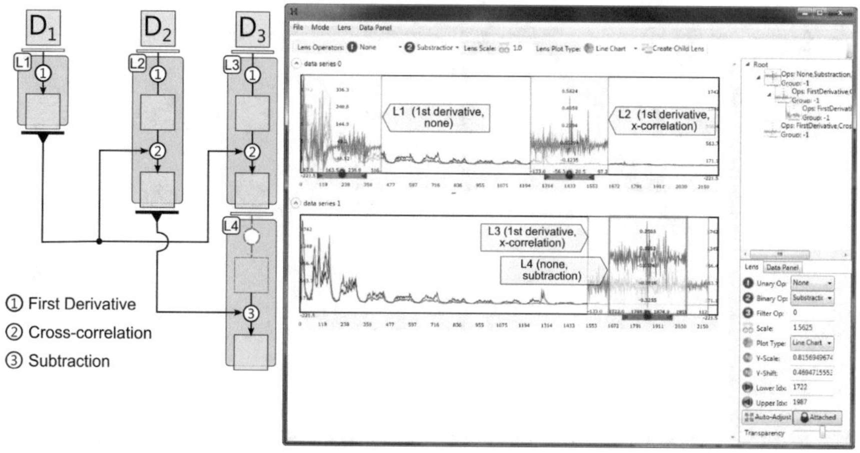

Fig. A.72: ChronoLenses define pipelines of transformations (left) to be applied to selected intervals of time-series data (right) in order to support various exploratory visual analysis tasks. © *Courtesy of Jian Zhao.*

ChronoLenses by Zhao et al. (2011b) is domain-independent time-series visualization technique supporting users in exploratory visual analysis tasks, such as visualizing derived values, identifying correlations, or discovering anomalies beyond obvious outliers. Such tasks often require carrying out elaborate transformations on the original time series. Tightly integrating visual analysis with interaction based on direct manipulation, ChronoLenses perform such transformations on-the-fly on an existing line plot (↪ p. 233) of the data. Users can build pipelines (shown left in the figure) composed of lenses performing various transformation functions, effectively creating flexible and reusable time-series visual analysis interfaces. At any moment, users can change the parameters of already created lenses, with the modifications instantaneously propagating down through the pipeline, providing immediate visual feedback that supports the iterative exploration process. Applied to the visualization, ChronoLenses locally enhance the original time series with the results of the transformation (intervals marked in blue in the main interface) allowing users to gain a deeper insight into the data.

Relevant references: Zhao et al. (2011b)

---

time

primitives: **points**
arrangement: **linear**

data

number of variables: **multiple**
frame of reference: **abstract**

vis

mapping of time: **static**
dimensionality: **2D**

Connected Scatterplot

time

primitives: **points**
arrangement: **linear**

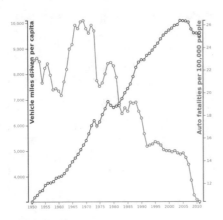 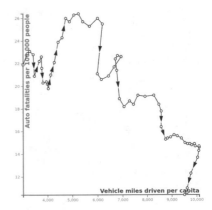

data

Fig. A.73: Comparison of a regular line plot (left) with a connected scatterplot (right). Both plots show the same data, but lead to different visual representations, which has an impact on how the data are interpreted. ⓒⓘ *The authors. Generated with the connected scatterplot software by Steve Haroz.*

number of variables: **multiple**
frame of reference: **abstract**

Connected scatterplots, also known as scatter line graphs (see Harris, 1999), are a classic visualization technique for time-oriented data and have been used widely in news media for the purpose of storytelling. Connected scatterplots are a variant of standard plot techniques such as point plot (↪ p. 232) and line plot (↪ p. 233). Traditional time-series plots show time on the horizontal axis and time-dependent variables on the vertical axes (see left-hand plot). In contrast, a connected scatterplot maps time-dependent variables to the vertical axis and the horizontal axis, which corresponds to classic scatterplots (which do not include time per se). For connected scatterplots, the dots in the plot are connected via lines or arrows based on the temporal order of the data points. That is, one can see the order of data points (qualitative information), but one cannot tell the temporal distance between them (quantitative information). The figure compares the line plot (left) and connected scatterplot (right) of two variables related to driving safety. Haroz et al. (2016) conducted several studies to test how well connected scatterplots can be interpreted. They found that plots of data with low complexity can be interpreted quite easily. Yet they also found issues with misinterpretations of connected scatterplots and suggest guidelines to circumvent or mitigate them. One such guideline is to clearly communicate the order of time, for example, by including further visual cues in addition to arrows.

vis

mapping of time: **static**
dimensionality: **2D**

Relevant references: Harris (1999) • Haroz et al. (2016)

DimpVis

time

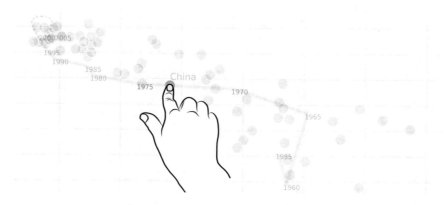

Fig. A.74: The DimpVis technique makes it possible to explore the temporal dimension even when time is not present in the original visual representation. This is achieved by dragging data items along hint paths that indicate the data item's position at different points in time. ⓒⓘ *The authors. Generated with the DimpVis software by Kondo and Collins (2014).*

DimpVis by Kondo and Collins (2014) is special in that it is not really a visualization technique for time-oriented data, but an interaction technique that makes the time dimension explorable even when time is not present in the visual representation. To this end, data items are visualized as graphical objects that can be interacted with via direct manipulation. Once a data item is picked, visual hint paths are displayed to indicate where the data item would be located at different points in time. The hint path can be a timeline connecting time points in sequence or a flashlight providing links to spatially adjacent time points. The figure shows how a timeline path can look like for a scatterplot visualization. Performing a drag gesture along the hint path allows the user to explore the dimension of time while the entire visualization (not just the dragged data item) is updated automatically. In this sense, the DimpVis approach is similar to animated scatter plots (↪ p. 330) where, however, the animation is replaced by the interactive dragging. DimpVis is applicable not only to scatterplots but to a variety of visualization techniques. Kondo and Collins (2014) demonstrate DimpVis for bar charts, pie charts, and heatmap visualizations. Their user study found that DimpVis is a suitable solution for a variety of visualization tasks and may even be preferable over standard solutions such as time sliders or small multiples (↪ p. 359).

Relevant references: Kondo and Collins (2014)

primitives: **points**
arrangement: **linear**

data

number of variables: **multiple**
frame of reference: **abstract**

vis

mapping of time: **static**
dimensionality: **2D**

FluxFlow

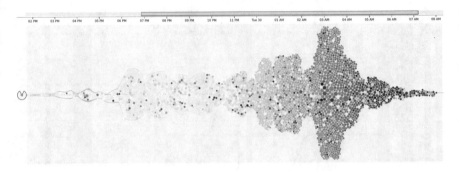

Fig. A.75: FluxFlow's thread timeline visualization (called volume circle view) represents the top 100 ranked anomalous retweeting threads during the Hurricane Sandy in 2012. The background colors illustrate the eight hidden state variables generated by the model. The circles indicate the participating Twitter users in the thread using a circular glyph to visually summarize important aspects of a retweeting thread. ⓒ① *The authors. Generated with the FluxFlow software by Jian Zhao.*

FluxFlow by Zhao et al. (2014) is an interactive visual analysis system for revealing and analyzing anomalous information spreading in social media. The challenge is to distinguish anomalous information behavior, such as the spread of rumors or misinformation, from normal behavior, such as popular topics and newsworthy events. FluxFlow incorporates machine learning algorithms to detect such behavior and novel visualization techniques for in-depth analysis. Consequently, FluxFlow provides (1) a thread glyph, which visually aggregates important aspects of a retweeting thread, (2) a volume chart, (3) a linear circle view, and (4) a volume circle view. Various interactively coordinated user interface components ease the exploration process. Through quantitative evaluation of the model and qualitative interviews with three domain experts, study results indicated that FluxFlow's anomaly detection algorithm is efficient in identifying misinformation, and the visualization is useful for analysts to discover insights and comprehend the model.

Relevant references: Zhao et al. (2014)

Line Density Plot

time

primitives: **points**
arrangement: **linear**

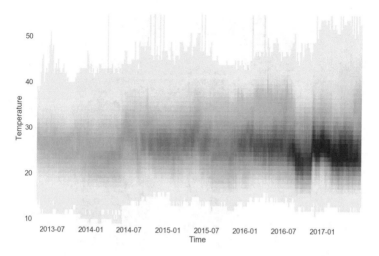

Fig. A.76: Line density plot of the temperature of more than 100.000 hard drives over a period of four years. Yellow to violet colors indicate low to high density of the underlying time series. ©① *Courtesy of Dominik Moritz.*

data

number of variables: **multiple**
frame of reference: **abstract**

Datasets that consist of very many time series are difficult to visualize using the standard line plot (↪ p. 233) technique. The reason is that for each time series a separate curve must be drawn in the plot. For large numbers of time series, this not only drains the rendering performance, but also leads to cluttered visual representations that are difficult to decipher. The idea of line density plots is to visualize not individual lines, but the density of lines on the display. Moritz and Fisher (2018) describe the steps that are necessary to construct a line density plot. First, the time axis and the value axis of the plot are subdivided into bins. As a result, the plot area is defined as grid of discrete cells. Then a time series is rendered as a polyline onto the grid, but only virtually so. Instead of actually drawing onto the grid, the grid cells just store the value of 1 if the polyline passes through them. The stored values are then normalized on a per column basis. This is done for all time series. In the end, one obtains a grid where cells that are passed by many lines contain a large density value, whereas cells passed by only a few lines have a lower value stored in them. The density values are then mapped to colors using an appropriate color scheme. The figure shows an example where low density is indicated by yellow, medium density is shown in green, and high density values are represented in violet. From the density plot, one can easily see where the majority of data values lie. Also, trends can be discerned, for example, the increasing accumulation of high density around the mid-twenty degrees temperature toward the right end of the time axis.

vis

mapping of time: **static**
dimensionality: **2D**

Relevant references: Moritz and Fisher (2018)

Matrix-Based Comparison

primitives: **points**
arrangement: **linear**

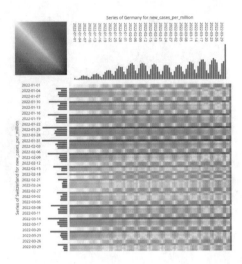

Fig. A.77: Matrix-based visual comparison of time-series data. The example shows data about CoViD cases over three months in Germany (horizontal axis) and Switzerland (vertical axis). ⓒ① *Courtesy of Sarah Clavadetscher.*

number of variables: **multiple**
frame of reference: **abstract**

Behrisch et al. (2012) and Beck et al. (2016) introduce a matrix-based visualization approach for comparing two time series. The idea is to map the two time series to the vertical and horizontal axes of a matrix. The figure shows an example of CoViD data for three months from 2022. We see the number of cases of CoViD in Germany along the horizontal axis, whereas the vertical axis shows the data of Switzerland. Each cell of the matrix corresponds to a pair of time points and their associated data values. The cells are color-coded to communicate two pieces of information: a difference measure δ representing the difference between two time points and an aggregation measure σ representing the combined value of two data points. In the example, δ is encoded by varying the hue from red via neutral hue-less gray to blue. Color brightness encodes σ, where brighter colors correspond to smaller aggregated values and darker colors stand for larger values. A non-linear mapping function is used to make colors better distinguishable. The resulting two-dimensional color map is shown in the top-left corner. The advantage of this color-coding is that it can directly visualize the difference between two values and also provide an indication of the overall magnitude of two values. In the figure, dark and saturated red cells represent pairs of time points where Switzerland had a higher number of cases, while the cases in both Switzerland and Germany were generally high. Brighter less saturated blue cells indicate that Germany had a slightly higher number of cases, while the numbers were generally low in both countries.

mapping of time: **static**
dimensionality: **2D**

Relevant references: Behrisch et al. (2012) • Beck et al. (2016)

MultiStream

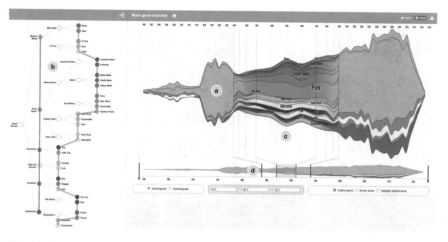

Fig. A.78: MultiStream is a streamgraph-based visualization approach (a) that incorporates the hierarchical structure of multiple time series (b) and adds a focus+context view (c) as well as overview+detail navigation (d). The example shows the evolution of different music genres over time. ⓒⓘ *The authors. Generated with the MultiStream prototype by Erick Cuenca.*

Streamgraph ($\hookrightarrow$ p. 286) representations are well suited for representing multiple time series that can be stacked, i.e., with a meaningful sum. However, they do not scale very well when a large number of time series or longer time intervals need to be displayed. MultiStream by Cuenca et al. (2018) has been designed to address both of these challenges. The approach makes use of the hierarchical organization of the time series, e.g., music genres and sub genres, in order to provide less cluttered overviews that still convey the most important information. The MultiStream interface is split into multiple components: a streamgraph (a), a hierarchy manager (b), a multi-resolution focus+context view (c), and an overview+detail view with navigation (d). The hierarchy manager on the left allows navigating through the hierarchical structure of the data. The multi-resolution view on the top right represents the data at three different levels of detail – full temporal and hierarchical information in the central focus area as well as less detailed and more compressed information in the two context areas to the left and right. Finally, the overview+detail view on the bottom right shows a course overview of the whole dataset while allowing one to navigate in time and select the time interval shown in the multi-resolution view.

Relevant references: Cuenca et al. (2018)

time

primitives: **points**
arrangement: **linear**

data

number of variables: **multiple**
frame of reference: **abstract**

vis

mapping of time: **static**
dimensionality: **2D**

netflower

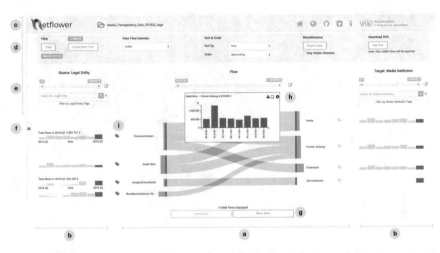

Fig. A.79: Netflower allows exploring the temporal dynamics of flows in bipartite networks. Its central visual metaphor is based on Sankey diagrams. The example shows quarterly money flows between government institutions (nodes on the left) and media outlets (nodes on the right). ⓒⓘ *The authors. Generated with the netflower prototype by Stoiber et al. (2019).*

Netflower supports the visual exploration of flows in dynamic networks, such as money flows between organizations or migration flows between countries. The central visualization (a) is a Sankey diagram (↪ p. 313) whereas the left side shows the originating nodes and the right side the destination nodes of a bipartite network. The width of each connecting line is proportional to the flow quantity as sum over the selected time interval. Netflower particularly aims for scalability to large graphs and supporting the exploration of the temporal dynamics of flows using details-on-demand, filters and tags, as well as auxiliary views. To keep the initial amount of elements shown on the screen manageable when loading new data, only the first dozen of elements are represented in full detail while more elements are shown upon user request (g). To signify that only a part of the total flows for each node is shown, a node's area is split into a filled and a hatched area, whereas the latter represents the portion of flows not currently displayed on the screen. Netflower provides functions for filtering, sorting and ordering, as well as a provenance notebook (d-f, i). The time interval to be shown in the Sankey view can be selected using a filter (d). Upon selection of a flow, a bar graph (↪ p. 234) shows the flow values for each time point in the data (h). To the left and right of the nodes, bar graph sparklines (↪ p. 235) on both sides (b) give a temporal overview of outgoing (source nodes) and incoming flows (target nodes).

Relevant references: Stoiber et al. (2019)

Optimized Stream Graphs

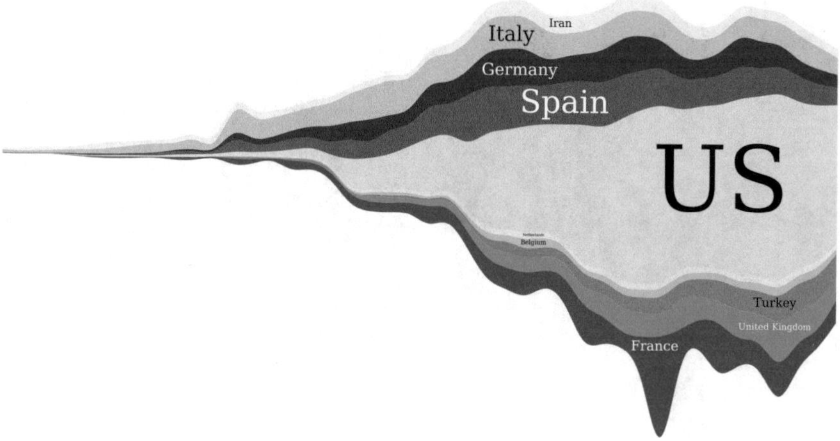

Fig. A.80: Optimized stream graphs adjust the baseline and the stacking order of stacked graphs to reduce sine illusion when interpreting the data. ⓒⓘ *The authors. Generated with the SineStream software by Chuan Bu.*

Visualization techniques that represent individual time-dependent variables as layers (↪ p. 289) and also stack these on top of each other like ThemeRiver (↪ p. 293), 3D ThemeRiver (↪ p. 317), and generally stacked graphs (↪ p. 286) inevitably suffer from sine illusion. While said visualization techniques map data values by varying the height of the layers, human perception tends to interpret not the height of layers with respect to the vertical y-axis only, but the "width" of layers with respect to their main direction of flow, more exactly perpendicular to the perceived main direction. This perceptual "illusion" can have adverse consequences for the extraction of individual data values from the visualization. Therefore, optimization techniques have been developed to reduce these adverse effects by appropriately arranging the layers in the visualization. This can be achieved by adjusting the baseline of the visualization and the stacking order of layers. Byron and Wattenberg (2008), Bartolomeo and Hu (2016), and Bu et al. (2021) proposed several optimization strategies with different trade-offs and suitability for different types of data. Their studies show that optimized stream graphs improve the perception of data and make it easier to extract data values from the visualization.

Relevant references: Byron and Wattenberg (2008) • Bartolomeo and Hu (2016) • Bu et al. (2021)

time

primitives: **points**
arrangement: **linear**

data

number of variables: **multiple**
frame of reference: **abstract**

vis

mapping of time: **static**
dimensionality: **2D**

time

RankExplorer

primitives: **points**
arrangement: **linear**

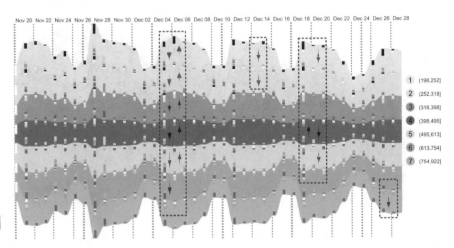

data

Fig. A.81: RankExplorer visualization of a subset of the top 2000 search keywords on Bing. Each colored layer summarizes multiple keywords. Small bars and glyphs indicate how keywords switch layers from one time step to the next. ⓒ⊕ *Courtesy of Conglei Shi.*

number of variables: **multiple**
frame of reference: **abstract**

When it comes to understanding the ranking of many time series, the RankExplorer by Shi et al. (2012) offers a useful solution. The ranking per time point depends on which time series have the highest, second highest, third highest value, and so on. In other words, the sorted data values of a time point define the ranking order for that time point. It is now interesting to analyze how the ranking order changes over time. As this would be too complex on the level of individual time series, the RankExplorer first groups the time series into a smaller number (seven in the figure) of ranking segments or layers. These ranking layers are represented as differently colored, layered bands of varying width along a horizontal time axis, which is analog to the ThemeRiver (↪ p. 293) approach. Changes in the ranking occur when time series leave one layer and enter another one. These changes are represented by small colored bars within the bands. The bars indicate (by the proportion of colors per bar) how many time series enter a layer from the previous time point and how many leave a layer toward the next time point. Additionally, small glyphs (see arrow-like shapes in dashed frames) visualize the content changes between ranking layers. For detailed inspection, users can zoom in to be provided with more fine-grained ranking layers, bars, and glyphs.

vis

mapping of time: **static**
dimensionality: **2D**

Relevant references: Shi et al. (2012)

Sankey Diagram, Alluvial Diagram

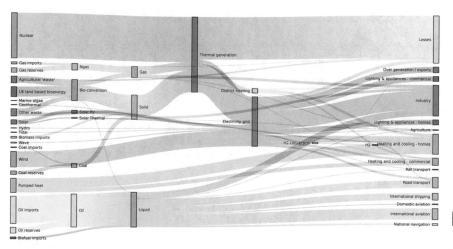

Fig. A.82: Sankey diagram showing how energy in the UK is converted or transmitted before being consumed or lost. Supplies are on the left and demands are on the right. The connecting lines visualize energy flows between stages and line thickness is proportional to the flow quantity. ⓒⓘ *The authors. Generated with the Sankey Diagram notebook by Mike Bostock.*

primitives: **points**
arrangement: **linear**

Sankey diagrams are used to represent flows of data such as money, people, energy, or material in a system by means of their input and output flow distributions. The width of the lines or arrows connecting different nodes or stages is proportional to the flow quantity. In most cases, Sankey diagrams do not reflect time via absolute position but as a sequence of stepwise changes over time, i.e., they use an ordinal time scale. Originally developed for depicting energy flows, its basic principle is very versatile and can be applied to all kinds of flow data in different application domains. A historical example of the application of this approach is Minard's graphical representation of Napoleon's Russian campaign that was created in 1869 (see Chapter *Historical Background*). In addition to changes in line thickness, Minard also used color hue as an additional visual variable for representing the direction of movement. Interestingly, this form of graphic representation is named after Irish Captain Matthew Henry Phineas Riall Sankey who only used this visualization method once in 1898 to represent energy flows in steam engines in Kennedy and Sankey (1898). *Alluvial diagrams* by Rosvall and Bergstrom (2010) are a special form of Sankey diagrams intended to represent structural changes in large complex networks over time, that is, they can be used to show changes in group composition over time. Brinton (1939), the representation concept is called *Cosmograph* and focuses on the comparison of source and destination flow distributions.

number of variables: **multiple**
frame of reference: **abstract**

Relevant references: Kennedy and Sankey (1898) • Rosvall and Bergstrom (2010) • Brinton (1939)

mapping of time: **static**
dimensionality: **2D**

TACO

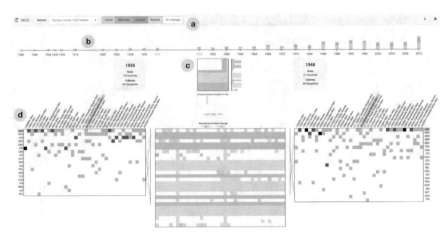

Fig. A.83: Visualization of changes in tables over time using TACO. Multiple levels of detail are employed along an overview+detail concept: (a) selection of types of changes to be displayed, (b) aggregated changes between two consecutive time points, (c) more detailed comparison of two selected time points showing changes and their distribution for rows and columns, (d) a detailed view of raw data table versions (bottom left and bottom right) as well as a diff heatmap (bottom center). The example shows differences between Summer Olympic Games medal tables over time between 1896 and 2012. ⓒⓘ *The authors. Generated with the TACO demo by Niederer et al. (2018).*

Tables may change over time in both structure and content, resulting in multiple versions. A challenging task is to understand what exactly has changed between versions (additions/deletions, reorder, merge/split, and content changes). TACO addresses this by visualizing the differences between multiple tables at various levels of detail. For this, it allows users to gradually add more focused and detailed views by interacting with the interface along three levels of detail. The interface contains a change overview timeline for aggregated differences between multiple table versions over time (level 1), allows for aggregated pairwise comparisons with a more detailed view of the distribution of changes along rows and columns (level 2), and shows a detailed pairwise comparison as well as the raw data (level 3). Across all views, color hue is used to differentiate the four different types of changes. On level 1, stacked bar graphs (↪ p. 234) on a timeline provide an overview of when changes to a table occurred along with the amount and types of changes between consecutive points in time. For the second level, a 2D ratio chart and additional histograms for rows and columns provide more details about changes between two selected time points. On the third level, a detailed view shows heatmap representations of the two selected raw data table versions (left and right) as well as a diff heatmap in the center. Reordering changes between rows are shown with rank charts.

Relevant references: Niederer et al. (2018)

WireVis

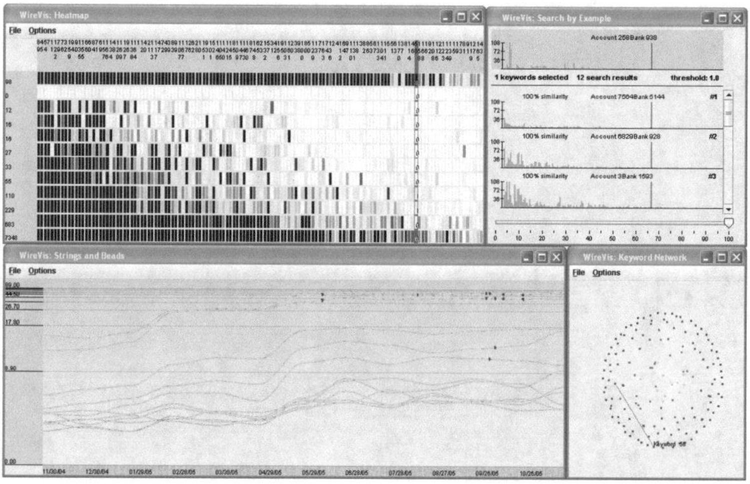

time

primitives: **points**
arrangement: **linear**

Fig. A.84: WireVis's multiple coordinated view interface capturing four views to explore complex numerical and categorical time-oriented transaction data: heatmap, search-by-example, strings-and-beads view, and keyword network view. © *2007 IEEE. Reprinted, with permission, from Chang et al. (2007).*

data

Financial transaction data capture complex numerical and categorical time-oriented data, which need to be investigated for suspicious behavior (fraudulent activities, which are changing over time). WireVis by Chang et al. (2007) is a visual analytics approach that utilizes four tightly coordinated views of transaction activity. A heatmap view shows relationships between accounts and keywords. A search-by-example approach supports analysts to discover accounts with similar activities. A strings-and-beads view depicts the transactions over time. And finally, the keyword network view is used to represent the relationships between keywords. All four views rely on high interactivity along with the ability to see global trends and capabilities to drill down into specific transaction records. The usefulness of WireVis was demonstrated in case studies using transaction data of the Bank of America.

number of variables: **multiple**
frame of reference: **abstract**

vis

Relevant references: Chang et al. (2007)

mapping of time: **static**
dimensionality: **2D**

MOSAN

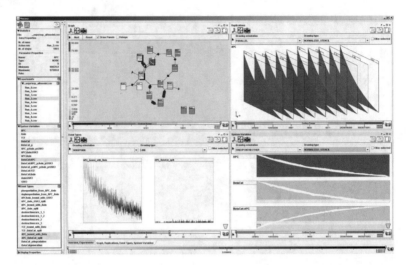

Fig. A.85: The top-left view shows a simulated reaction network. An overview of the time-dependent simulation data is given by the small line plots in the boxes of the network nodes. Three coordinated linked views are provided for the comparison of simulation runs and variables. © *Courtesy of Andrea Unger.*

MOSAN is a tool for visualizing multivariate time-oriented data that result from simulation of reaction networks. Due to the stochastic multi-run simulation, each variable comprises multiple time series. In order to facilitate the understanding of the complex dependencies in the data, it is necessary to jointly visualize structural information and stochastic simulation data together. To this end, Unger and Schumann (2009) combine different views within a single interactive interface. In an overview, time-oriented data are shown along with the structural relations among the variables in the reaction network. The structural relations are shown by a graph layout, where boxes correspond to variables, and simulation data are visualized by small line plots within the boxes (top-left view). The small line plots provide a highly aggregated view of the stochastic simulation data, thus focusing on the communication of the general temporal trends. Furthermore, advanced color-coding is applied to the plots to support the comparison of heterogeneous value ranges among variables. In addition to the overview, coordinated linked views support the inspection of individual time series of the same variable (top right) as well as the detailed inspection and comparison of temporal developments of different variables selected from the overview (bottom). Filter sliders further support the drill down into the data.

Relevant references: Unger and Schumann (2009)

3D ThemeRiver

Fig. A.86: Distinctly colored currents form the overall shape of the 3D ThemeRiver. The width, and additionally the height of currents is varied to visualize time-oriented data. In this figure, width encodes the overall distribution of 17 clusters of aerosol data and height indicates the incidence of zinc. © *2003 IEEE. Reprinted, with permission, from Imrich et al. (2003).*

Imrich et al. (2003) propose a 3D variant of the ThemeRiver technique (↪ p. 293). The 3D approach inherits the basic visual design from its 2D counterpart: multiple time-oriented variables are encoded to the widths of individually colored currents that form a river flowing through time along a horizontal time axis. In the 2D variant, only one data variable can be visualized per current, namely by varying the current's width. Imrich et al.'s extension addresses this limitation. By extending the design to the third dimension it is possible to use an additional visual encoding: the height (in 3D) of a current can be varied to encode further information. This design is particularly suited to visualizing ternary covariate trends in the data. Imrich et al. conducted user tests to evaluate the usefulness of the 3D encoding, and indeed got positive results that indicate that the 3D variant has advantages over the 2D variant. Specifically, the availability of appropriate interactive 3D navigation tools is highlighted as an important factor contributing to the success of the 3D ThemeRiver.

Relevant references: Imrich et al. (2003)

time

primitives: **points**
arrangement: **linear**

data

number of variables: **multiple**
frame of reference: **abstract**

vis

mapping of time: **static**
dimensionality: **3D**

Data Tube Technique

time

primitives: **points**
arrangement: **linear**

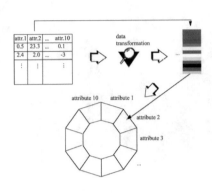

exceptional stock   crash

data

number of variables: **multiple**
frame of reference: **abstract**

Fig. A.87: Data are mapped onto the inside of a 3D tube using a tabular layout. Each slice of the tube represents a time point and each cell represents a data parameter by color. Left: visual mapping schema; right: exploring 50 different stocks. © *Courtesy of Mihael Ankerst.*

In the data tube technique by Ankerst (2001) multiple time-oriented variables are mapped to bands that follow the inside of a 3D tube. The basic mapping procedure is depicted in the left part of the figure. Each slice of the tube represents a time point and each cell represents a data value by color. The tube is viewed from above and time is flowing to or from the center of the tube. The user is able to explore the data by interactively moving through the 3D tube. Because of the 3D perspective distortion, cells that are further away appear to be smaller in size, much like in a focus+context display. As a result of this, the number of displayed variables and the number of displayable time points can be quite large. Later, Ankerst et al. (2008) also developed a comprehensive temporal data mining architecture called DataJewel that is closely integrated with pixel-oriented visualization techniques. Further visual and interactive extensions have been developed by Sureau et al. (2009) and Bouali et al. (2016) in DataTube2. One such extension detailed by Devaux et al. (2014) is to include 3D link arcs into the tube to support the visualization of time-oriented log data.

Relevant references: Ankerst (2001) • Ankerst et al. (2008) • Sureau et al. (2009) • Devaux et al. (2014) • Bouali et al. (2016)

vis

mapping of time: **static**
dimensionality: **3D**

Kiviat Tube

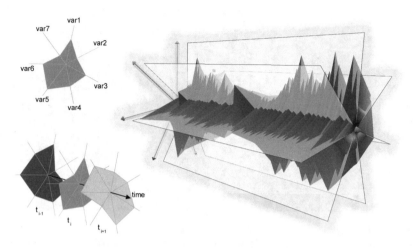

Fig. A.88: Construction of a three-dimensional Kiviat tube representing seven time-dependent variables. Peaks and valleys indicate ups and downs in the evolution of the data over time. Wings assist in associating features of the Kiviat tube to particular variables in the data. ⓒ① *The authors. Generated the VisAxes3D software by Clemens Holzhüter.*

The Kiviat tube described by Tominski et al. (2005a) visualizes multiple time-dependent variables. The construction of a Kiviat tube is as simple as stacking multiple Kiviat graphs (see Kolence and Kiviat, 1973) along a shared time axis. Each Kiviat graph represents the data for multiple variables for a specific point in time. But instead of drawing individual Kiviat graphs, a three-dimensional surface is constructed. This way, multiple, otherwise separated time points are combined to form a single 3D body that represents the dataset as a whole. The spatial characteristics of a Kiviat tube can be recognized easily, as it allows users to identify peaks or valleys in the data over time. Additional semi-transparent wings assist in relating identified patterns to particular variables. Similar wings were also employed by Akase and Okada (2015). Common interaction methods can be used for zooming and rotation around arbitrary axes. Rotation specifically around the time axis enables users to quickly access variables on all sides of the Kiviat tube. Interactive axes allow users to navigate back and forth in time to visit different portions of a possibly large time series.

Relevant references: Tominski et al. (2005a) • Kolence and Kiviat (1973) • Akase and Okada (2015)

time

primitives: **points**
arrangement: **linear**

data

number of variables: **multiple**
frame of reference: **abstract**

vis

mapping of time: **static**
dimensionality: **3D**

Temporal Star

primitives: **points**
arrangement: **linear**

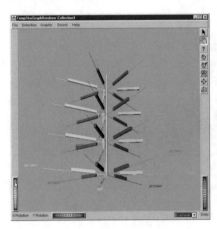 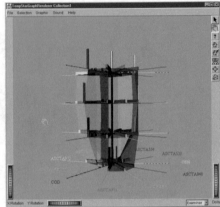

data

Fig. A.89: 3D representation of circular column graphs that are arranged in a row to represent each time step. A transparent veil can be displayed to enhance the perception of the dataset's evolution. © *Courtesy of Monique Noirhomme.*

number of variables: **multiple**
frame of reference: **abstract**

The temporal star technique by Noirhomme-Fraiture (2002) visualizes multivariate data structures in 3D. For each point in time, a circular column graph is drawn that represents each variable's value as a bar in a circular arrangement. These graphs are aligned in a row to represent the development of the dataset over time (left figure). A unique color is assigned to each variable to aid the recognition of variables across time. Moreover, a transparent veil can be displayed to enhance the perception of the dataset's evolution as a whole (green parts in the right figure). The concept used is similar to that of the Kiviat tube ($\hookrightarrow$ p. 319), which uses Kiviat graphs instead of circular column graphs. In the temporal star technique, difference plots are also integrated, showing the relative differences between variables rather than absolute values. The rendering parameters, the shown time intervals, and the configuration of the axes can be adjusted interactively. Furthermore, the temporal star technique is integrated with a data warehousing application that provides rich data manipulation features.

mapping of time: **static**
dimensionality: **3D**

Relevant references: Noirhomme-Fraiture (2002)

Time-tunnel

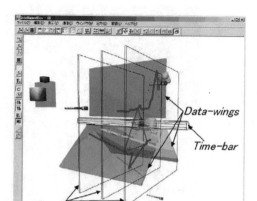

time

primitives: **points**
arrangement: **linear**

data

number of variables: **multiple**
frame of reference: **abstract**

vis

mapping of time: **static**
dimensionality: **3D**

Fig. A.90: Time-tunnel visualization with three time-dependent variables. Left: individual plots are put onto semi-transparent planes (data wings) that are positioned around a central time bar; top right: selected time points can be viewed as radar charts; bottom right: multivariate perspective on the data. © *2004 IEEE. Reprinted, with permission, from Akaishi and Okada (2004).*

Akaishi and Okada (2004) developed time-tunnel as a data analysis technique for visualizing a number of time-series plots in a 3D virtual space. The individual plots are put onto semi-transparent planes (data wings) that are positioned around a central time bar in a fan-like manner (left figure). In the example figure, line plots are used for visualizing the time-dependent data but any other linear time visualization might also be used. Multiple planes can be overlapped and compared thanks to their transparency. Furthermore, single time slices or selected time intervals can be viewed as radar charts (top right) and provide a multivariate point of view (bottom right). Additionally, not only time series, but any kind of information can be put onto a plane, making the time-tunnel a multimedia presentation tool.

Relevant references: Akaishi and Okada (2004) • Akase and Okada (2015)

Worm Plots

primitives: **points**
arrangement: **linear**

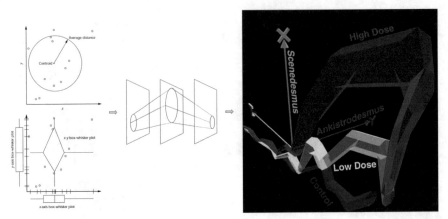

number of variables: **multiple**
frame of reference: **abstract**

Fig. A.91: Worm plots are generated by creating visual abstractions of point groups at multiple time steps and by assembling a 3D surface that resembles a worm. The worm plot on the right shows three groups (control, high dose, low dose) of data points of toxicology experiments plotted against the two variables Scenedesmus and Ankistrodesmus. © *1997 IEEE. Reprinted, with permission, from Matthews and Roze (1997).*

Worm plots have been developed to help scientists gain qualitative insights into the temporal development of groups of points in scatter plots. The initial step necessary to construct a worm plot is to generate a visual abstraction of multiple points. One way to do this is to compute the centroid of a group of points and the average distance of points to the centroid. The visual abstraction is then a circle that is located at the centroid and has a radius equal to the average distance. Alternatively, a 2D generalization of box-whisker plots can be used to form a diamond-shaped visual abstraction. Such abstractions are computed for each time step (i.e., each scatter plot). Subsequently, a three-dimensional surface (worm) is assembled from the visual abstractions of each group. This procedure is illustrated in the left part of the figure. Presented in an interactively manipulatable virtual world, worm plots allow users not only to see where in the variable space point groups are located, but also to discern the compactness of point groups, and to understand the development of these characteristics over time.

Relevant references: Matthews and Roze (1997)

mapping of time: **static**
dimensionality: **3D**

Software Evolution Analysis

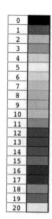 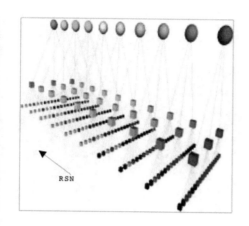 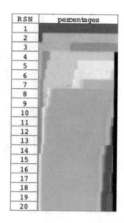

Fig. A.92: 3D visualization to analyze software systems and product families. Left: color legend for different versions; center: hierarchical decomposition by modules, packages, and files (color represents current version); right: individual file where each row represents a single version colored by percentage of code originating from a particular (previous) version. © *1999 IEEE. Reprinted, with permission, from Gall et al. (1999).*

The software evolution analysis technique by Gall et al. (1999) uses 3D representations to depict software systems or product families respectively. The information is decomposed hierarchically into modules, packages, and files or similar concepts. This hierarchy is depicted as a three-dimensional tree structure in which the leaf nodes represent individual files. Multiple such trees are aligned in layers in the 3D space, with one layer for each revision of the software. Color is used to distinguish different versions and to show changes over time. Furthermore, individual files might be inspected in more detail to explore the evolution of changes over time using proportionally colored version lines (right in the figure). This way, patterns are formed that can be used to identify, for example, stable parts of a system, frequently changed parts, similarities, and more.

Relevant references: Gall et al. (1999)

time

primitives: **points**
arrangement: **linear**

data

number of variables: **multiple**
frame of reference: **abstract**

vis

mapping of time: **static**
dimensionality: **3D**

3D TimeWheel

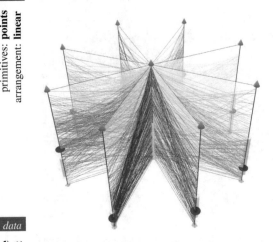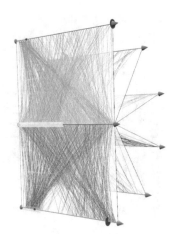

Fig. A.93: The 3D TimeWheel uses a central axis to represent time. Several axes are arranged around the time axis to visualize time-dependent variables. Left: radial layout; right: page-flip layout. ⓒ① *The authors. Generated with the VisAxes3D software by Clemens Holzhüter.*

The 3D TimeWheel by Tominski et al. (2005a) is a three-dimensional variant of the TimeWheel technique (↪ p. 298). Analogous to the original TimeWheel, the 3D TimeWheel arranges multiple axes of time-dependent variables around a central axis representing the dimension of time. Lines between axes connect time points to data values. Different arrangements of the axes are possible. The central time axis can be surrounded by evenly distributed time-dependent axes (left). This is similar to the classic Parallel Coordinates technique, with the only difference being that for Parallel Coordinates lines are drawn between adjacent axes, while for the 3D TimeWheel lines are always linked to the central time axis. For easier comparison of two selected time-dependent variables, it is also possible to create a distribution of axes that grants a full view on the axes of interest (right). Using a page-flip metaphor, the user can flip the axes panes like the pages of a book. Handles on the axes further support filtering and navigation in time. A study by Hassan et al. (2019) found that 3D representations can be beneficial for the visual exploration of time-oriented data.

Relevant references: Tominski et al. (2005a) • Hassan et al. (2019)

Vanishing-Point Plot

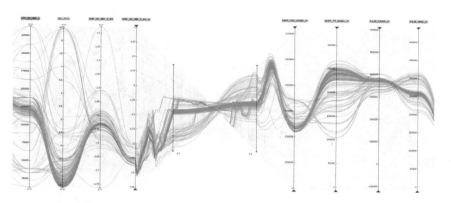

Fig. A.94: Vanishing-point plots add the time dimension to Parallel Coordinates plots by embedding additional 3D perspective time plots for selected data attributes. The example shows a crash simulation dataset containing design parameters and the simulated responses for each simulation run as axes. © *2016 Eurographics Association and John Wiley & Sons. Reprinted, with permission, from Gruendl et al. (2016).*

Parallel Coordinates by Inselberg and Dimsdale (1990) are one of the most well-known visualization techniques for exploring multivariate datasets. However, a remaining challenge is the integration of the time dimension in addition to showing multivariate relationships. One option is to add time as an additional axis to the Parallel Coordinates plot. But this leads to clutter as well as a loss of the temporal order for all axes not adjacent to the time axis. A second option is to use animation to dynamically visualize temporal developments. Yet, this makes comparisons between points in time difficult. As a third option, the vanishing-point plot adds time on demand between any two neighboring data variables. This is in contrast to the related TimeWheel (↪ p. 298) and 3D TimeWheel (↪ p. 324) techniques, where different axes layouts are used to allow for having all variable axes in direct relation to the time axis. Integration of time via the vanishing-point plot is achieved by adding a pseudo-perspective view of two line plots over time that are integrated between two adjacent axes of a Parallel Coordinates plot (see figure). To connect two variables, a translucent parallel coordinates panel is shown that depicts the relationship of the two variables according to the selected time points. Time extends from the foreground into the background toward a vanishing point. Users can move forward and backward in time by moving the panel back and forth. This allows for the analysis of trends over time as well as the exploration of changes in the relationship between two variables over time.

Relevant references: Gruendl et al. (2016) • Inselberg and Dimsdale (1990)

time

primitives: **points**
arrangement: **linear**

data

number of variables: **multiple**
frame of reference: **abstract**

vis

mapping of time: **static**
dimensionality: **3D**

InfoBUG

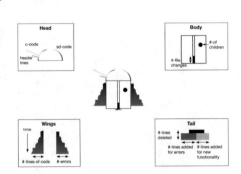

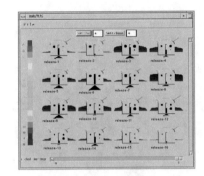

Fig. A.95: The InfoBUG encodes data about software to the wings, head, tail, and body of a glyph (left). The wings show lines of code and number of errors over time, whereas body, head, and tail show further data for a selected time point. A small multiples view can be used to compare different software releases over time (right). © *1998 IEEE. Reprinted, with permission, from Chuah and Eick (1998).*

Chuah and Eick (1998) developed InfoBUG for visualizing changes in software projects. The InfoBUG is an information-rich graphic that combines a multitude of different heterogeneous data values. The glyph resembles an insect with wings, head, tail, and body. The different parts of the glyph are used to represent four different classes of information about software projects as shown to the left in the figure: code lines and errors (wings), types of code (head), added and deleted lines of code (tail), and number of file changes and children (body). The wings represent the lines of code (left wing) and the number of errors (right wing) over time as vertical silhouette graphs. While the wings show data over time, the other parts of the glyph show only the data of a user-selected time point, which is indicated as a red line at the wings. Antennas on the InfoBUG's head represent different types of code, where color indicates the type of code, and the relative sizes of different types are encoded by antenna length. The bug's tail represents the number of deleted and added lines. Finally, the InfoBUG's body visualizes information about the number of altered files via a bar in the middle of the body and the number of child objects via filled circles. Small multiples (↪ p. 359) can be used to compare the different releases of a software product over time as shown in the figure to the right. Furthermore, the representation can be animated to follow the course of time.

Relevant references: Chuah and Eick (1998)

Gravi++

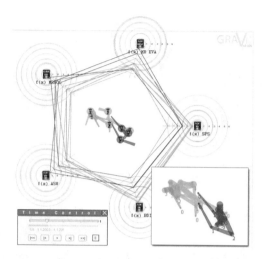

Fig. A.96: Patient icons in the center of the display are positioned relative to the surrounding parameters (in this case items of a questionnaire) following a spring-based model. Individual answers to questionnaire items are shown as concentric rings, and star glyphs show sets of answers as polygonal lines. The user can step through time manually or can use animation, which can be steered via the control panel on the lower left. Furthermore, traces might be displayed to convey information about the evolution of values over time as shown in detail on the lower right. © *Courtesy of Klaus Hinum.*

Gravi++ has been developed as a tool for finding predictors for the treatment planning of anorexic girls. It represents patients and data gathered from questionnaires during treatment over the course of several weeks or months. Patients are represented by icons that are laid out according to a spring-based model relative to the surrounding icons, which represent items of a questionnaire. This leads to the formation of clusters of persons who gave similar answers. Animations are used to visualize the change in values over time. The position of each person's icon changes over time, making it possible to trace, compare, and analyze the changing values. Alternatively, the change over time can be represented by traces. The size and path of the person's icon are shown corresponding to all time steps or only to a restricted subset like the previous and the next time step. To visualize the exact values of each question, rings around the question's icon can be drawn and star glyphs might be shown.

Relevant references: Hinum et al. (2005)

time

primitives: **points**
arrangement: **linear**

data

number of variables: **multiple**
frame of reference: **abstract**

vis

mapping of time: **static, dynamic**
dimensionality: **2D**

CircleView

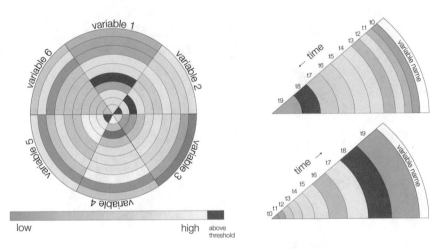

Fig. A.97: Multiple variables are shown as segments of a circle. Each segment is further subdivided into time slots that represent data values using color. Left: six variables over ten time steps; right: different time axes arrangements (outside–in vs. inside–out) and space assignment that emphasizes more recent values. ⓒ ⓘ *The authors. Adapted from Keim et al. (2004).*

Keim et al. (2004) describe the CircleView as a technique for visualizing multi-variate streaming data as well as static historical data. The basic idea is to divide a circle into a number of segments, each representing one variable. The segments are further divided into slots covering periods of time, and color shows the (aggregated) data value for the corresponding interval. Thus, time is mapped linearly along the segments. The user can interactively adjust the number of time slots, the time span per slot, different layouts for the time axis, and might emphasize more recent slots by assigning increasingly more space to them (bottom-right figure). Since the or-der of segments is important for the visual appearance and the comparison of the data, segments can be reordered interactively by the user or automatically based on similarity measures. For streaming data, the segments of the circle are shifted automatically from the center to the edge (or vice versa). Keim and Schneidewind (2005) also presented a multi-resolution approach on top of CircleView, where time slots for coarser granularities are shown besides detail values.

Relevant references: Keim et al. (2004) • Keim and Schneidewind (2005)

CloudLines

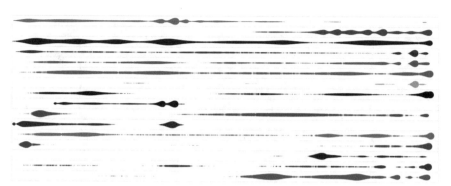

time

primitives: **points**
arrangement: **linear**

Fig. A.98: CloudLines provide a compact overview of event episodes from multiple time series. Individual events are shown as semi-transparent circles, whereas an importance function is mapped to both opacity and size for improved visibility. The figure shows a dataset of key politicians' appearances in the news on February 2011, whereas each politician is shown in an individual row. © *2011 IEEE. Reprinted, with permission, from Krstajic et al. (2011).*

data

number of variables: **multiple**
frame of reference: **abstract**

Application areas that deal large amounts of data, like news publishing, network security, or financial services, are challenging in terms of data analysis as they require scalable solutions. CloudLines is an incremental time series visualization technique that uses interactive distortion to handle time-based representations of large datasets in limited space. It focuses on recent events while providing context on the past, and makes relevant patterns salient at any scale. In order to better focus on recent events, a logarithmic time scale can be used that allows for direct interaction with recent events. The method uses an importance function, which allows the visualization to adjust the opacity and the size of the events according to their age and data density based on a decay function and kernel density estimates. Each row represents a variable, e.g., weekly amount of tweets of a certain person, and qualitative color-coding is used to make row distinction easier. Access to details is provided through a magnifying lens, which takes the distortions in size and opacity into account. This increases readability in selected areas of interest. In order to be applicable for streaming data such as online news streams, incremental kernel density estimate algorithms and smoothly animated transitions for display updates can be used. In the work of Krstajic et al. (2011), CloudLines have been applied for visualizing news appearances of politicians and news event monitoring. They showed that CloudLines help to detect important event episodes in the time series, show the detailed structure of event episodes, and interact with time series on atomic level.

vis

mapping of time: **static, dynamic**
dimensionality: **2D**

Relevant references: Krstajic et al. (2011)

Animated Scatter Plot

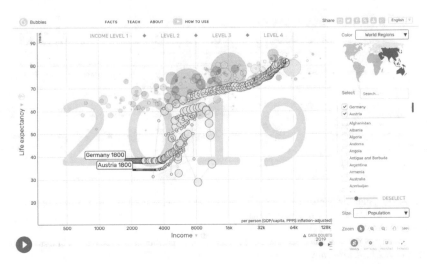

Fig. A.99: Two data variables (life expectancy and income) are mapped to the horizontal and vertical axes, symbol size represents a third variable (population), and animation is used to step through time. Additionally, trails are activated for the selected countries, Austria and Germany, which help viewers track countries over time. ⓒ⊕ *The authors. Generated with Bubbles by the Gapminder Foundation.*

Animated scatter plots show the data in a Cartesian coordinate system similar to point plots (↪ p. 232). For point plots, one axis of the coordinate system represents time, the other axis encodes one time-dependent data variable. For animated scatter plots, the coordinate axes represent two time-dependent variables, while the dimension of time is visualized by means of animation. Animated scatter plots have been popularized by Hans Rosling, who used them extensively to explain facts about the world. The Bubbles tool (formerly known as Trendalyzer) by the Gapminder Foundation (2021) makes animated scatter plots available to a wide audience. The figure shows an example where the two data variables *life expectancy* and *income* are mapped onto the axes of the coordinate system. The dot size represents a third variable, namely *population*, and different dot colors mark groups of data items. An animation visualizes how the dots (and the underlying data) change over time. The animation can be controlled via a time slider, a play/pause button, and a slider for adjusting animation speed. Furthermore, trails can be displayed, which means that dots stay visible and are connected over time. This helps users to better see the path of a variable through time. Robertson et al. (2008) evaluated the use of animation in conveying trends over time and compared animated scatter plots with a modified version of trails, and small multiples (↪ p. 359). The results show that animation is both slower and less accurate than the other representations but is well suited as a presentation aid.

Relevant references: Gapminder Foundation (2021) • Robertson et al. (2008)

Process Visualization

time

primitives: **points**
arrangement: **linear**

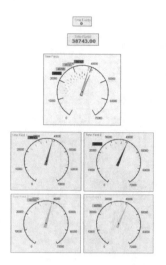 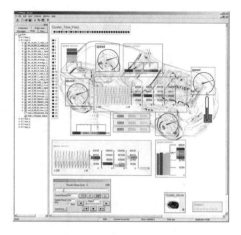

Fig. A.100: Process visualization can be supported by providing virtual instruments at different levels of detail (left). This helps users stay focused on important variables of an automotive process, while less relevant information is presented at a higher level of abstraction only (right). © *2002 IEEE. Reprinted, with permission, from Matković et al. (2002).*

Process visualization, for instance in automotive environments, has to deal with a multitude of time-varying input variables to be monitored. Matković et al. (2002) suggest a focus+context approach to help users keep track of the important changes in a process. The key idea is to provide virtual instruments that represent monitored variables at different levels of detail. Instruments representing focused variables provide more detailed information, for example, a brief view on a variable's history, which is not possible with classic gauges. On the other hand, less relevant variables are visualized using heavily abstracted virtual instruments that might show just the numeric value or even only a colored dot. Multiple such instruments are arranged in a virtual environment that is used as visual reference for the monitoring scenario. Focus and context within the environment can change dynamically during monitoring, either upon detection of certain events in the data or via user interaction. The approach of Matković et al. (2002) is an example of a visualization for dynamic temporal data where only the current state of the data is available without any history.

Relevant references: Matković et al. (2002)

data

number of variables: **multiple**
frame of reference: **abstract**

vis

mapping of time: **dynamic**
dimensionality: **2D**

TimeRider

time

primitives: **points**
arrangement: **linear**

data

number of variables: **multiple**
frame of reference: **abstract**

vis

mapping of time: **dynamic**
dimensionality: **2D**

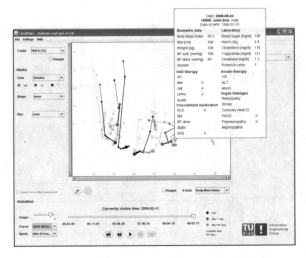

Fig. A.101: Patients are represented as animated marks in a scatter plot. Body mass index is mapped to the horizontal axis, HbA1c to the vertical axis, and mark color shows whether a patient smoked. Time controls for animation and synchronization settings are visible at the bottom. Additionally, traces are displayed that connect values over time. A detail-on-demand window showing further patient data is displayed when hovering over a patient's mark. ⓒⓘ *The authors. Generated with the TimeRider software.*

TimeRider by Rind et al. (2011a) is an enhanced animated scatter plot (↪ p. 330) for exploring multivariate trends in cohorts of diabetes patients. The enhancements tackle three challenges of medical data: irregular sampling, data wear (i.e., decreasing validity over time), and patient records covering different portions of time. Animation of irregularly sampled data is achieved via interpolation of individual values along a linear trajectory. To account for data wear and to maintain temporal context, transparency and traces are used to enrich the visual encoding of time. For comparing patient histories that cover different portions of time, TimeRider provides four synchronization modes: by calendar date, patient age, start of treatment, and end of treatment. To take better advantage of animation, TimeRider is highly interactive; apart from common interactions to select, pan, zoom, filter, and show details on demand, the user can change the visual mapping of axes, color, shape, and size. Other task-specific features are value ranges that can be highlighted in the background of the scatter plot and dynamic queries on data variables.

Relevant references: Rind et al. (2011a)

Flocking Boids

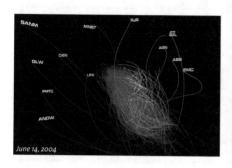

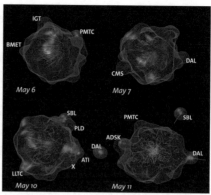

Fig. A.102: Stock market data are represented as flocking boids that move in a three-dimensional presentation space. Left: boids leaving the flock indicate that the corresponding stock price behaves differently than the majority of prices; right: implicit surfaces surrounding boids help users to recognize the spatial structure of the flock. © *2004 IEEE. Reprinted, with permission, from Vande Moere (2004).*

Stock market data change dynamically during the day as prices are constantly up-dated. Vande Moere (2004) proposes to visualize such data by means of information flocking boids. The term boids is borrowed from the simulation of birds (bird objects = boids) in flocks. In order to visualize stock market prices, each stock is considered to be a boid with an initially random position in a 3D presentation space. Upon ar-rival of new data, boid positions are updated dynamically according to several rules. These rules attempt to avoid collisions of boids, to move boids at the same speed as their neighbors in the flock, to move boids toward the flock's center, to keep similar boids close to each other, and to let boids stay away from boids that are dissimilar. The visual representation is inherently dynamic and aims at the users' capability to perceive emergence of patterns as the visualization updates. To this end, boids and corresponding traces are visualized as animated curves, as shown in the left figure. This 3D visual representation is enhanced by enclosing boids within implicit surfaces, which helps users recognize the spatial structure of the flock (right). The flocking boids visualization can be useful for detecting various patterns in the data such as the emergence of clusters, the separation of boids from the main flock, or a general chaotic behavior of boids.

Relevant references: Vande Moere (2004) • Vande Moere and Lau (2007)

time

primitives: **points**
arrangement: **linear**

data

number of variables: **multiple**
frame of reference: **abstract**

vis

mapping of time: **dynamic**
dimensionality: **3D**

Time Line Browser

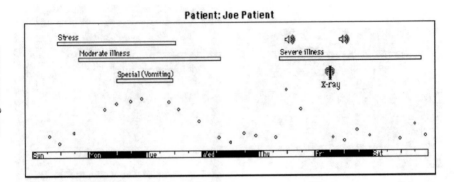

Fig. A.103: Heterogeneous patient information is visualized along a common horizontal time axis. Intervals are displayed as labeled bars and events are displayed as icons. The small circles form a point plot that shows the patient's blood glucose over time. © *1991 Elsevier. Reprinted, with permission, from Cousins and Kahn (1991).*

Cousins and Kahn (1991) developed the time line browser for visualizing heterogeneous time-oriented data. The time line browser integrates qualitative and quantitative data as well as point and interval data into a single coherent view. The time line browser distinguish simple events, complex events, and intervals. Simple events are represented as small circles, whereas complex events are shown as icons. Bars are used to indicate the location and duration of intervals. These depictions are aligned with respect to a common horizontal time axis (↪ p. 258), where textual labels might be used to display further details. In addition to the visualization, a formal system for timeline elements and timeline operations has been developed. It defines five basic operations (i.e., slice, filter, overlay, add, new) for manipulating timelines and also supports composite operations. These operations are useful for addressing the issues of different temporal granularities and the calendar mapping problem.

Relevant references: Cousins and Kahn (1991)

PatternFinder

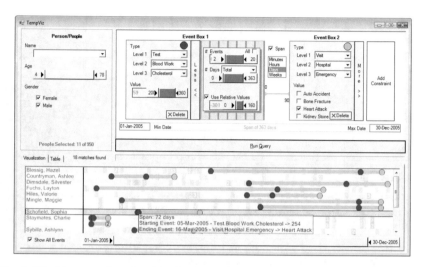

Fig. A.104: The query formulated in the visual interface (top) relates a cholesterol test to a subsequent emergency visit to a hospital. The resulting visualization (bottom) shows several patient records matching the query as vertically stacked ball-and-chain representations. © *Courtesy of Jerry Alan Fails.*

PatternFinder by Fails et al. (2006) can be used for constructing queries to find temporal patterns in medical record databases. The temporal patterns consist of events that are associated with data, and time spans between events. Users formulate queries by imposing constraints on events and time spans. Events can be selected from a hierarchically structured vocabulary, and constraints for associated variables can be specified in a visual interface along with temporal constraints. This way, users can build queries for the existence of events (e.g., persons with heart attack), temporally ordered events (e.g., heart attack followed by stroke), temporally ordered value changes (e.g., BMI of 25 or higher followed by BMI of 20 or lower), and trends over time (e.g., BMI decreasing). Event sequences might not only be specified in terms of temporal order, but also in terms of temporal distance (e.g., time span of 28 days or less between heart attack and stroke). Moreover, all of the mentioned query types can also be combined. For visualizing query results, a so-called ball-and-chain representation is used: results are shown as vertically stacked timelines, where colored circles represent matched events and bars stand for matched time spans.

Relevant references: Fails et al. (2006)

---

time

primitives: **points, intervals**
arrangement: **linear**

data

number of variables: **multiple**
frame of reference: **abstract**

vis

mapping of time: **static**
dimensionality: **2D**

FacetZoom

primitives: points, intervals
arrangement: linear

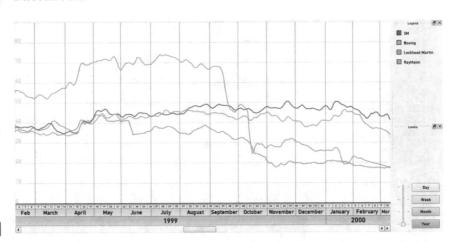

Fig. A.105: The hierarchical structure of time is shown as an interactive horizontal time axis widget that has a data view attached to it, in this case, a visualization of stock market data. © *Courtesy of Raimund Dachselt.*

number of variables: multiple
frame of reference: abstract

FacetZoom is a technique that enables users to navigate hierarchically structured information spaces (see Dachselt et al., 2008). The hierarchical structure of time is a natural match for this technique. What Dachselt and Weiland (2006) originally called TimeZoom is a visual navigation aid for time-oriented data. The basic idea is to display a horizontal time axis that represents different levels of temporal granularity as stacked bars (e.g., decades, years, months, weeks, days). The time axis is an interactive widget that can be used to access data from different parts of the time domain at different levels of abstraction. In addition to continuous zooming and panning via mouse, it is also possible to simply select discrete intervals from the time axis. Depending on the user's selection, the time axis display is altered to accommodate the selected part of the time axis with more display space. Accordingly, the data view, which is attached to the time axis, can use the extra space to represent more data items in greater detail. While the actual mapping of time is static, the navigation steps of the user, including the visual adjustment of the time axis, are smoothly animated.

mapping of time: static
dimensionality: 2D

Relevant references: Dachselt et al. (2008) • Dachselt and Weiland (2006)

KNAVE II

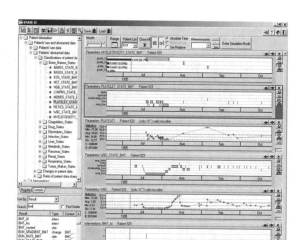

primitives: **points, intervals**
arrangement: **linear**

Fig. A.106: KNAVE II supports the visualization and knowledge-based navigation of patient data and abstractions thereof. Left: tree view of clinical domain ontology for navigation; right: visualization panels for raw data and abstractions; bottom left: search panel. © *2006 Elsevier. Reprinted, with permission, from Shahar et al. (2006).*

KNAVE II enables visual browsing and exploring of patient's data (raw measured values and external interventions such as medications). The system focuses mainly on the visual display of temporal abstractions of the data and shows domain-specific concepts and patterns. In order to abstract the raw data, a predefined knowledge base is used that defines three types of interpretations: classification of data (e.g., low–normal–high), change of data (e.g., increasing–decreasing), and rate of change (e.g., slow–fast). Colored timelines ($\hookrightarrow$ p. 258) depict abstracted intervals as bars where a bar's vertical position within a panel encodes its qualitative value. For example, the second panel (blue frame) in the figure shows the platelet state abstracted to the qualitative values: very low, low, moderately low, normal, and high (from bottom to top). KNAVE II also allows users to view the raw data as line plots ($\hookrightarrow$ p. 233). Moreover, statistics for parameters and corresponding abstractions can be superimposed as bar graphs ($\hookrightarrow$ p. 234) as shown in the topmost panel, or can be given as text labels, as shown in the third and fifth panels. A granularity-based zoom (bottom of each panel) allows users to quickly navigate in time by simply clicking individual granules.

Relevant references: Shahar et al. (2006)

number of variables: **multiple**
frame of reference: **abstract**

mapping of time: **static**
dimensionality: **2D**

Continuum

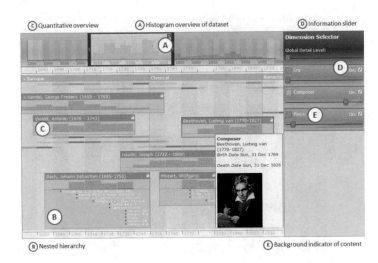

Fig. A.107: Continuum showing a music dataset with the three variables era, composer, and piece. Scalable histograms provide a complete representation of the data and quantify the focal data item. Top left: timeline overview; bottom left: timeline detail view; right: dimension selector panel. © *Courtesy of Paul André.*

Collections of small events often constitute larger, more complex events, like for example talks at conferences or legs of a race. Moreover, events might also be related to other events at other points in time (e.g., a paper written at some point in time and referenced later). Continuum by André et al. (2007) is a timeline visualization tool to represent large amounts of hierarchically structured temporal data and their relationships. It addresses the three problems of scale, hierarchy, and relationships by using scalable histogram overviews, flattening high-dimensional data into dynamically adjustable hierarchies, and arching connection lines for representing non-hierarchical relationships. The interface consists of three main panels that show overview, detail, and dimension configuration. The timeline overview always represents the complete timespan of the dataset using scalable histograms where the vertical axis quantifies the user-selected focal data item. The timeline detail view shows hierarchical relationships as nested elements and applies semantic zooming depending on the amount of information to be displayed. With the dimension selector, users can interactively control the hierarchical buildup and the level of detail to be shown.

Relevant references: André et al. (2007)

VisuExplore

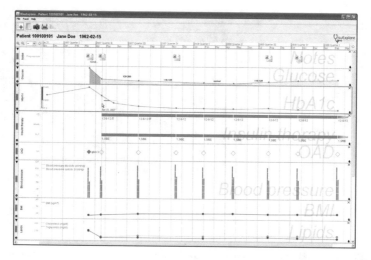

Fig. A.108: Visualization of heterogeneous medical parameters of a diabetes patient. Eight visualization panels show progress notes in a document browser, glucose and HbA1c as line plots with semantic zoom, insulin therapy as timelines, OAD as event chart, blood pressure as bar graphs, and BMI as well as lipids as line plots (top to bottom). ⓒⓘ *The authors. Generated with the VisuExplore software.*

VisuExplore by Rind et al. (2011b) is an interactive visualization system for exploring a heterogeneous set of medical parameters over time. It uses multiple views along a common horizontal time axis to convey the different medical parameters involved. VisuExplore provides an extensible environment of pluggable visualization techniques and its primary visualization techniques are deliberately kept simple to make them easily usable in medical practice: line plots (↪ p. 233), timeline charts (↪ p. 258), bar graphs (↪ p. 234), event charts, line plots with semantic zoom, and document browsers (notes panel on the top). Furthermore, data can also be presented as textual tables to augment the visual representations. VisuExplore's interactive features allow physicians to get an overview of multiple medical parameters and focus on parts of the data. Users may add, remove, resize, and rearrange visualization views. A measurement tool can be employed to determine the duration of time spans between user-selected points of interest. This works not only within one but also across different views.

Relevant references: Rind et al. (2011b)

Midgaard

primitives: **points, intervals**
arrangement: **linear**

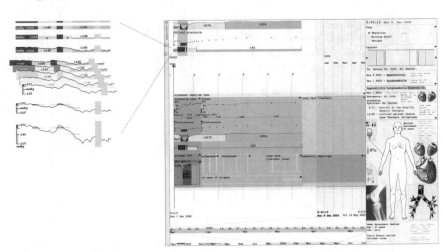

number of variables: **multiple**
frame of reference: **abstract**

Fig. A.109: Midgaard employs different steps of semantic zooming of a time series from a broad overview (top left) to a detailed view with details of fine structures (bottom left). The main user interface shows different measurements (e.g., blood gas measurements, blood pressure), treatment plans, and additional patient information. ⓒ① *The authors.*

mapping of time: **static**
dimensionality: **2D**

Several tightly integrated visualization techniques have been developed in the Midgaard project by Bade et al. (2004) to enhance the understanding of heterogeneous patient data. To support the user in exploring the data and to capture as much qualitative and quantitative information as possible on a limited display space, Midgaard supports different levels of abstractions for time-oriented data (left). Switching between these levels is achieved via a smoothly integrated semantic zoom functionality. These methods were designed to allow users to interact with data and time. Navigation in time is done using three linked time axes (bottom right). The first one (bottom) provides a fixed overview of the underlying time interval covering its full range. Selecting a subrange in that time axis defines the temporal bounds for the main display area and the second time axis (middle). Selecting a further subrange in the middle time axis defines detail and surrounding context areas in time. By interactively adjusting the subranges, users can easily zoom and pan in time.

Relevant references: Bade et al. (2004)

LifeLines

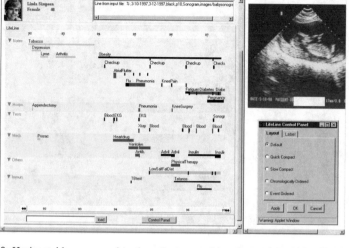

Fig. A.110: Horizontal bars are used to show the temporal location and duration of health-related incidents. The example shows several facets of patient information and an additionally linked sonogram on the right. © *Courtesy of Catherine Plaisant and University of Maryland Human-Computer Interaction Lab.*

A simple and intuitive way of depicting incidents is by drawing a horizontal line on a time scale for the time span the incident took. This form of visualization is called timeline (↪ p. 258). Plaisant et al. (1996) apply and extend this concept for visualizing health-related incidents in personal histories and patient records. Consequently, they call their approach LifeLines. Horizontal bars are used to show the temporal location and duration of incidents, treatments, or rehabilitation. Additional information can be encoded via the height as well as the color of individual bars. In order to structure the displayed information in groups, the so-called facets are introduced. Multiple such facets are stacked vertically. Depending on the information sought by the user, facets can be expanded and collapsed. When collapsed, only a very small, geometrically and semantically reduced visual representation without textual labels is displayed. When expanded, a facet shows full detail. External information related to certain incidents are provided on demand in a linked view, for example, x-ray images or sonograms.

Relevant references: Plaisant et al. (1996) • Plaisant et al. (1998)

EventRiver

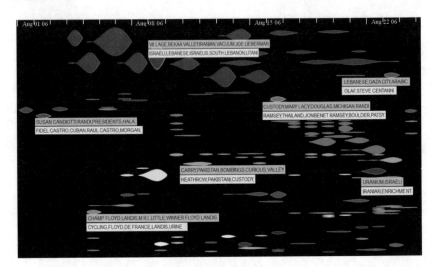

Fig. A.111: EventRiver visualization of CNN news data from August 2006. Event bubbles flow in a horizontal river of time, where important events are highlighted in red in the top part of the river. © *2011 IEEE. Reprinted, with permission, from Luo et al. (2012).*

Text collections such as news corpora or email archives often contain temporal references, which embed the text's information into a temporal context. Luo et al. (2012) describe a technique, called EventRiver, for exploring such text collections interactively in terms of important events and the stories that these events constitute. In the first step, events are extracted from the data using a number of analytical methods, including keyword identification and temporal locality clustering. This analysis yields a set of events that are characterized by their position in time, by their duration, and by several other measures (e.g., temporal influence, strength, and co-strength). The visual design of EventRiver is based on the so-called event bubbles that flow in a horizontal river of time, that is, along a horizontal time axis. The bubbles are placed horizontally where events are located in time. A bubble's shape illustrates how an event has emerged and disappeared over time. Colors and the bubbles' vertical positions in the river are chosen so as to highlight important interconnected events that constitute long-term stories in the text documents. While tooltip labels show the important keywords of events, document details are provided on demand in separate views. Analysts can adjust the EventRiver by using various interaction techniques including dynamic filtering, semantic and temporal zooming, and manual relocation of event bubbles.

Relevant references: Luo et al. (2012)

Story Curves

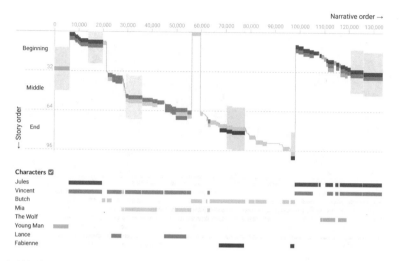

Fig. A.112: Story curves representation of the movie "Pulp Fiction". The narrative order is mapped to the horizontal axis, whereas the chronological order of events is mapped to the vertical axis. The curve visualizes how the movie's story unfolds and which events are told at what time. ⓒⓘ *The authors. Generated with the Story Curves software by Nam Wook Kim.*

The temporal order in which stories are told and the actual chronological order of the events being told are two distinct things. Movies often make use of this fact. They tell a story deliberately disrupting the chronological order by flashbacks, reversing time, and switching between past, present, and future. Story curves by Kim et al. (2018) is a technique for visualizing such nonlinear narratives. The core idea is to map the narrative order to the horizontal axis and the chronological order of story events to the vertical axis. A curve then visualizes how the story unfolds and which events are told at what point during the movie. The curve can be enhanced with color-coded bars representing additional information, such as the characters appearing in a scene or the location in which a scene takes place. The figure shows the story curve of the movie "Pulp Fiction". We can see that the initial scene (leftmost cyan bar) is actually almost in the middle of the chronological order of the story. Moreover, we see that the actual first events of the story are only told in the middle of the movie (topmost yellow bar). From the paired bars that run in parallel, we can further learn when key characters of the movie co-occur. With such insights, story curves offer new and interesting perspectives on movies (or other narratives) in terms of how they twist the dimension of time to tell a compelling and fascinating story.

Relevant references: Kim et al. (2018)

time

primitives: **points, intervals**
arrangement: **linear**

data

number of variables: **multiple**
frame of reference: **abstract**

vis

mapping of time: **static**
dimensionality: **2D**

TextFlow

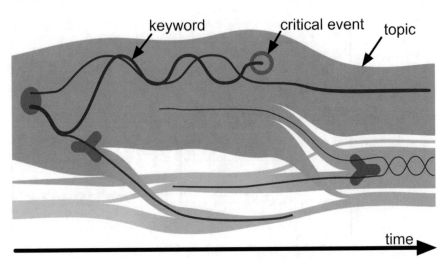

keyword   critical event   topic

time

time

primitives: **points**
arrangement: **linear**

data

number of variables: **multiple**
frame of reference: **abstract**

vis

mapping of time: **static**
dimensionality: **2D**

Fig. A.113: The TextFlow technique uses bands of varying widths to represent the evolution of topics over time. Critical events in the evolution are marked with glyphs. So-called threads (blue lines) show details about selected keywords. ⓒⓘ *Courtesy of Weiwei Cui.*

The TextFlow technique by Cui et al. (2011) has been designed to support the analysis of text over time. The general idea is similar to the ThemeRiver (↪ p. 293) approach. Text topics are represented as colored bands whose height varies over time to represent topic importance. Yet, TextFlow also allows bands to split and merge, and to change their vertical positions to represent important changes in the topic evolution. As such, TextFlow bears resemblance to storyline visualizations (↪ p. 264). Split and merge events as well as source (emergence of a topic) and sink (disappearance of a topic) events are additionally marked in the visualization by small glyphs. For a more detailed analysis of topic evolution, so-called threads are used to visualize the co-occurrence of user-selected keywords. This allows users to understand how individual keywords contribute to the formation of topics and their overall relevance in comparison to other topics. The TextFlow technique provides interactive filtering to focus the analysis on selected topics and also provides recommendations on what keywords are worth investigating in detail based on correlation scores.

Relevant references: Cui et al. (2011)

Event-Flow Visualization

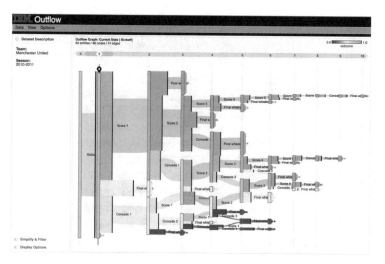

Fig. A.114: Event-flow visualization via Outflow. Outflow processes multiple temporal event data and visualizes aggregate event-flow pathways together with associated statistics. This screenshot shows a visualization of Manchester United's 2010-2011 soccer season. Green and red show pathways with good and bad outcomes (i.e., wins and losses), respectively. © *2012 IEEE. Reprinted, with permission, from Wongsuphasawat and Gotz (2012).*

Event-flow visualization is concerned with the question of how linked events literally flow through time. This question is relevant in various domains, ranging from electronic medical records to sports events, where events can be linked to time points, time intervals, or combinations thereof. The general design of event-flow visualizations is similar to Sankey diagrams (↪ p. 313). To overcome visual clutter, methods to reduce edge crossing and straighten unnecessarily curvy edges while preventing overlaps are usually applied. A concrete example of an event-flow visualization is Outflow by Wongsuphasawat and Gotz (2012). It supports the visual analysis of multiple event progression pathways and their associated properties (timing, cardinality, and outcomes). The EventFlow technique by Monroe et al. (2013b) tackles the complexity of point- and interval-based event sequences by providing user-driven data simplification and search techniques to support the users to discover temporal patterns in electronic health records. The DecisionFlow approach by Gotz and Stavropoulos (2014) directly supports the visual analysis of high-dimensional temporal event sequence data with orders-of-magnitude more event types than previously achieved. Various interaction techniques are usually provided, including navigation, simplification, and correlated factor analysis.

Relevant references: Wongsuphasawat and Gotz (2012) • Monroe et al. (2013b) • Gotz and Stavropoulos (2014)

time

primitives: **points,intervals**
arrangement: **linear**

data

number of variables: **multiple**
frame of reference: **abstract**

vis

mapping of time: **static**
dimensionality: **2D**

Co-Bridges

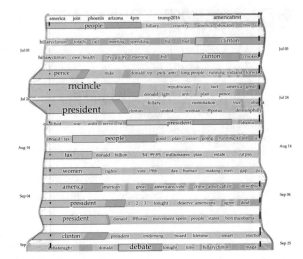

Fig. A.115: The Co-Bridges visualization is based on the metaphor of a river flowing top to bottom, whose river banks represent two social media streams by Hillary Clinton (left, pink) and Donald Trump (right, blue). Bridges cross the river and their segments represent the frequency of keywords in the two data streams. © *Courtesy of Siming Chen.*

The Co-Bridges technique supports the visual comparison of two time-varying data streams with multiple items. Chen et al. (2021) designed Co-Bridges with two metaphors in mind: the river metaphor and the bridge metaphor. The river part of the technique depicts a quantitative characteristic (e.g., stream volume) about the two data streams being compared via two distinctly colored vertical line plots (left and right in the figure), which correspond metaphorically to the river banks. Several bridges cross the river and connect corresponding parts of the two data streams. The bridges' width represents the accumulated frequency of data items. Thin curves at each bridge head mark the time interval covered by a bridge. The bridges themselves are paved with labeled item segments. There is one segment per item, and per segment, the proportion of colors indicates the frequencies of the associated item for the two data streams. The segments are sorted in such a way that they are closer to the river bank whose stream has the higher count of the associated item. Dynamic querying and semantic zooming allow users to refine their analysis and compare data streams in detail. The figure shows an example of Co-Brides being applied to social media data. One can see, for example, that the item "president" is more frequent in the left stream, which represents posts by former presidential candidate Hillary Clinton.

Relevant references: Chen et al. (2021)

LiveGantt

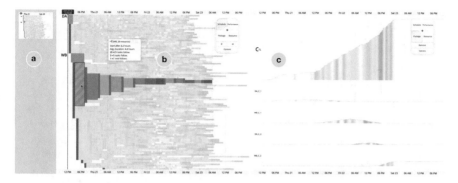

Fig. A.116: Interactive visualization of large manufacturing schedules. The interface shows (a) an exploration history, (b) the schedule view that shows tasks with their durations and orders, as well as (c), a package view that shows changes in packages' production. The example depicts a schedule containing thousands of tasks on hundreds of machines. ⓒⓘ *The authors. Generated with the LiveGantt demo by Jaemin Jo.*

LiveGantt aims for supporting the visual exploration of large manufacturing schedules from different perspectives. It extends traditional Gantt charts (↪ p. 253) which are limited in terms of scalability and interaction support. To achieve a scalability of the method to thousands of tasks and hundreds of machines, time-based task aggregation and resource reordering are applied. The LiveGantt approach consists of five different interactive views. First, the main view is the scheduling view in the center (b) which is based on timeline bars as used in Gantt charts. To tackle scalability, an interactive aggregation algorithm is used. It is controlled interactively by the user selection of a focus time that is shown by a thick black vertical line. Second, the performance view is used to monitor the performance of a schedule over time. It consists of a line plot (↪ p. 233) that shows resource utilization over time which is superimposed over a bar graph (↪ p. 234) that represents concurrent changeovers over time. Third, the resource view contains information about utilization and total changeover time for each resource in bar graphs. In addition, operation start and finish times are displayed in a Gantt-like chart. Fourth, the package view on the right (c) shows the number of completed packages and work-in-progress (WIP). It allows for stepwise drill down by expanding the contents of each package and showing individual charts. For each package, a line plot shows the development of WIPs over time. To emphasize sudden changes, a dual encoding is used by coloring the area below the line depending on the slope of the line plot. Color hue is representing the direction of the slope (orange for upwards and blue for downwards) and saturation represents the steepness of the slope (more saturated colors for steeper slopes). Fifth, the exploration history view on the left (a) visualizes the exploration sequence with thumbnails and allows users to interact with it.

Relevant references: Jo et al. (2014)

time

primitives: **points**
arrangement: **linear, cyclic**

PeopleGarden

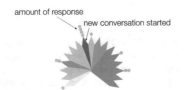

data

number of variables: **multiple**
frame of reference: **abstract**

Fig. A.117: This PeopleGarden shows about 1200 posts entered into a discussion board during a period of two months. Users are represented by flowers whose petals represent individual messages posted by a user. © *1999 ACM. Reprinted, with permission, from Xiong and Donath (1999).*

PeopleGarden is a graphical representation of users' interaction histories in discussion groups. PeopleGarden visualizes data about users and messages posted to an online interaction environment. It integrates information on the time of posting, amount of response, and whether a post starts a new conversation. For intuitive understanding, PeopleGarden uses the metaphor of a garden of flowers. The garden represents the whole environment and flowers represent individual users within the environment. The petals of a flower stand for the messages posted by a user, the time of posting is mapped to the ordering and saturation of the petals, the amount of response is represented by circles that are stacked on top of petals, and the color is used to depict whether a post starts a new conversation. Furthermore, the height of the flower gives information about how long a specific user has been a member of the discussion group. Using these visual representations, one can easily spot dominant voices, long-time participants, or very active groups.

Relevant references: Xiong and Donath (1999)

vis

mapping of time: **static**
dimensionality: **2D**

PostHistory

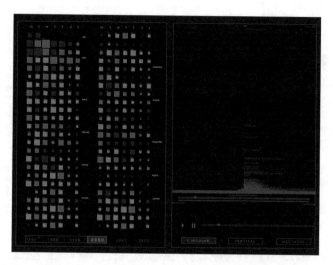

Fig. A.118: The calendar panel on the left shows e-mail activity on a daily basis, where the number of emails and their average directedness are mapped to box size and color, respectively. The contacts panel on the right displays the names of people who sent messages to the user. © *Courtesy of Fernanda B. Viégas.*

PostHistory is a visualization technique for visually uncovering different patterns of e-mail activity (e.g., social networks, e-mail exchange rhythms) and the role of time in these patterns. PostHistory is user-centric and focuses on a single user's direct interactions with other people through e-mail. The social patterns are derived from analyzing e-mail header information. So, not the content of messages, but the tracked traffic is used as the basis for the analysis of people's e-mail conversations over time. Basically, the user interface visualizes a full year of e-mail activity and is divided into two main panels: a calendar panel on the left and a contacts panel on the right. The calendar panel shows the intensity of e-mail activity on a daily basis: a square represents a single day and each row of squares represents a week. The size of a square is determined by the quantity of e-mail received on that day and its color represents the average directedness of messages, i.e., whether a mail was received via TO, CC, or BCC. The brighter the color of a square, the more directed the messages are on that day. The contacts panel is used for displaying the names of people who sent messages to the user.

Relevant references: Viégas et al. (2004a)

time

primitives: **points**
arrangement: **linear, cyclic**

data

number of variables: **multiple**
frame of reference: **abstract**

vis

mapping of time: **static**
dimensionality: **2D**

Pixel-Oriented Network Visualization

time

primitives: **points**
arrangement: **linear, cyclic**

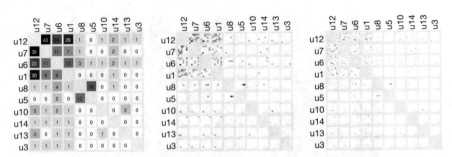

Fig. A.119: Wiki collaboration patterns. Left: adjacency matrix showing users as rows and columns and collaboration intensity by color brightness of cells that connect two users (darker means more collaboration); center: pixel-oriented view where collaboration dynamics are shown in a 6x6 pixel array laid out row by row and each pixel represents a four week period; right: more fine-grained configuration that shows weekly steps. © *Courtesy of Klaus Stein.*

data

number of variables: **multiple**
frame of reference: **abstract**

Social networks consist of actors and relationships between them. Unlike most static node-link representations of graph-like structures would suggest, these networks are dynamically changing over time. The two most common forms of visualizing time-varying networks are applying animation to node-link diagrams or applying the concept of small multiples (↪ p. 359) by showing snapshots of different points in time. An alternative display is suggested by Stein et al. (2010). They developed a pixel-oriented visualization of networks (PONV) that reveals interaction patterns between actors by integrating pixel-based representations (↪ p. 282) within the cells of an adjacency matrix. An adjacency matrix can be represented visually as a matrix table whose rows and columns represent the nodes of the network. The table cell at the intersection of a particular column and row visualizes information about the relationship between the corresponding nodes. The left figure shows an example of an adjacency matrix where the darkness of a cell represents the collaboration intensity of two individuals (the value 0 and a white cell background indicate that there is no relationship between two individuals). The figure in the center uses a different representation to show the temporal evolution of the relationships. Now each cell contains a 6x6 pixel glyph, where each pixel represents the aggregated collaboration intensity of a four-week period. The user can interactively control various parameters, including the color scale, the pixel pattern arrangement, and the time period to be covered by each pixel (e.g., daily values as in the right figure).

vis

mapping of time: **static**
dimensionality: **2D**

Relevant references: Stein et al. (2010)

Parallel Glyphs

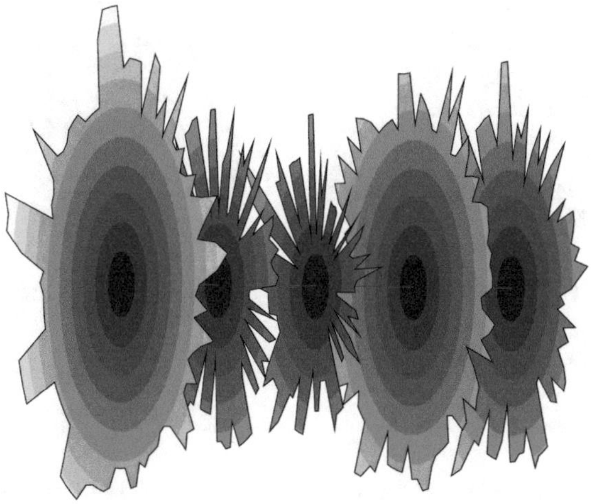

Fig. A.120: Parallel glyphs are used to visualize five different variables over time. Each radial glyph shows a single variable over time where each point on the outside of a glyph corresponds to a data value measured at a specific point in time. Differently colored rings assist in comparing data values. © *2005 IEEE. Reprinted, with permission, from Fanea et al. (2005).*

Multivariate time series can be visualized as parallel glyphs. Fanea et al. (2005) synthesized this technique as a combination of parallel coordinates and star glyphs. The visualization uses multiple star glyphs, each of which consists of as many radially arranged spikes as there are time points in the data. The length of a spike corresponds to the data value measured at the spike's associated time point. The tips of subsequent spikes are connected via a polyline, effectively creating a polygonal shape that visualizes the data of one variable in a radial fashion. As an alternative representation, the shape can be filled with differently colored rings (as in the figure) to make the data values easier to compare. Multiple such star glyphs are generated, one for each variable of the dataset. These glyphs are then arranged in the three-dimensional space along a shared axis in a parallel fashion. To assist users in identifying correlations among variables, polylines can be used to connect the same time step along the glyphs. In order to avoid clutter, it is possible to restrict this feature to a user-selected number of time steps. The technique offers various ways of manipulating the display, including switching the role of variables and data records, rotation and zooming in the 3D presentation space, and adjustment of colors.

Relevant references: Fanea et al. (2005)

time

primitives: **points**
arrangement: **linear, cyclic**

data

number of variables: **multiple**
frame of reference: **abstract**

vis

mapping of time: **static**
dimensionality: **3D**

Circos

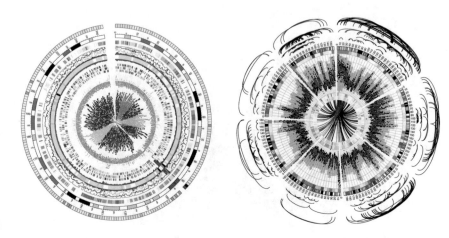

Fig. A.121: Display of multivariate data using data tracks in a radial layout. Images show human chromosome data using point plots, line plots, tiles, histograms, heatmaps, and connectors that link points on the circle. © *Courtesy of Martin Krzywinski.*

The Circos technique by Krzywinski et al. (2009) uses a circular design to generate multivariate displays. It uses concentric bands (data tracks) as display areas and is capable of displaying data as point plots (↪ p. 232), line plots (↪ p. 233), histograms, heat maps, tiles, connectors, and text. Time is mapped circularly to the circumference of data tracks. The configuration of a visual representation is based on rules that filter and format data elements based on position, value, or previous formatting. Circos was initially developed for genomics and bioinformatics data to visualize alignments, conservation, and intra- and inter-chromosomal relationships. Relationships between pairs of positions are represented by the use of ribbons that connect elements. In the same way, relational data encoded in tabular formats can be shown. Thanks to its flexible approach, Circos has also been applied to numerous other application areas, such as urban planning, and has been used for infographics in newspapers and ads to display complex relationships.

Relevant references: Krzywinski et al. (2009)

Kaleidomaps

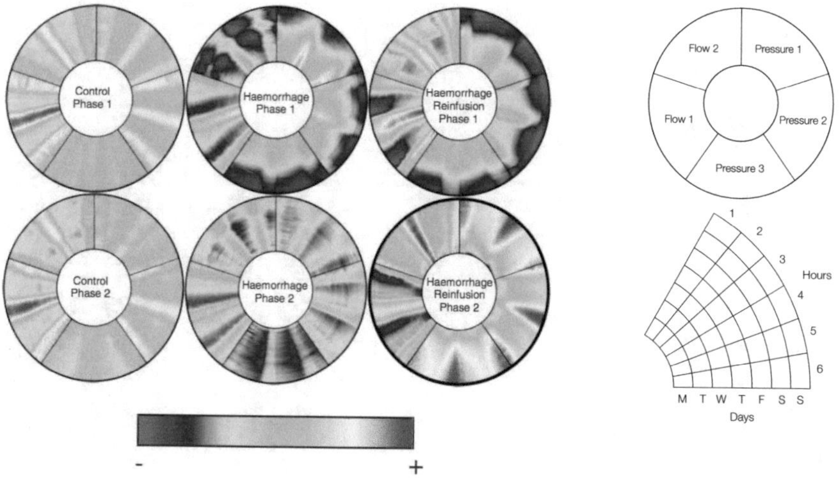

time

primitives: **points**
arrangement: **cyclic**

Fig. A.122: Left: six kaleidomaps show the morphology of blood pressure and flow waves over two experimental phases; top right: illustration of the layout of variables within a kaleidomap; bottom right: layout of time within a segment. © *2006 IEEE. Reprinted, with permission, from Bale et al. (2006).*

Kaleidomaps visualize multivariate time-series data and the results of wave decomposition techniques using the curvature of a line to alter the detection of possible periodic patterns. The overall idea of kaleidomaps is similar to the rendered output of a kaleidoscope for children, from whence the name comes. A base circle is broken into segments of equal angles for different variables. Each circle segment has two axes representing time, one along the radius and one along the arc of the segment. The data values and categories are represented using color. Due to the circular nature of kaleidomaps, the number of variables in one circle is limited to a maximum of six to eight. Interaction techniques within the kaleidomaps allow an analyst to drill down both in time and frequency domains in order to uncover potential relationships between time, space, and waveform morphologies. Kaleidomaps were developed in the domain of critical care medicine, but case studies have shown their usefulness in other domains as well, like in environment analysis.

Relevant references: Bale et al. (2006) • Bale et al. (2007)

data

number of variables: **multiple**
frame of reference: **abstract**

vis

mapping of time: **static**
dimensionality: **2D**

KAVAGait

time

primitives: **points**
arrangement: **linear**

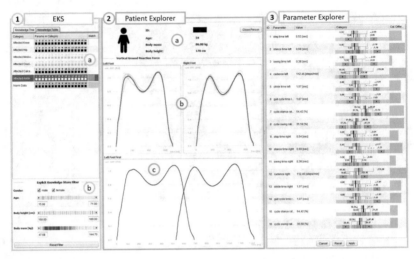

data

number of variables: **multiple**
frame of reference: **abstract**

Fig. A.123: KAVAGait's user interface, which combines three main components: (1) the explicit knowledge store (EKS) capturing an overview of the stored gait patterns, (2) the patient explorer combining patient's information and different visualizations of the ground reaction force, and (3) the parameter explore visualizing the 16 calculated spatio-temporal parameters of the patient to be explored in connection with various statistical measures. ⓒ① *Wagner et al. (2019)*.

Understanding the patient's gait performance is critical to provide appropriate diagnoses and treatment planning. KAVAGait by Wagner et al. (2019), which was designed in close cooperation with domain experts, provides a visual exploration environment incorporating principles of knowledge-assisted visual analytics to support clinicians inspecting complex data derived during clinical gait analysis. KAVAGait combines well-known visualization techniques data- and tasks-specifically, like interactive twin box plots to analyze the current patient with various statistical measures of the corresponding patient cohorts. Various interaction concepts as well as the modeling of explicit knowledge about various gait patterns ease the overall exploration process. The effectiveness of KAVAGait was validated by moderated expert reviews, user studies, and a case study with a national expert.

vis

mapping of time: **static**
dimensionality: **2D**

Relevant references: Wagner et al. (2019)

SentiCompass

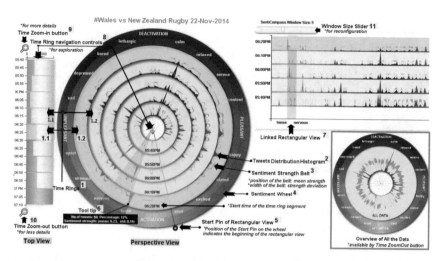

Fig. A.124: SentiCompass is used for exploring and comparing sentiments of Twitter data over time. The emotional spectrum of the sentiments is displayed on the outermost, colored ring. © *2015 IEEE. Reprinted, with permission, from Wang et al. (2015).*

SentiCompass collects tweets of specific events, such as sports events or elections, and performs text mining for sentiment analysis in combination with supervised classification along an affective dictionary. Thus, the approach does not limit sentiment to a one-dimensional variable but expands it to a 2D sentiment spectrum. The positive ends of the sentiment spectrum are pleasantness and activation, while the negative ends are deactivation and unpleasantness. SentiCompass contains three different views: a top view, a perspective view, and a linked rectangular view. For the main visual metaphor, the perspective view (center), time is shown as a cylindrical tunnel cylindrical tunnel (↪ p. 284), with different time periods represented by time rings with time advancing from the inside to the outside. The emotional spectrum of the sentiments is displayed on the outermost, colored ring. The distribution of the number of tweets is shown as circular histograms along the different sentiments within each ring. In addition, each ring contains a sentiment strengths belt, i.e, the mean and standard deviation of sentiment strengths as band graph (↪ p. 233) (line plot with varying line widths). As an alternative to the circular perspective view, a linked rectangular view can be shown (top right) which shows the same data in a linear, rectangular fashion. The top view (left) is used as an overview and navigation element for the perspective view. It splits a time interval into slices of same lengths that are displayed below each other. Each slice is proportionally filled to represent the total number of tweets collected in the respective time interval.

Relevant references: Wang et al. (2015)

time

primitives: **points**
arrangement: **cyclic**

data

number of variables: **multiple**
frame of reference: **abstract**

vis

mapping of time: **static**
dimensionality: **2D**

TreeRose

time

primitives: **points**
arrangement: **cyclic**

data

number of variables: **multiple**
frame of reference: **abstract**

vis

mapping of time: **static**
dimensionality: **2D**

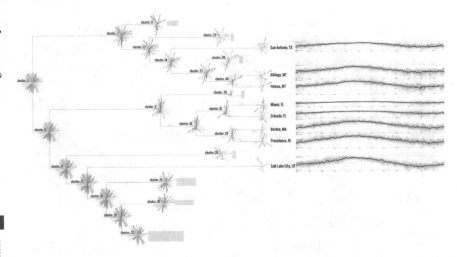

Fig. A.125: TreeRoses representing 15-year temperature data of major US cities. On the left-hand side, collapsible dendrogram clusters group cities with similar anomalous temperatures. The wind roses summarize the anomaly distributions and the gray rectangles beside the wind roses represent the depth and width of collapsed branches. On the right-hand side, the band curves visualize detailed temporal patterns and outliers. © *2019 Springer. Reprinted, with permission, from Tang et al. (2019a).*

Outlier detection in cyclic time series is an important, but challenging task. TreeRose by Tang et al. (2019a) is a visual analytics framework, which transforms a large number of multivariate time series into manageable visual representations that enable analysts to identify and monitor anomalies and trends. The time series are expected to be periodical with a known period (e.g., yearly, daily). Wind roses are used to represent the anomalies, which are glyph-based summaries of outliers for each time series. Hierarchical clustering is applied to group time series with similar anomalies (dendrogram clusters). The applicability and usability of TreeRose are illustrated using weather and building datasets as use cases.

Relevant references: Tang et al. (2019a)

Intrusion Detection

time

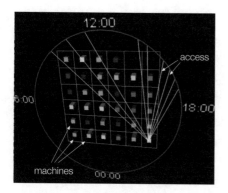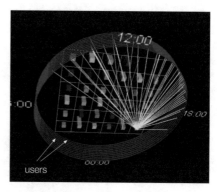

primitives: **points**
arrangement: **cyclic**

Fig. A.126: Machines in a network are represented by a matrix of 3D cubes in the center of the display. Time is mapped to the circumference of a circle enclosing the matrix of machines. When a particular machine is accessed, a line is drawn that links the particular machine with a point in time on the circular time axis. © *Courtesy of Kovalan Muniandy.*

data

A 3D visualization technique by Muniandy (2001) helps administrators analyze user access to computers in a network over time for intrusion detection. The different parameters time, users, machines, and access are mapped onto a 3D cylinder. In the figure, time is mapped onto the circumference of a circle showing the 24 hours of a day. The units along the circle can be configured to represent either hours, months, or years. Different users are represented by individual cylinder slices that are stacked upon each other and machines are represented as cubes that are arranged in a matrix. Access to a machine by a user is visualized by a line connecting the user slice at the corresponding access time with the accessed machine. This way, certain patterns of network access can easily be spotted visually and suspicious behavior can be revealed. Details are displayed when hovering with the mouse over an element of the visualization. To mitigate occlusion, the representation can be zoomed and rotated freely by the user. Moreover, filtering can be applied to remove clutter.

number of variables: **multiple**
frame of reference: **abstract**

vis

Relevant references: Muniandy (2001)

mapping of time: **static**
dimensionality: **3D**

KronoMiner

primitives: **points, intervals**
arrangement: **cyclic**

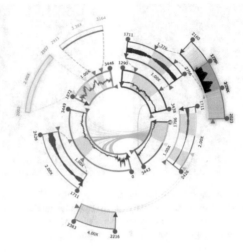

number of variables: **multiple**
frame of reference: **abstract**

Fig. A.127: KronoMiner shows the original data, four variables colored in red, blue, orange, and green, as radial line plots in the central ring and subintervals of interest on the outer rings for closer inspection. © *Courtesy of Jian Zhao.*

KronoMiner by Zhao et al. (2011a) is a multipurpose time-series exploration tool providing rich navigation capabilities and analytical support. It aims to make the analysis of selected regions of interest easier. To this end, the central view of KronoMiner is based on a hierarchical radial layout of line plots showing the raw data in the center and focused intervals in the periphery, allowing users to drill down into different parts of the data at different scales. The central ring contains all data, four time-dependent variables colored in red, blue, orange, and green in the figure. From there, users can select focus intervals, which are then extracted and shown in the outer rings. The focused intervals can be rotated, dragged, stretched or shrunken, supporting various kinds of time-series analysis and exploration tasks. KronoMiner also introduces two analytical techniques. One is a MagicAnalytics Lens, which shows the correlations between two focus intervals when overlapped. The second is the Best Match mode in which an arch shape is displayed indicating the matching parts of two intervals under a specific similarity measure.

mapping of time: **static**
dimensionality: **2D**

Relevant references: Zhao et al. (2011a)

A.3 Techniques Supporting a Spatial Frame of Reference

Small Multiples

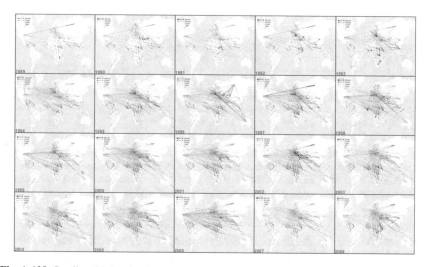

Fig. A.128: Small multiples showing migration data. Each miniature map visualizes the migration of people as links between countries. Links start in red to mark the origin of the migration, they end in green at the destination, and their line width indicates the volume of the migration. ⓒⓘ *The authors. Generated with the JFlowMap software by Ilya Boyandin.*

Small multiples are more of a general concept than a specific technique. They are described as sets of miniature visual representations (see Tufte, 1983; Tufte, 1990). For time-oriented data, each miniature visualizes a selected time point. The concrete depiction may show a single variable or multiple variables in an abstract or spatial context using a 2D or 3D presentation space. Particularly relevant is the arrangement of the small multiples as it dictates how the time axis is perceived. Linear or circular arrangements can be used, or specific arrangement patterns can be applied to account for different granularities of the time axes. Small multiples provide an overview of the data and allow users to visually compare the data at different time points. Another advantage of small multiples is that the concept can be applied to virtually any existing visualization technique. The only thing to do is to create a thumbnail from an existing visual representation for each time step. Depending on the amount of screen space occupied by each thumbnail, however, the number of representable time steps could be rather small. Or, if the images are shrunk to fit more time steps, fewer details are visible.

Relevant references: Tufte (1983) • Tufte (1990)

time

primitives: **points**
arrangement: **linear, cyclic**

data

number of variables: **single, multiple**
frame of reference: **abstract, spatial**

vis

mapping of time: **static**
dimensionality: **2D, 3D**

time

EventViewer

primitives: **points, intervals**
arrangement: **linear, cyclic**

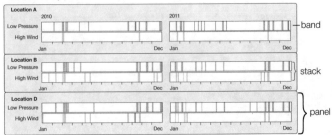

data

number of variables: **single, multiple**
frame of reference: **abstract, spatial**

Fig. A.129: Visual exploration of event data via the EventViewer. Spatial, temporal, and thematic dimensions of events can be assigned to configurations of bands, stacks, and panels. Left: low pressure and high wind events are shown along three locations and two years; right: display configuration to reveal temporal patterns along hours of a day. ©① *The authors. Adapted from Beard et al. (2008).*

vis

mapping of time: **static**
dimensionality: **2D**

The EventViewer by Beard et al. (2008) is a framework that has been developed to visualize and explore spatial, temporal, and thematic dimensions of sensor data. The system supports queries on events that have been extracted from such data and are stored in an events database. The spatial, temporal, and thematic categories of selected events can flexibly be assigned to three kinds of nested display elements called bands, stacks, and panels. Bands are the primary graphic object and act as display containers for a set of events. The horizontal dimension of a band represents time and bars within a band represent instances of events. The length of a bar corresponds to the event's duration and color can be used to encode other data values. Furthermore, missing data are shown by using gray bars to make a clear visual distinction to areas without events (shown as empty areas). Stacks consist of event bands that are placed on top of each other and panels are collections of stacks. Each of the three data dimensions space, time, and theme can be modeled along hierarchies or lattices. For time, calendric systems consisting of time granularities like hours, days, weeks, and years are used. The configuration of display elements is broken down along these hierarchical and lattice structures and form small multiples (↪ p. 359). The assignments can be changed interactively by the user via direct manipulation, thus revealing different kind of patterns, for example, periodic patterns, spatial and temporal trends, or event-event relationships.

Relevant references: Beard et al. (2008)

Ring Maps

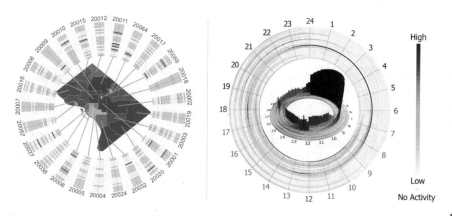

Fig. A.130: Ring maps representing health-related alert levels (green, yellow, orange, red) for various zip code regions during a period of 24 weeks (left) and degree of activity during the course of a day for 96 human activities, one shown per ring (right). *Left:* © *Courtesy of Guilan Huang. Right:* © *Courtesy of Jinfeng Zhao.*

The basic idea of ring maps is to create multiple differently sized rings, each of which is subdivided into an equal number of ring segments (see Zhao et al., 2008b; Huang et al., 2008). The rings and their segments as well as the center area of the overall visual representation can be used in various ways. One can utilize ring maps to visualize spatio-temporal data. To this end, a map is shown in the center and the ring segments of a particular angle are associated with a specific area of the map. This is depicted in the left figure, where different angles show the data for different zip code regions. A time series for each region can then be represented by the rings, for instance, by assigning the first series entry to the innermost ring and the last one to the outermost ring. The actual data visualization is done by color-coding. There are other ways of mapping information to rings and segments. The right figure shows an application of ring maps where the hours of the day are mapped to the ring segments and the rings represent different activities a person can be busy with during the course of a day. The degree of activity is encoded by color. This time the center of the display is used to show a complementary 3D representation of the same data to assist users in spotting highly active regions.

Relevant references: Zhao et al. (2008b) • Huang et al. (2008)

ThermalPlot

time

primitives: **points**
arrangement: **linear**

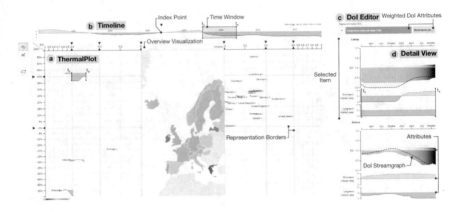

data

Fig. A.131: ThermalPlot summarizes combinations of multiple attributes over time using user-defined degree-of-interest (DoI) functions. The example shows the development of short-term and long-term interest rates for OECD countries between January 2000 and July 2015. ⓒ① *The authors. Generated with the ThermalPlot demo by Holger Stitz and Samuel Gratzl.*

number of variables: **multiple**
frame of reference: **abstract, spatial**

Visual exploration of large and multivariate time-series datasets to identify interesting items and patterns is an important analysis task in many application domains. However, this poses significant challenges to users as particularly the identification of important attributes and comparison of temporal developments among large numbers of items are complex tasks. Stitz et al. (2016) tackle this challenge in their ThermalPlot technique by mapping a user-specified degree-of-interest (DoI) value on the x-axis and the change of the DoI value (ΔDoI) on the y-axis. That is, the DoI and its first derivative are plotted against each other. Moreover, the DoI value is composed of a weighted combination of one or multiple attributes over time. Users can interactively compose the resulting DoI value and adjust individual weights to reflect the relevance of each component. The interface consists of three main parts: the ThermalPlot view (a) depicting the development of DoI vs. ΔDoI relative to a selected index point including a multi-level display, the overview timeline (b) which shows the overall DoI over time using a silhouette graph ($\hookrightarrow$ p. 281), and the DoI editor (c) that uses silhouette graphs and streamgraphs ($\hookrightarrow$ p. 286) to visualize the contribution of each component to the overall DoI. For high absolute or relative DoI ranges in the ThermalPlot that contain potentially more interesting items, a more detailed representation based on silhouette graphs is used. Toward the center of the representation, less details are shown (label + point or point only) and optionally, an overview representation that summarizes the less interesting items according to the DoI specification might be displayed (e.g., a choropleth map for spatial data or a treemap for abstract data).

mapping of time: **dynamic**
dimensionality: **2D**

vis

Relevant references: Stitz et al. (2016)

Circular

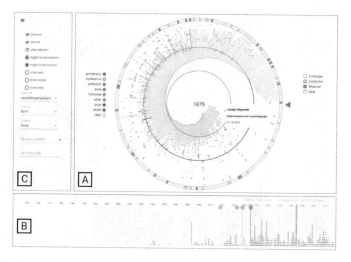

Fig. A.132: An overview of Circular. The main view (A) represents people, locations, events, and thematic changes by means of radially arranged rays. The rays are ordered according to user-selected criteria, in this case, by date of birth. People of interest in this dataset have many events occurring after their death. A timeline (B) shows historic events in a temporally ordered view. Several controls (C) allow users to modify the visualization. ⓒⓘⓢⓔ *Filipov et al. (2021).*

Circular by Filipov et al. (2021) is a technique to visualize a large amount of linked historical events and entities that are embedded in spatial and temporal frames of reference. Circular depicts entities as rays that are arranged in a radial fashion and sorted depending on a user-selected data attribute. Along the rays, one can see the events related to the temporal development of an entity represented as colored dots. The colors represent different types of event themes. On the circle's exterior, a selected categorical attribute can be visualized. While the data are based on time points, they can also be grouped according to a particular context, like artists' lifelines and related events. Circular provides a straightforward and engaging interface to present and get an overview of the data. Various interaction techniques support data exploration. The data can be sliced in various ways based on analysis tasks and user interests. The grouping and ordering can be changed to allow users to extract different insights from the visualization. Domain experts from the digital humanities tested Circular with data containing historical events and entities related to public music festivities in Vienna. Through the visualization they could better understand their data and answer questions like "Can you name exiled musicians that do not have a street in the city named after them?", "Which location was used most often for political stagings during the Second Republic?", or "Which exiled musicians never returned home?".

Relevant references: Filipov et al. (2021)

time

primitives: **points, intervals**
arrangement: **linear**

data

number of variables: **multiple**
frame of reference: **abstract, spatial**

vis

mapping of time: **static**
dimensionality: **2D**

ViDX

time

primitives: **points, intervals**
arrangement: **linear, cyclic**

data

number of variables: **multiple**
frame of reference: **abstract, spatial**

vis

mapping of time: **static, dynamic**
dimensionality: **2D, 3D**

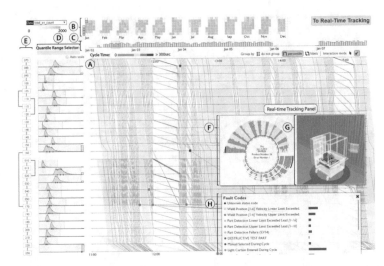

Fig. A.133: The ViDX system combines several views (e.g., Marey's graph, calendar view, bar chart, histograms, and radial graph) to facilitate the visual analysis of assembly lines. © *2017 IEEE. Reprinted, with permission, from Xu et al. (2017).*

The ViDX system by Xu et al. (2017) is a comprehensive solution for visually analyzing and monitoring assembly line performance in smart factories. As such, the system is related to the Industry 4.0 trend, which is concerned with increasing interconnectivity and smart automation. ViDX is based on the concept of coordinated multiple views (CMV) according to which the visual representation of complex data is divided into several dedicated views that are tightly linked via interaction mechanisms. The figure shows several such linked views, including a calendar view (↪ p. 269) and a bar graph (↪ p. 234) timeline at the top, a Marey's graph (center) and histograms and the assembly line schema (right). These views show historical data about the assembly line, its efficiency, and failure rates. For example, the Marey's graph shows the assembly line's stations along the vertical axis and time along the horizontal axis. Vertical lines in the Marey's graph represent products being processed simultaneously on the assembly line. A smoothly operating assembly line can be recognized by the lines forming parallel patterns without any gaps. Failures, interruptions, and delays are visible as line crossings and gaps in Marey's graph. Further information about such problems is available in a separate on-demand view. In addition to supporting the investigation of historical data, ViDX also allows users to monitor the current state of the assembly line. For this purpose, ViDX offers a radial graph and a 3D representation of the assembly line (center). While the radial graph represents the current status of all the products being processed at the stations of the assembly line, the 3D view provides the spatial context necessary to understand the overall operation of the assembly line.

Relevant references: Xu et al. (2017)

Value Flow Map

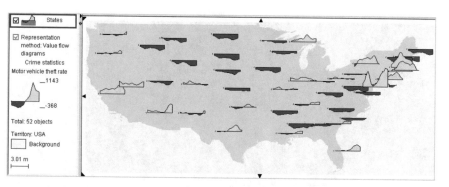

Fig. A.134: Univariate spatio-temporal data are represented by embedding multiple miniature silhouette graphs into regions of a map. The graphs use a specific encoding where the yellow color corresponds to a positive deviation of the variable from the data's mean and the blue color indicates a negative deviation. © *Courtesy of Gennady Andrienko.*

What Andrienko and Andrienko (2004) call value flow map is a technique to visualize variation in spatio-temporal data. A value flow map is an instance of the icons on maps approach (↪ p. 379), where a value flow map shows miniature silhouette graphs (↪ p. 281) for each map area to represent the temporal behavior of one data variable per area. Typically, temporal smoothing is carried out by replacing the values of a point-based time scale with the mean values of an interval-based time scale. In this way, small fluctuations are disregarded and major trends become visible. Moreover, a number of data transformations can be applied to define the mapping of the graphs. An example of such a transformation is to show the variable's deviation from the data's mean, rather than the raw data, that is, data values are replaced by their differences from the mean in order to represent positive and negative variations. This way, the silhouette graphs visualize quite well how the data values flow in time and space (hence the value flow map). This is a necessary requirement to enable analysts to detect patterns, and thus to support exploring spatial distributions, comparing data evolution at different locations, as well as finding similarities and outliers.

Relevant references: Andrienko and Andrienko (2004)

Flowstrates

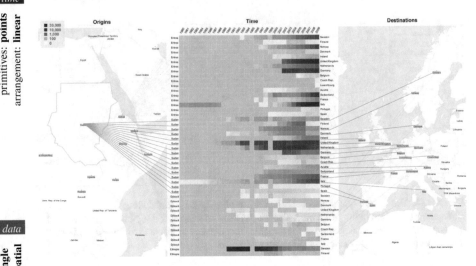

Fig. A.135: Flowstrates visualization showing migration of refugees from eastern Africa to European countries. Currently highlighted is the flow of refugees from Sudan. © *Courtesy of Ilya Boyandin.*

Flowstrates by Boyandin et al. (2011) extend the idea of flow maps (↪ p. 371) to the temporal dimension and allow the user to analyze changes in the flow magnitudes over time. In Flowstrates, the origins and the destinations of the flows are displayed in two separate maps, and the temporal changes of the flow magnitudes are displayed between the two maps in a heatmap in which the columns represent time periods. As in most flow maps that focus on representing the flow magnitudes, the exact routes of the flows are not accurately represented in Flowstrates. Instead, the flow lines are rerouted so that they connect the flow origins and destinations with the corresponding rows of the heatmap as if the flows were going through it. The flow lines help viewers see in the geographic maps the origin and the destination corresponding to each of the heatmap rows. To allow users to explore the whole data in every bit of detail, Flowstrates provide interactive support for performing spatial visual queries, focusing on different regions of interest for the origins and destinations, zooming and panning, sorting and aggregating the heatmap rows.

Relevant references: Boyandin et al. (2011)

Traffigram

primitives: **points**
arrangement: **linear**

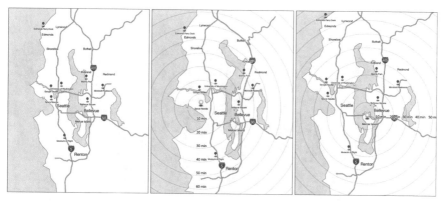

Fig. A.136: Regular maps (left) show spatial distances. The Traffigram distorts a map so that distances correspond to travel times from a selected origin. The distorted map in the center shows the Seattle Space Needle as the origin, the example to the right has the origin set to Mercer Island. ⓒⓘ *Courtesy of Sungsoo (Ray) Hong.*

number of variables: **single**
frame of reference: **spatial**

Traffigram is a technique for visualizing time, more concretely the time it takes to travel from a selected origin in space to other locations on the map. This technique is also called distance cartogram. Hong et al. (2014) designed their Traffigram as a distortion-based approach where a map is distorted so that distances between one origin to the rest of the points on the map correspond to travel time. The first step is to pinpoint a user-selected location with respect to which the Traffigram is to be generated. Secondly, isochronal contours are computed. These contours connect points in the map that are reachable from the focus location in the same amount of time. In other words, traveling from the focus point to any point on an isochronal contour takes constant time. The third step in generating a Traffigram is to distort the map by warping. Hong et al. (2014) use the thin-plate spline (TSP) algorithm for this purpose. The effect of this warping is that the isochronal contours become a circular shape so that readers can visually compare travel time between multiple locations at a glance. Two such distorted maps (center) and (right) and the original undistorted map (left) are shown in the figure. It can easily be seen from the figures that much of the distortion takes place in west/east directions away from the major highways, which go in north/south directions. So traveling in areas further away from highways takes more time. More recent work related to distance cartograms includes the design of an advanced visualization technique for creating the outcome without introducing excessive distortion that can hamper readers' understanding of geographical relationships (see Hong et al., 2017) and understanding the user-side effect of using distance cartograms in a more realistic situation through a two-weeks deployment study (see Hong et al., 2018).

mapping of time: **static**
dimensionality: **2D**

Relevant references: Hong et al. (2014) ● Hong et al. (2017) ● Hong et al. (2018)

Time-Varying Hierarchies on Maps

primitives: **points**
arrangement: **linear**

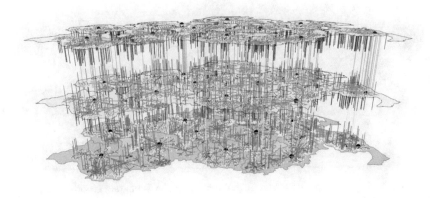

number of variables: **single**
frame of reference: **spatial**

Fig. A.137: Hierarchy layouts are embedded into areas of a map, where each map layer corresponds to one time step. Colored links and spikes between layers indicate significant changes from one time step to the other. ⓒⒾ *The authors. Generated with the LandVis system by Christian Tominski.*

Hierarchical structures can be found in many application areas. A technique for visualizing hierarchies that change over time in a geo-spatial context is described by Hadlak et al. (2010). This technique follows the idea of using the third dimension of the presentation space to represent the dimension of time, which is analog to the space-time cube approach (↪ p. 377). For a series of time steps, individual map layers are constructed, where each map region shows an embedded hierarchy layout and each node's color visualizes a data value. To facilitate the identification of changes between two layers, visual cues are added. Differently colored links between subsequent layers are used to indicate nodes that have moved or whose attribute values have changed significantly. Significance is determined by a user-selectable threshold. Positive attribute changes are shown as red links and negative changes are shown in blue. Links representing node movements are colored with a shade of gray. Addition or deletion of nodes and edges is indicated by spikes. Spikes that represent deletion leave a layer in the direction of the time axis and are shown in blue. Those that mark addition enter a layer and are shown in red. The layering approach in combination with the described visual cues allows users to compare successive time steps more closely.

mapping of time: **static**
dimensionality: **3D**

Relevant references: Hadlak et al. (2010)

Great Wall of Space-Time

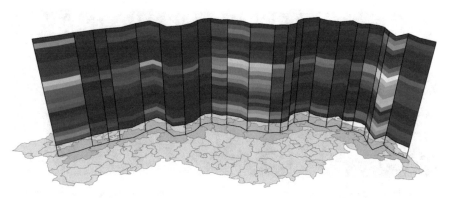

time

primitives: **points**
arrangement: **linear**

Fig. A.138: A Great Wall of Space-Time showing human health data. A path through space is extruded along the third dimension to create a slice through space-time, the wall, onto which the actual data representation is projected. ⓒⓘ *The authors. Generated with the LandVis system by Christian Tominski.*

data

number of variables: **single**
frame of reference: **spatial**

Tominski and Schulz (2012) introduce a visualization technique for spatio-temporal data that refers to 2D geographical space and 1D linear time. The idea is to construct a non-planar slice – called the Great Wall of Space-Time – through the 3D (2D+1D) space-time continuum. As such, the technique is based on the concept of the space-time cube (↪ p. 377). The construction of the wall is based on topological and geometrical aspects of the geographical space. First, a topological path is established automatically or interactively based on a neighborhood graph of the map regions. The topological path is transformed into a geometrical path that respects the geographic properties of the areas of the map. The geometrical path is then extruded to a 3D wall, whose 3rd dimension can be used to map the time domain. Different visual representations can be projected onto the wall in order to display the data. Examples illustrate data visualizations based on color-coding (as shown in the figure) and parallel coordinates. The wall has the advantage that it shows a closed path through space with no gaps between the information-bearing pixels on the screen.

vis

mapping of time: **static**
dimensionality: **3D**

Relevant references: Tominski and Schulz (2012)

MoSculp

Fig. A.139: MoSculp-rendered sculpture of an Olympic runner. The sculpture captures the 3D characteristics of the motion of the runner's limbs and torso. ⓒⓘ *The authors. Generated with the MoSculp software by Xiuming Zhang.*

Zhang et al. (2018) describe their MoSculp approach as a novel way of visualizing motion and time. MoSculp is similar to stroboscopic photography and shape-time photography in that it makes intermediate states of motion visible. In fact, MoSculp creates a three-dimensional sculpture of the motion of complex objects, such as human bodies or animals. While existing approaches can represent only one or a few discrete states of an object's motion at a time, a MoSculp sculpture creates a contiguous visual and tangible representation that captures all the motion's states. The figure shows MoSculp applied to the movement of an Olympic runner. More specifically, the sculpture (in blue) captures the movement of several parts of the runner's body, including the lower legs, the lower arms, and the torso. The creation of a MoSculp sculpture takes several steps. First, key points are extracted from the frames of a video clip. These 2D key points are then mapped to a sequence of 3D body models, which are then used to estimate the 3D motion for the entire video. The skeletons of the 3D body models form the basis for the creation of a MoSculp sculpture. The final visualization shows either the sequence of 3D models or a rendering of the sculpture, where different rendering styles can be selected by the user. It is even possible to create a physically 3D-printed version of the sculpture. In this sense, the MoSculp approach not only makes the key characteristics of motions visible, but also physically explorable.

Relevant references: Zhang et al. (2018)

Flow Map

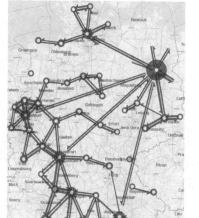

Fig. A.140: The flow map shows characteristic movements of photographers over time extracted from the metadata of more than 590,000 geo-referenced and time-stamped photographs. © *Courtesy of Gennady Andrienko.*

Flow maps show movements of objects over time, that is, they show a change of positions over time, rather than a change of data values. Usually, such movements form directed (optionally segmented) trajectories connecting the starting point of a movement and its end point. Such trajectories can be represented visually as more or less complex arrows or curves, where width, color, and other attributes can be used to encode additional information (see Kraak and Ormeling, 2020). A famous example is Minard's flow map of Napoleon's Russian campaign. A high number of flows, however, leads to overlapping trajectories and thus to visually cluttered flow maps. In order to represent massive data, flow maps can show abstractions of movements, rather than individual movements (see Andrienko and Andrienko, 2011). To faithfully communicate the underlying data, characteristic movements need to be extracted. First, time points are aggregated to larger time intervals and individual places are substituted with larger regions so as to arrive at abstracted trajectories that show mean trends. Secondly, the trajectories are grouped based on a similarity search, e.g., by applying cluster analysis or self-organizing maps (SOM). In this way, places with similar dynamics are merged, and individual trajectories are replaced by trajectories associated with the groups.

Relevant references: Kraak and Ormeling (2020) • Andrienko and Andrienko (2011)

Visits

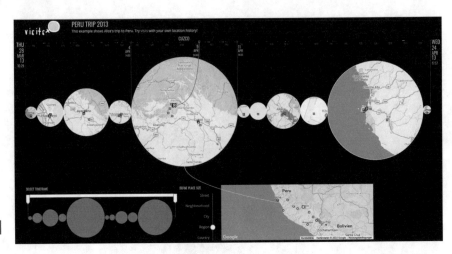

Fig. A.141: Location history of a trip to Peru visualized using the Visits approach. Map insets (center) with location markers are temporally arranged along a time axis (top). The insets have been generated automatically to match a region-level spatial resolution. Sliders (bottom left and center) can be used to adjust the spatial resolution and filter the time axis. An overview map (bottom right) provides spatial context. ⓒ① *The authors. Generated with the Visits software by Alice Thudt.*

Location histories contain information about a person's spatial and temporal where-abouts. That is, they describe at what time (when) a person was at certain locations (where). The Visits approach by Thudt et al. (2013) makes location histories visually accessible with a combined spatial and temporal visualization. The dimension of time is represented as a horizontal time axis (top in the figure). Each visited location is marked as a small dot at the time axis. More details about the locations' spatial context are represented in circular map insets (center) that are aligned along with the time axis. A map inset represents a temporally contiguous group of locations, which are marked by small crosses. The specific time point when a location was visited can be interactively highlighted via an arc drawn between the location and the time point at the time axis. The diameter of a map inset corresponds to the time interval covered in the data. The number of map insets depends on the chosen spatial resolution. Different resolutions (e.g., street, neighborhood, city, region) can be selected via a vertical slider (bottom center). Choosing street-level resolution will lead to more and smaller map insets, and region-level resolution leads to fewer and larger map insets. To keep users oriented, an overview map (bottom right) indicates where the map insets are located on a global scale. Moreover, a dual-handle range slider can be used to filter the time axis.

Relevant references: Thudt et al. (2013)

Time-Oriented Polygons on Maps

time

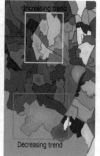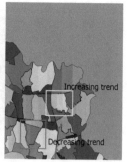

Fig. A.142: Time-oriented polygons are sub-dived according to time using three different schemes: wedges (left), rings (center), and time slices (right). Data values are encoded by color. The maps show the development of the high school population over the course of three years. © *2005 IEEE. Reprinted, with permission, from Shanbhag et al. (2005).*

Shanbhag et al. (2005) present three time-oriented visualization methods to analyze and support the effective allocation of resources in a spatio-temporal context. Wedges, rings, and time slices are the three basic layouts used to display changes in data values over time on a map. For all three variants, data values and categories are represented using color components (hue, saturation, and brightness). In the layout of the wedges, the area of a polygon is partitioned into radial sectors in a clock-like manner (left figure). The ring layout is inspired by the concentric rings of a tree trunk where the innermost ring corresponds to the earliest time point and the outermost ring corresponds to the latest time point (center figure). The time slices layout divides a polygon into vertical slices that are ordered from left to right according to the progress in time (right figure). The wedges, rings, and time slice layouts are applied to polygonal areas of a map. Time-oriented polygons on maps were used, for example, to plan the future allocation of resources for schools based on time-dependent variables such as student population by grade, number of students requiring free meals, and test scores needed.

Relevant references: Shanbhag et al. (2005)

primitives: **points**
arrangement: **linear, cyclic**

data

number of variables: **single**
frame of reference: **spatial**

vis

mapping of time: **static**
dimensionality: **2D**

Growth Ring Maps

time

primitives: **points**
arrangement: **linear, cyclic**

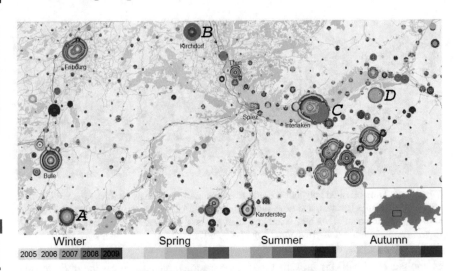

data

number of variables: **single**
frame of reference: **spatial**

Fig. A.143: Growth Ring Maps visualize spatio-temporal events as colored pixels accumulating in space around locations where the events occurred. This map shows information from a photo database, more concretely, it shows where and when photos were taken. © *Courtesy of Peter Bak.*

Growth Ring Maps by Bak et al. (2009) are a technique for visualizing the spatio-temporal distribution of events. Every spatio-temporal event is represented by one pixel, which makes the technique highly scalable with the number of events. Each location (for example the centroid of spatial clusters of events) is taken as the center point for the computation of growth rings. The pixels (i.e., events) are placed around this center point in an orbital manner resulting in the so-called Growth Ring representations. The pixels are sorted by the date and time the event occurred: the earlier an event happened, the closer the pixel is to the central point. Additional color-coding is used to visualize the association of a pixel with a time primitive. When two or more neighboring growth rings are about to overlap, the layout algorithm displaces the pixels in such a way that none of them is covered by another pixel. Hence, when big clusters of events are close in space, the corresponding growth rings will not have perfectly circular shapes but will be distorted. The resulting Growth Ring Maps are overlaid over a cartographic map to capture their spatial context. The figure illustrates Growth Ring Maps where events correspond to photos taken by tourists.

vis

mapping of time: **static**
dimensionality: **2D**

Relevant references: Bak et al. (2009) ● Andrienko et al. (2011)

Trajectory Wall

Fig. A.144: Visualization of trajectories of migrating storks. Colored trajectory bands represent the speed of migration in km/d. A radial time lens (bottom right) represents the distribution of speeds for the months of a year. ⓒⓘ *The authors. Generated with the TrajectoryVis software by Christian Tominski.*

The Trajectory Wall by Tominski et al. (2012b) shows temporal, spatial, and attribute information of movement data as a hybrid 2D/3D visual representation. Starting with an appropriate grouping of movement trajectories, the trajectories are mapped to 3D bands that are stacked above a map display to erect a wall of trajectories. A user-selected data attribute associated with the movement (e.g., speed, acceleration, sinuosity) is color-coded along the individual bands. The 3D representation is suitable for being watched from an oblique perspective. When looking at the map from a bird's-eye view, colored 2D lines appear on the map and provide an overall visualization of all movement trajectories. An interactive time lens (bottom right) enables the user to access temporarily aggregated information about a selected spatial region of the data. While the trajectory bands communicate the spatial information and linear temporal character of the data, the time lens emphasizes the cyclic temporal components. In combination, the interactive visualizations enable users to explore trajectory attributes with regard to their spatial and temporal dependencies. Movement patterns such as general commuting behavior, unexpected deviations, or trends in the development of trajectory attributes can be discerned.

Relevant references: Tominski et al. (2012b) • Andrienko et al. (2014)

primitives: **points**
arrangement: **linear, cyclic**

data

number of variables: **single**
frame of reference: **spatial**

vis

mapping of time: **static**
dimensionality: **2D, 3D**

time

Spatio-Temporal Event Visualization

primitives: **points**
arrangement: **linear**

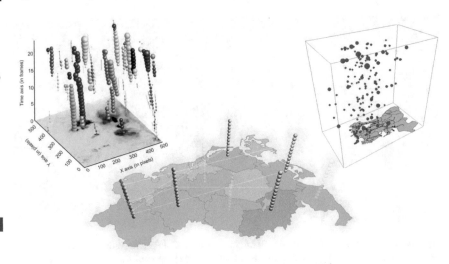

data

number of variables: **single, multiple**
frame of reference: **spatial**

Fig. A.145: Events in space and time are visualized by embedding graphical objects of varying size and color into space-time cubes. From left to right, the cubes show events related to convective clouds, human health data, and earthquakes. *Left:* © *2007 Elsevier. Reprinted, with permission, from Turdukulov et al. (2007).; Center:* ⓒⓘ *The authors. Generated with the LandVis system by Christian Tominski; Right:* © *2004 IEEE. Reprinted, with permission, from Gatalsky et al. (2004).*

vis

mapping of time: **static**
dimensionality: **3D**

Events usually describe happenings of interest. In order to analyze events in their spatial and temporal context, one can make use of the space-time cube concept (↪ p. 377), where space is mapped to the x-y plane and time is mapped to the z-axis of a 3D presentation space. The actual events are visualized by placing graphical objects in the space-time cube at those positions where events are located in time and space. Attributes associated with events can be encoded, for example, by varying the size, color, shape, or texture of the graphical objects. Marking events in a space-time cube is a general concept with a wide range of applications: Turdukulov et al. (2007) explore events related to the development of convective clouds, Tominski et al. (2005b) consider maxima in human health data as events of interest, and Gatalsky et al. (2004) visualize earthquake events.

Relevant references: Turdukulov et al. (2007) • Gatalsky et al. (2004) • Tominski et al. (2005b)

Space-Time Cube

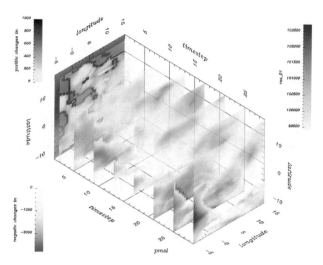

Fig. A.146: This space-time cube represents two spatial dimensions (latitude and longitude) along the y-axis and the z-axis, and time along the x-axis. Multiple color-coded layers are embedded into the cube to visualize spatio-temporal climate data. © *Courtesy of Thomas Nocke.*

A classic concept combining the visualization of space and time is the space-time cube, which is attributed to the pioneering work of Hägerstrand (1970). The basic idea is to map two spatial dimensions to two axes of a virtual three-dimensional cube and to use the third axis for the mapping of time. The spatial context is often represented as a map that constitutes one face of the space-time cube. The three-dimensional space inside the cube is used to represent spatio-temporal data, where manifold visual encodings are possible. One can place graphical objects in the cube in order to mark points of interest, or one can construct trajectories that illustrate paths of objects paths of objects (↪ p. 378). Associated data can be encoded to the properties of graphical objects and trajectories, where color and size are common candidates. Another technique is to place multiple layers along the time axis, each of which encodes the data for a specific time point. Space-time cubes usually rely on appropriate interaction to allow users to view the data from different perspectives. A contemporary review of the concept can be found in the work by Kraak (2003). User studies found that the 3D nature of space-time cubes matches quite well with the conditions in immersive environments (see Filho et al., 2020).

Relevant references: Hägerstrand (1970) • Kraak (2003) • Filho et al. (2020)

Space-Time Path

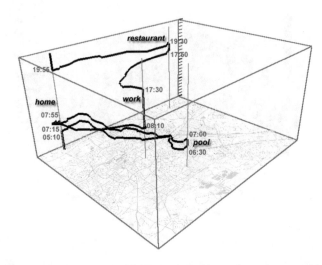

Fig. A.147: A space-time path is embedded into a space-time cube to show a person's movement through space and time. For better orientation, important places are marked by vertical lines and annotations. © *2003 International Cartographic Association. Reprinted, with permission, from Kraak (2003).*

The space-time path is a specific representation of data in a space-time cube ($\hookrightarrow$ p. 377). The roots of the concept of space-time paths can be found in the work by Lenntorp (1976). Kwan (2009) describes contemporary visual representations that are based on the classic concept. A space-time path is constructed by considering the location of an object as a three-dimensional point in space and time. Multiple such points ordered by time describe the path that an object has taken. The path can be rendered as a polyline that connects successive points. In order to encode data along a space-time path, one can vary the line's color, use differently dashed line segments, or employ other visual attributes. Alternatively, a space-time path can be represented as a three-dimensional tube, where the tube's radius can be varied to encode additional data values. Today's implementations usually offer interaction to allow for virtual movements through space and time, or for rotation and zoom.

Relevant references: Kraak (2003) • Lenntorp (1976) • Kwan (2009)

Icons on Maps

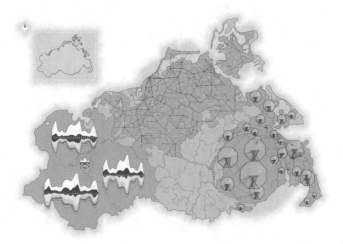

Fig. A.148: The figure illustrates the embedding of time-representing icons into a map in order to visualize spatio-temporal data. This illustration shows ThemeRiver icons and TimeWheel icons in the left and the right part of the map, respectively. The northern part of the map illustrates a conflict graph as used for local optimization of icon positions. ⓒⓘ *The authors. Generated with the LandVis system by Christian Tominski.*

When time-oriented data contain additional spatial dependencies, it is necessary to visualize both the temporal and the spatial aspects of the data. A sensible approach to achieving this is to adapt and combine existing solutions. Maps are commonly applied to represent the spatial context of the data. In order to combine maps with existing visualization techniques for representing the temporal context, they must be made compatible with the map display. First and foremost, this implies a reduction in size, which effectively means creating icons from otherwise full-size visual representations. In a second step, it is then possible to place icons on the map, exactly where the data are anchored in space. Tominski et al. (2003) and Fuchs and Schumann (2004b) demonstrate the integration of the ThemeRiver (↪ p. 293) and the TimeWheel (↪ p. 298) into a map display as shown in the figure. If there are too many icons on a map, they might occlude each other. Therefore, an additional computational step can be used to determine suitable overlap-free icon positions, a problem that is very much related to the cartographic map labeling problem. This problem can be tackled by global or local optimization of icon positions (see Fuchs and Schumann, 2004a; McNabb and Laramee, 2019).

Relevant references: Tominski et al. (2003) • Fuchs and Schumann (2004b) • Fuchs and Schumann (2004a) • McNabb and Laramee (2019)

time

primitives: **points, intervals**
arrangement: **linear, cyclic**

data

number of variables: **single, multiple**
frame of reference: **spatial**

vis

mapping of time: **static**
dimensionality: **2D, 3D**

VIS-STAMP

Fig. A.149: Multiple views show multivariate spatio-temporal crime data that have been clustered by a self-organizing map (SOM). An enhanced color-coding schema is used consistently across all views to visualize cluster affiliation. ⓒⓘ *The authors. Generated with the VIS-STAMP system by Diansheng Guo.*

Spatio-temporal data can be complex and multi-faceted. Guo et al. (2006) developed a system called VIS-STAMP that integrates computational, visual, and cartographic methods for visual analysis and exploration of such data. At the heart of the system is a self-organizing map (SOM) that is used for multivariate clustering, sorting, and coloring. The visual ensemble comprises a matrix view (top left), a map view (top right), a parallel coordinates view (bottom right), and a SOM view (bottom left). The matrix view's columns represent time points and its rows stand for geographic regions. Cluster affiliation of the matrix cells is visualized by means of an enhanced color-coding schema. The color-coding is consistent across all views. The map view follows the small multiples approach (↪ p. 359) and shows color-coded map thumbnails, one for each time point. The parallel coordinates view addresses the multivariate character of the data. Finally, the SOM view offers a detailed view and control interface of the underlying self-organizing map. A number of automatic and interactive manipulation techniques (e.g., reordering and sorting) facilitate the data analysis.

Relevant references: Guo et al. (2006)

Time-Ray Maps

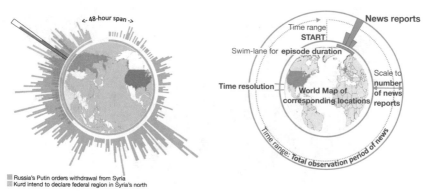

time

primitives: **points**
arrangement: **linear**

Fig. A.150: Time-Ray Maps allow users to assess the temporal evolution along with the geographic origins of news items at a glance. A circular time axis around the circumference of the world map is used to depict the observation period starting from the top. Bar height represents the amount of news items for a particular topic while color hue is used to distinguish between different news report types. Upon selection of a time interval in the bar graph, corresponding spatial information is highlighted in the world map. ⓒⓘ *Courtesy of Julia Sheidin.*

data

number of variables: **multiple**
frame of reference: **spatial**

Time-Ray Maps represent the temporal and spatial evolution of news episodes. Moreover, news stories can be compared to each other, as well as a quick analysis of a specific story in terms of influence is made possible. The circular bar graphs (↪ p. 234) along the outer circle represent the amount of news reports over time, starting at the top and continuing clockwise. Inside the inner circle, highlighted regions are displayed on a circular world map. For representing multiple news episodes, different color hues are overlaid semi-transparently to allow for direct comparison of temporal and spatial patterns. Interactive brushing by selecting individual bars or time intervals along the circular time axis highlights corresponding spatial regions in the center of the representation. The highlighted regions and bars are linked by using the same color hues. Inside the world map, the user can pan/rotate the visible area and zoom to regions of interest.

Relevant references: Sheidin et al. (2017)

vis

mapping of time: **static**
dimensionality: **2D**

Wakame

time

primitives: **points**
arrangement: **linear**

data

number of variables: **multiple**
frame of reference: **spatial**

vis

mapping of time: **static**
dimensionality: **3D**

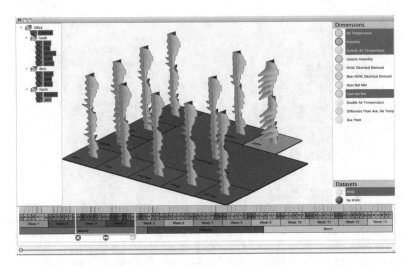

Fig. A.151: Three-dimensional visualization objects, so-called wakame, show multiple variables along time. Placing multiple wakame on a map allows analysts to make sense of spatio-temporal data. A time scale widget on the bottom can be used to select intervals of interest. © *Courtesy of Clifton Forlines.*

Forlines and Wittenburg (2010) describe an interactive system for visualizing multivariate spatio-temporal data. The temporal aspects are encoded to multivariate glyphs, so-called wakame. A single wakame basically corresponds to a radar chart that has been extruded along the third dimension. In a radar chart, different variables are represented on radially arranged axes that are connected to form a polyline. Wakame are constructed as solid three-dimensional objects whose shapes indicate temporal trends and relations among time-dependent variables, similar to the Kiviat tube (↪ p. 319). Embedding multiple wakame into a map display facilitates the understanding of spatial aspects. Noteworthy about the wakame approach are its interaction and animation facilities. An intelligent camera positioning mechanism supports users in finding perspectives on the wakame that most likely bear interesting information. A hierarchical time axis widget (↪ p. 336) denotes by color how different each time primitive is from its neighbors. This allows users to pick interesting time primitives easily. Upon interaction, an animation is used to smoothly interpolate between views. Additional animation schemes are offered to switch between the three-dimensional wakame view and traditional two-dimensional radar charts and line plots (↪ p. 233), which might be better suited for certain analysis tasks. Given their 3D nature, wakame can also be useful in the context of immersive analytics as demonstrated by Reski et al. (2020), who refer to their technique as 3D radar charts.

Relevant references: Forlines and Wittenburg (2010) • Reski et al. (2020)

GeoTime

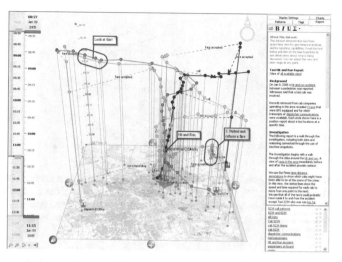

Fig. A.152: The visualization shows taxis involved in a hit and run accident as colored points and paths. Patterns of interest are annotated in the scene, and a narrative can be authored on the right. © *Courtesy of William Wright.*

Kapler and Wright (2005) describe GeoTime®, which is a registered trademark of Oculus Info Inc., as a system to visualize data items (e.g., objects, events, transactions, flows) in their spatial and temporal context. It provides a dynamic, interactive version of the space-time cube concept (↪ p. 377), where a map plane illustrates the spatial context and time is mapped vertically along the third display dimension. Items and tracks are placed in the space-time cube at their spatial and temporal coordinates. GeoTime provides a variety of visual and interactive capabilities. Time intervals of interest can be selected by the user and events are smoothly animated along the time axis. Alternative projections of the display allow users to focus more on either temporal or spatial aspects. Notable about GeoTime are its annotation, storytelling, and pattern recognition features (see Eccles et al., 2008). They enable automatic as well as user annotation of the representation with findings, as well as the creation of stories about the data for analytic exploration and communication. Additional functionality allows the analysis of events and transactions in time above a network diagram (see Kapler et al., 2008).

Relevant references: Kapler and Wright (2005) • Eccles et al. (2008) • Kapler et al. (2008)

time
primitives: **points**
arrangement: **linear**

data
number of variables: **multiple**
frame of reference: **spatial**

vis
mapping of time: **static, dynamic**
dimensionality: **3D**

Multiple Temporal Axes Model

time

primitives: **points, intervals**
arrangement: **linear**

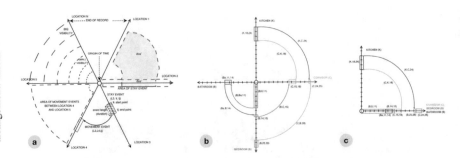

Fig. A.153: Multiple Temporal Axes Model. (a) Using the model to represent movement and activity events across different locations in the home. (b) A person's movements between the locations kitchen, corridor, bedroom, and bathroom in her home. (c) Axes representing different locations might also be collapsed on top of each other. © *2013 IEEE. Reprinted, with permission, from Juarez et al. (2013).*

data

number of variables: **multiple**
frame of reference: **spatial**

Tracking a person's movements between a set of known locations over time is an important task in a number of domains. In their work in the context of ambient assisted living (AAL), Juarez et al. (2013) designed a multiple temporal axes (MTA) model to detect accidents at home, which are still one of the major risk factors for the elderly. The data are collected via sensors and automated preprocessing is applied to detect stay as well as movement events. Using a specialized visualization model helps users to identify common risk scenarios via visual mining. In the visualization model, axes represent different locations and are arranged in a star-like layout. Time extends from the center to the outside on each axis and distinct colors are used for each location. Thus, each axis represents both, a spatial and a temporal perspective. When a person stays in one location (stay event), a colored rectangle is shown along the time axis. A movement event is represented by an arc that connects the origin and

vis

mapping of time: **static**
dimensionality: **2D**

destination location. As time evolves from the inside to the outside of the axes, arcs are getting longer the later the activity events occur. For summarizing movements in the house, a number of axes can be collapsed on top of each other (see (c) in figure). This can also be used to alleviate problems with arcs crossing multiple axes in case they cannot be arranged next to each other. The MTA model supports users in gaining an overview of the stay and activity sequences in terms of intervals and interval relations and identifying densities in the activities in order to monitor and detect accidents.

Relevant references: Juarez et al. (2013)

Data Vases

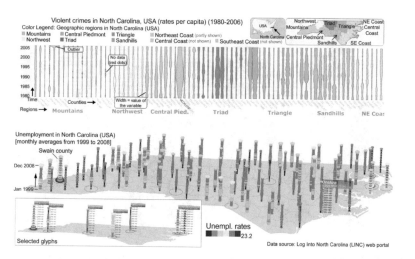

Fig. A.154: Shape and color of a data vase encode time-varying data values. The spatial component of the data can be communicated by a vertical alignment of 2D data vases (top), or by embedding 3D data vases into a space-time cube (bottom). © *2010 IEEE. Reprinted, with permission, from Thakur and Hanson (2010).*

The data vases technique has been designed to visualize multiple time-varying variables. Thakur and Rhyne (2009) describe two alternative designs: a 2D and a 3D variant. A 2D data vase is basically a graph constructed by mirroring a line plot (↪ p. 233) against the time axis, effectively creating a symmetric shape that can be filled (segment-wise) with a data-specific color. Such data vases can then be arranged on the screen to create a meaningful visualization. For spatio-temporal data, one can use multiple vertically aligned data vases, each of which represents an individual geographic region. Thakur and Hanson (2010) further elaborate on the idea of extending data vases to the third dimension. In 3D, data vases are constructed by stacking discs along a vertical time axis, where each disc maps the data for a particular time primitive to disc size and color. Such 3D data vases can then be embedded into a space-time cube (↪ p. 377), i.e., a virtual 3D world where two dimensions are used to show a geographic map and the third dimension encodes time.

Relevant references: Thakur and Rhyne (2009) • Thakur and Hanson (2010)

time

primitives: **points, intervals**
arrangement: **linear**

data

number of variables: **multiple**
frame of reference: **spatial**

vis

mapping of time: **static**
dimensionality: **2D, 3D**

Pencil Icons

Fig. A.155: Multiple time-dependent variables are mapped onto the faces of pencil icons to visualize temporal dependencies in the data. By placing the icons on a map, the spatial dependencies are communicated. ⓒ① *The authors. Generated with the LandVis system by Christian Tominski.*

Pencil icons visualize multivariate spatio-temporal data. The technique is based on the space-time cube concept (↪ p. 377), where the spatial frame of reference is represented as a map in the x-y plane of a virtual three-dimensional cube. The dimension of time is mapped along the cube's z-axis. Within the cube, pencil icons are positioned where data are available. This way, the spatial context is communicated. Each pencil icon represents the temporal context and multiple time-dependent variables by mapping time along the pencil, starting at the tip, and associating each face of the pencil with an individual time-dependent variable. Color-coding is applied to visualize the data. Color lightness is varied according to data values, and different hues are used to help users identify particular variables. The linear shape of the pencil is suited to represent the linear characteristics of the underlying time axis. Heterogeneous data can be depicted by using appropriate color scales. In order to deal with occlusion and information displayed at the pencils' back faces, several interaction techniques are provided, including navigation in the virtual world as well as the individual and linked rotation of pencils.

Relevant references: Tominski et al. (2005b)

Temporal Focus+Context

primitives: **intervals**
arrangement: **linear**

Fig. A.156: Using temporal focus+context, phases of a building's history are visualized using different graphical properties (e.g., rendering style, transparency) to emphasize and attenuate different time periods. © *Courtesy of Alexandre Carvalho.*

number of variables: **multiple**
frame of reference: **spatial**

Carvalho et al. (2008) introduce a temporal focus+context visualization model to meaningfully display several time points simultaneously. In this model, focus+context is applied to time rather than, more typically, to attributes or space. Underlying the proposed technique is the calculation of a temporal degree of interest (TDoI), which is driven by the valid time attribute, by specific analysis, exploration, or presentation goals as well as by user-defined visualization requirements. The TDoI is used to convey the temporal aspects of the data via adjusting graphical properties, such as transparency, color, sketchiness, or other non-photorealistic enhancements. This makes it possible to meaningfully compress information about distinct temporal states of the data into the same visualization display. The concept of temporal focus+context is generally applicable to different types of visualization of time-oriented data.

Relevant references: Carvalho et al. (2008)

mapping of time: **static**
dimensionality: **3D**

Chro-Ring

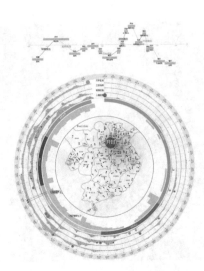

Fig. A.157: The time ring of Chro-Ring illustrates the life of the writer Su Shi. The upper part shows the life bar, illustrating the various events. The lower part shows the composition histogram, summarizing his life in different rings. Red arcs are successful periods and blue arcs indicate setbacks. In the middle of the ring, the map illustrates the various locations. © *2016 Springer Nature. Reprinted, with permission, from Zhu et al. (2016).*

Visualizing and exploring the personal histories of writers according to temporal and geo-spatial dimensions is a challenging task. Chro-Ring by Zhu et al. (2016) is a visual exploration environment that captures the individual histories of Chinese writers according to their life stories, geographic information, and relationships. The chronology graph is represented as ring, and various facets, views, and interaction techniques are provided to visually explore the writers' history. The most important views are (1) time ring, (2) life river, (3) histograms, (4) cloud map, (5) line chart, (6) tag cloud, and (7) radial spikes. Color-coding (red and blue) annotates successful and unsuccessful life periods. A usage scenario and user study (with 24 participants) illustrate the usability of Chro-Ring.

Relevant references: Zhu et al. (2016)

Helix Icons

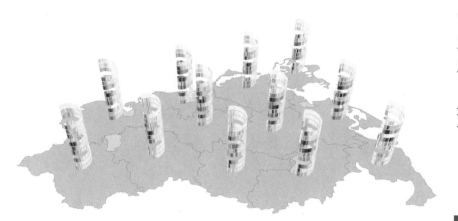

Fig. A.158: Helix icons use color-coding to visualize multivariate spatio-temporal data along helix ribbons, which emphasize the data's cyclic temporal character. The spatial aspect of the data is illustrated by embedding helix icons in a space-time cube. ©① *The authors. Generated with the LandVis system by Christian Tominski.*

Helix icons by Tominski et al. (2005b) are useful for emphasizing the cyclic character of spatio-temporal data. The underlying model of this technique is the space-time cube (↪ p. 377), which maps the spatial context to the x-axis and the y-axis, and the dimension of time to the z-axis of a virtual three-dimensional cube. The actual data visualization is embedded into the cube. Helix icons use a helix ribbon to roll up the time domain along the z-axis. Each segment of the helix ribbon visualizes a specific time point (or interval) in time by means of color-coding. Multiple time-dependent variables can be visualized by subdividing the helix ribbon into narrower sub-ribbons, each of which represents a different variable. Using unique hues for each sub-ribbon helps the user distinguish variables. As for spiral graphs (↪ p. 284), interaction techniques help users in finding an appropriate number of segments per cycle so that periodic patterns in the data are revealed. The inherent 3D representation problems (i.e., information displayed on helix back faces or inter-icon occlusion) are dealt with by offering 3D navigation through the space-time cube and rotation of helix icons. It is worth noting that helical representations are useful for time-oriented data with and without spatial context (see Gautier et al., 2016; Weber et al., 2001).

Relevant references: Tominski et al. (2005b) • Gautier et al. (2016) • Weber et al. (2001)

time

primitives: **points, intervals**
arrangement: **cyclic**

data

number of variables: **multiple**
frame of reference: **abstract, spatial**

vis

mapping of time: **static**
dimensionality: **3D**

Appendix B
Examples of Data Quality Issues

In Section 3.4, we briefly discussed a taxonomy of data quality problems. This appendix provides a list of concrete examples for all categories of problems from the taxonomy. We start with single-source problems followed by multi-source problems.

B.1 Single-Source Problems

B.1.1 Missing Data

Missing Value

- Missing time/interval and/or missing value
  (Date: NULL, items-sold: 20)
- Dummy entry
  (Date: 1970-01-01); (duration: -999)

Missing Tuple

- Missing time/interval + values
  (The whole tuple is missing)

B.1.2 Duplicate Data

Unique Value Violation

- Exact same time/interval although time/interval is defined as unique value
  (Holidays: 2012-04-09; 2012-04-09)

© The Editor(s) (if applicable) and The Author(s) 2023
W. Aigner et al., *Visualization of Time-Oriented Data*, Human–Computer
Interaction Series, https://doi.org/10.1007/978-1-4471-7527-8

Exact Duplicates

- Same time/interval and same values
  (Date: 2012-03-29, items-sold: 20 is in table twice)

Inconsistent Duplicates

- Same real entity with different times/intervals or values
  (patient: A, admission: 2012-03-29 8:00) vs.
  (patient: A, admission: 2012-03-29 8:30)
- Same real entity of time/interval (values) with different granularities (rounding)
  (Time: 11:00 vs. 11:03); (Weight: 34,67 vs. 35)

B.1.3 Implausible Data

Implausible Range

- Very early date/time in the future
  (Date: 1899-03-22); (date: 2099-03-22); (date: 1999-03-22, duration: 100y)

Unexpected Low/High Values

- Deviations from daily/weekly... profile or implausible values
  (Average sales on Monday: 50) vs. (this Monday: 500)
- Changes of subsequent values implausible
  (Last month: 4000 income) vs. (this month: 80000 income)
- Too long/short intervals between start–start/end–end
  Below one second at the cash desk
- Too long/short intervals between start–end/end–start
  Off-time between two shifts less than 8h
- Too long/short overall timespan (first to last entry)
  Continuous working for more than 12 hours
- Same value for too many succeeding records
  17 customers in every interval of the day

B.1.4 Outdated Data

Outdated Temporal Data

- Only old versions available
  Sales values from last year
- New version replaced by the old version
  Project plan tasks overwritten by prior version

B.1.5 Wrong Data

Wrong Data Type

- No time/interval
  Date: AAA; duration: *

Wrong Data Format

- Wrong date/time/datetime/duration format
  (Date: YYYY-MM-DD) vs. (date: YY-MM-DD); (duration: 7.7h) vs. (duration: 7h42')
- Times outside raster (e.g., for denoting end of the day)
  1-hour-raster but time is 23:59:00 for the end of the last interval

Misfielded Values

- Time in date field, date in time field/duration field
  (Time in date field: 14-03, date in timefield: 12:03:08)
- Values attached to the wrong/adjacent time/interval
  GPS data shows sprints followed by slow runs although the velocity was constant

Embedded Values

- Date+time in date field, time zone in time field/duration field
  (Time: 22:30) vs. (time: 22:30 CET)

Coded Wrongly or not Conform to Real Entity

- Wrong time zone
  UTC data in stead of local time
- Valid time/interval but not conform to the real entity
  (Admission: 2012-03-04) vs. (real admission: 2012-03-05)

Domain Violation (Outside Domain Range)

- Outliers in % of concurrent values (attention with small values) for a given point in time/interval
  On average (median) 30 customers in a shop in a given hour – in a 10' interval within that hour, a value of 200 is present
- Uneven or overlapping intervals
  Turnover data for 8:00–9:00, 9:00–11:00, 11:00–12:00
- Minimum/Maximum violation for given time/interval/type of day
  Sales at night even though no employees were present
- Sum of subintervals impossible
  Seeing the doctor + working hours longer than regular working hours
- Start, end, or duration do not form a valid interval
  (End ≤ start); (duration ≤ 0)
- Circularity in a self-relationship
  Interval A ⊂ interval B, interval B ⊂ interval A, A ≠ B

Incorrect Derived Values

- Error in computing duration
  Error computing sum of employees present within two intervals: (interval: 8:00–8:30, employees: 3), (interval: 8:30–9:00, employees: 3) → (interval: 8:00–9:00, employees: 6); no proper dealing with summer time-change; computing the number of work hours per day without deducting the breaks

B.1.6 Ambiguous Data

Abbreviations or Imprecise Unusual Coding

- Ambiguous time/interval/duration due to short format
  (Date: 06-03-05) vs. (date: 06-05-03); 5' interval encoded as '9:00': (interval: 8:55–9:00) vs. (interval: 9:00–09:05); average handling time per given interval: 3' – not clear: (average of completed interactions) vs. (average of started interactions) within this interval

- Extra symbols for time properties
  *+ or * or 28:00 for next day*

B.2 Multi-Source Problems

B.2.1 Heterogeneous Syntaxes

Different Data Formats/Synonyms

- Different date/duration formats
  (Date: YYYY-MM-DD) vs. (date: DD-MM-YYYY); (Date: 03-05 (March 5)) vs. (date: 03-05 (May 3))

Different Table Structure

- Time separated from date vs. date+time or start+duration in one column
  (Table A: start-date, start-time) vs. (table B: start-timestamp)

B.2.2 Heterogeneous Semantics

Heterogeneity of Scales (Measure Units/Aggregation)

- Different granularities; different interval length
  (Table A: whole hours only) vs. (table B: minutes)

Information Relates to Different Times/Intervals

- Different times/intervals
  (Table A: current sales as of yesterday) vs. (table B: sales as of last week)

B.2.3 References

Referential Integrity Violation/Dangling Data

- No reference to a given time/interval in another source
  (Table A: sales per day), (table B: sales assistants per day), problem: table B

does not contain a valid reference to a given day from table A or table A does not contain any referencing time

Incorrect Reference

- Reference exists in other sources but does not conform to real entity
  Sales of one day (table A) are assigned to certain sales assistants (from table B) because they reference the same day, however, in reality, a different crew was working on that day

B.3 Summary

With the help of the above examples, one can systematically check whether data quality problems exist. If some problems are overlooked, it is often the case that problems will pop up anyway when one tries to visualize incorrect data. This can often be recognized by artifacts in the visualization or algorithms failing to execute properly. Yet, verifying data correctness before the actual visualization is always preferable to stumble upon data problems during the data exploration.

References

Agarwal, S. and F. Beck (2020). "Set Streams: Visual Exploration of Dynamic Overlapping Sets". In: *Computer Graphics Forum* 39.3, pp. 383–391. DOI: `10.1111/cgf.13988`.

Aigner, W., C. Kainz, R. Ma, and S. Miksch (2011). "Bertin Was Right: An Empirical Evaluation of Indexing to Compare Multivariate Time-Series Data Using Line Plots". In: *Computer Graphics Forum* 30.1, pp. 215–228. DOI: `10.1111/j.1467-8659.2010.01845.x`.

Aigner, W., S. Miksch, W. Müller, H. Schumann, and C. Tominski (2007). "Visualizing Time-Oriented Data – A Systematic View". In: *Computers & Graphics* 31.3, pp. 401–409. DOI: `10.1016/j.cag.2007.01.030`.

Aigner, W., S. Miksch, W. Müller, H. Schumann, and C. Tominski (2008). "Visual Methods for Analyzing Time-Oriented Data". In: *IEEE Transactions on Visualization and Computer Graphics* 14.1, pp. 47–60. DOI: `10.1109/TVCG.2007.70415`.

Aigner, W., S. Miksch, B. Thurnher, and S. Biffl (2005). "PlanningLines: Novel Glyphs for Representing Temporal Uncertainties and Their Evaluation". In: *Proceedings of the International Conference Information Visualisation (IV)*. IEEE Computer Society, pp. 457–463. DOI: `10.1109/IV.2005.97`.

Akaishi, M. and Y. Okada (2004). "Time-tunnel : Visual Analysis Tool for Time-series Numerical Data and its Aspects as Multimedia Presentation Tool". In: *Proceedings of the International Conference Information Visualisation (IV)*. IEEE Computer Society, pp. 456–461. DOI: `10.1109/IV.2004.1320184`.

Akase, R. and Y. Okada (2015). "Web Based Time-Tunnel: An Interactive Multidimensional Data Visualization Tool Using Genetic Algorithm". In: *Proceedings of the International Conference Information Visualisation (IV)*. IEEE Computer Society, pp. 95–100. DOI: `10.1109/iV.2015.93`.

Albers, D., M. Correll, and M. Gleicher (2014). "Task-Driven Evaluation of Aggregation in Time Series Visualization". In: *Proceedings of the SIGCHI Conference on Human Factors in Computing Systems (CHI)*. ACM Press, pp. 551–560. DOI: `10.1145/2556288.2557200`.

© The Editor(s) (if applicable) and The Author(s) 2023
W. Aigner et al., *Visualization of Time-Oriented Data*, Human–Computer Interaction Series, https://doi.org/10.1007/978-1-4471-7527-8

Ali, M., A. Alqahtani, M. W. Jones, and X. Xie (2019). "Clustering and Classification for Time Series Data in Visual Analytics: A Survey". In: *IEEE Access* 7, pp. 181314–181338. DOI: `10.1109/ACCESS.2019.2958551`.

Allen, J. F. (1983). "Maintaining Knowledge about Temporal Intervals". In: *Communications of the ACM* 26.11, pp. 832–843. DOI: `10.1145/182.358434`.

André, P., M. L. Wilson, A. Russell, D. A. Smith, A. Owens, and m. schraefel m.c. (2007). "Continuum: Designing Timelines for Hierarchies, Relationships and Scale". In: *Proceedings of the ACM Symposium on User Interface Software and Technology (UIST)*. ACM Press, pp. 101–110. DOI: `10.1145/1294211.1294229`.

Andrienko, G., N. Andrienko, P. Bak, D. A. Keim, S. Kisilevich, and S. Wrobel (2011). "A Conceptual Framework and Taxonomy of Techniques for Analyzing Movement". In: *Journal of Visual Languages & Computing* 22.3, pp. 213–232. DOI: `10.1016/j.jvlc.2011.02.003`.

Andrienko, G., N. Andrienko, H. Schumann, and C. Tominski (2014). "Visualization of Trajectory Attributes in Space-Time Cube and Trajectory Wall". In: *Cartography from Pole to Pole*. Edited by Buchroithner, M., Prechtel, N., and Burghardt, D. Lecture Notes in Geoinformation and Cartography. Springer, pp. 157–163. DOI: `10.1007/978-3-642-32618-9_11`.

Andrienko, N. and G. Andrienko (2004). "Interactive Visual Tools to Explore Spatio-Temporal Variation". In: *Proceedings of the Conference on Advanced Visual Interfaces (AVI)*. ACM Press, pp. 417–420. DOI: `10.1145/989863.989940`.

Andrienko, N. and G. Andrienko (2006). *Exploratory Analysis of Spatial and Temporal Data*. Springer. DOI: `10.1007/3-540-31190-4`.

Andrienko, N. and G. Andrienko (2011). "Spatial Generalization and Aggregation of Massive Movement Data". In: *IEEE Transactions on Visualization and Computer Graphics* 17.2, pp. 205–219. DOI: `10.1109/TVCG.2010.44`.

Andrienko, N., G. Andrienko, G. Fuchs, A. Slingsby, C. Turkay, and S. Wrobel (2020). *Visual Analytics for Data Scientists*. Springer. DOI: `10.1007/978-3-030-56146-8`.

Andrienko, N., G. Andrienko, S. Miksch, H. Schumann, and S. Wrobel (2021). "A Theoretical Model for Pattern Discovery in Visual Analytics". In: *Visual Informatics* 5.1, pp. 23–42. DOI: `10.1016/j.visinf.2020.12.002`.

Angelini, M., G. Santucci, H. Schumann, and H.-J. Schulz (2018). "A Review and Characterization of Progressive Visual Analytics". In: *Informatics* 5.3, p. 31. DOI: `10.3390/informatics5030031`.

Ankerst, M. (2001). "Visual Data Mining with Pixel-oriented Visualization Techniques". In: *Proceedings of ACM SIGKDD Workshop on Visual Data Mining*. ACM Press.

Ankerst, M., A. Kao, R. Tjoelker, and C. Wang (2008). "DataJewel: Integrating Visualization with Temporal Data Mining". In: *Visual Data Mining*. Edited by Simoff, S., Böhlen, M., and Mazeika, A. Vol. 4404. Lecture Notes in Computer Science. Springer, pp. 312–330. DOI: `10.1007/978-3-540-71080-6_19`.

Antunes, C. M. and A. L. Oliveira (2001). *Temporal Data Mining: An Overview*. Workshop on Temporal Data Mining at the ACM SIGKDD International Conference on Knowledge Discovery and Data Mining (KDD). URL: https://www.researchgate.net/publication/284602094.

Arbesser, C., F. Spechtenhauser, T. Mühlbacher, and H. Piringer (2017). "Visplause: Visual Data Quality Assessment of Many Time Series Using Plausibility Checks". In: *IEEE Transactions on Visualization and Computer Graphics* 23.1, pp. 641–650. DOI: 10.1109/tvcg.2016.2598592.

Bach, B., P. Dragicevic, D. W. Archambault, C. Hurter, and S. Carpendale (2017). "A Descriptive Framework for Temporal Data Visualizations Based on Generalized Space-Time Cubes". In: *Computer Graphics Forum* 36.6, pp. 36–61. DOI: 10.1111/cgf.12804.

Bach, B., N. H. Riche, T. Dwyer, T. M. Madhyastha, J.-D. Fekete, and T. J. Grabowski (2015). "Small MultiPiles: Piling Time to Explore Temporal Patterns in Dynamic Networks". In: *Computer Graphics Forum* 34.3, pp. 31–40. DOI: 10.1111/cgf.12615.

Bach, B., C. Shi, N. Heulot, T. Madhyastha, T. Grabowski, and P. Dragicevic (2016). "Time Curves: Folding Time to Visualize Patterns of Temporal Evolution in Data". In: *IEEE Transactions on Visualization and Computer Graphics* 22.1, pp. 559–568. DOI: 10.1109/TVCG.2015.2467851.

Bade, R., S. Schlechtweg, and S. Miksch (2004). "Connecting Time-oriented Data and Information to a Coherent Interactive Visualization". In: *Proceedings of the SIGCHI Conference on Human Factors in Computing Systems (CHI)*. ACM Press, pp. 105–112. DOI: 10.1145/985692.985706.

Bai, Z., Y. Tao, and H. Lin (2020). "Time-varying Volume Visualization: A Survey". In: *Journal of Visualization* 23.5, pp. 745–761. DOI: 10.1007/s12650-020-00654-x.

Bak, P., F. Mansmann, H. Janetzko, and D. A. Keim (2009). "Spatiotemporal Analysis of Sensor Logs using Growth Ring Maps". In: *IEEE Transactions on Visualization and Computer Graphics* 15.6, pp. 913–920. DOI: 10.1109/TVCG.2009.182.

Baldonado, M. Q. W., A. Woodruff, and A. Kuchinsky (2000). "Guidelines for Using Multiple Views in Information Visualization". In: *Proceedings of the Conference on Advanced Visual Interfaces (AVI)*. ACM Press, pp. 110–119. DOI: 10.1145/345513.345271.

Bale, K., P. Chapman, N. Barraclough, J. Purdy, N. Aydin, and P. Dark (2007). "Kaleidomaps: A New Technique for the Visualization of Multivariate Time-Series Data". In: *Information Visualization* 6.2, pp. 155–167. DOI: 10.1057/palgrave.ivs.9500154.

Bale, K., P. Chapman, J. Purdy, N. Aydin, and P. Dark (2006). "Kaleidomap Visualizations of Cardiovascular Function in Critical Care Medicine". In: *Proceedings of International Conference on Medical Information Visualisation - BioMedical Visualisation (MediVis)*. IEEE Computer Society, pp. 51–58. DOI: 10.1109/MEDIVIS.2006.19.

Barbeu-Dubourg, J. (1753). *Carte chronographique*. Paper roll in scroll case, 40 cm x 16.5 m. Princeton University Library.

Bartolomeo, M. D. and Y. Hu (2016). "There is More to Streamgraphs than Movies: Better Aesthetics via Ordering and Lassoing". In: *Computer Graphics Forum* 35.3, pp. 341–350. DOI: 10.1111/cgf.12910.

Baur, D., B. Lee, and S. Carpendale (2012). "TouchWave: Kinetic Multi-touch Manipulation for Hierarchical Stacked Graphs". In: *Proceedings of the International Conference on Interactive Tabletops and Surfaces (ITS)*. ACM Press, pp. 255–264. DOI: 10.1145/2396636.2396675.

Beard, K., H. Deese, and N. R. Pettigrew (2008). "A Framework for Visualization and Exploration of Events". In: *Information Visualization* 7.2, pp. 133–151. DOI: 10.1145/1466620.1466623.

Beck, F., M. Burch, S. Diehl, and D. Weikopf (2017). "A Taxonomy and Survey of Dynamic Graph Visualization". In: *Computer Graphics Forum* 36.1, pp. 133–159. DOI: 10.1111/cgf.12791.

Beck, F., M. Burch, and D. Weiskopf (2016). "A Matrix-Based Visual Comparison of Time Series Sports Data". In: *Proceedings of the Symposium on Vision, Modeling & Visualization (VMV)*. Eurographics Association. DOI: 10.2312/vmv.20161342.

Becker, R. A. and W. S. Cleveland (1987). "Brushing Scatterplots". In: *Technometrics* 29.2, pp. 127–142. DOI: 10.2307/1269768.

Bederson, B. B., A. Clamage, M. P. Czerwinski, and G. G. Robertson (2004). "DateLens: A Fisheye Calendar Interface for PDAs". In: *ACM Transactions on Computer-Human Interaction (TOCHI)* 11.1, pp. 90–119. DOI: 10.1145/972648.972652.

Been, K., E. Daiches, and C.-K. Yap (2006). "Dynamic Map Labeling". In: *IEEE Transactions on Visualization and Computer Graphics* 12.5, pp. 773–780. DOI: 10.1109/TVCG.2006.136.

Behrisch, M., J. Davey, T. Schreck, D. Keim, and J. Kohlhammer (2012). "Matrix-based Visual Correlation Analysis on Large Timeseries Data". In: *Proceedings of the IEEE Conference on Visual Analytics Science and Technology (VAST)*. IEEE Computer Society, pp. 209–210. DOI: 10.1109/VAST.2012.6400549.

Bergman, L., B. E. Rogowitz, and L. A. Treinish (1995). "A Rule-based Tool for Assisting Colormap Selection". In: *Proceedings of the IEEE Visualization Conference (Vis)*. IEEE Computer Society, pp. 118–125. DOI: 10.1109/VISUAL.1995.480803.

Bernard, J., C. Bors, M. Bögl, C. Eichner, T. Gschwandtner, S. Miksch, H. Schumann, and J. Kohlhammer (2018). "Combining the Automated Segmentation and Visual Analysis of Multivariate Time Series". In: *Proceedings of the EuroVis Workshop on Visual Analytics (EuroVA)*. Eurographics Association, pp. 49–53. DOI: 10.2312/eurova.20181112.

Bernard, J., T. Ruppert, O. Goroll, T. May, and J. Kohlhammer (2012). "Visual-Interactive Preprocessing of Time Series Data". In: *Proceedings of the Annual Conference of the Swedish Computer Graphics Association (SIGRAD)*. 81. Linköping University Electronic Press, pp. 39–48. URL: https://ep.liu.se/ecp/081/006/ecp12081006.pdf.

Bernard, J., M. Steiger, S. Mittelstädt, S. Thum, D. A. Keim, and J. Kohlhammer (2015). "A Survey and Task-based Quality Assessment of Static 2D Colormaps". In: *Proceedings of the Conference on Visualization and Data Analysis (VDA)*. Vol. 9397. SPIE Proceedings. SPIE. DOI: 10.1117/12.2079841.

Berry, L. and T. Munzner (2004). "BinX: Dynamic Exploration of Time Series Datasets Across Aggregation Levels". In: *Poster Compendium of IEEE Symposium on Information Visualization (InfoVis)*. IEEE Computer Society, pp. 5–6. DOI: 10.1109/INFVIS.2004.11.

Bertin, J. (1981). *Graphics and Graphic Information-Processing*. Translated by William J. Berg and Paul Scott. de Gruyter.

Bertin, J. (1983). *Semiology of Graphics: Diagrams, Networks, Maps*. Translated by William J. Berg. University of Wisconsin Press.

Bettini, C., S. Jajodia, and X. S. Wang (2000). *Time Granularities in Databases, Data Mining, and Temporal Reasoning*. 1st edition. Springer. DOI: 10.1007/978-3-662-04228-1.

Bier, E. A., M. C. Stone, K. Pier, W. Buxton, and T. D. DeRose (1993). "Toolglass and Magic Lenses: The See-Through Interface". In: *Proceedings of the Annual Conference on Computer Graphics and Interactive Techniques (SIGGRAPH)*. ACM Press, pp. 73–80. DOI: 10.1145/166117.166126.

Borland, D. and R. M. Taylor (2007). "Rainbow Color Map (Still) Considered Harmful". In: *IEEE Computer Graphics and Applications* 27.2, pp. 14–17. DOI: 10.1109/mcg.2007.323435.

Börner, K., A. Bueckle, and M. Ginda (2019). "Data Visualization Literacy: Definitions, Conceptual Frameworks, Exercises, and Assessments". In: *Proceedings of the National Academy of Sciences* 116.6, pp. 1857–1864. DOI: 10.1073/pnas.1807180116.

Börner, K., A. Maltese, R. N. Balliet, and J. Heimlich (2016). "Investigating Aspects of Data Visualization Literacy Using 20 Information Visualizations and 273 Science Museum Visitors". In: *Information Visualization* 15.3, pp. 198–213. DOI: 10.1177/1473871615594652.

Bors, C. (2020). "Facilitating Data Quality Assessment Utilizing Visual Analytics: Tackling Time, Metrics, Uncertainty, and Provenance". PhD thesis. Institute of Visual Computing and Human-Centered Technology, TU Wien.

Bors, C., J. Bernard, M. Bögl, T. Gschwandtner, J. Kohlhammer, and S. Miksch (2019). "Quantifying Uncertainty in Multivariate Time Series Pre-Processing". In: Eurographics Association. DOI: 10.2312/eurova.20191121.

Bors, C., C. Eichner, S. Miksch, C. Tominski, H. Schumann, and T. Gschwandtner (2020). "Exploring Time Series Segmentations Using Uncertainty and Focus+Context Techniques". In: *Proceedings of the Eurographics / IEEE Conference on Visualization (EuroVis) - Short Papers*. Eurographics Association, pp. 7–11. DOI: 10.2312/evs.20201040.

Bouali, F., S. Devaux, and G. Venturini (2016). "Visual Mining of Time Deries Using a Tubular Visualization". In: *The Visual Computer* 32.1, pp. 15–30. DOI: 10.1007/s00371-014-1052-0.

Boyandin, I., E. Bertini, P. Bak, and D. Lalanne (2011). "Flowstrates: An Approach for Visual Exploration of Temporal Origin-Destination Data". In: *Computer Graphics Forum* 30.3, pp. 971–980. DOI: 10.1111/j.1467-8659.2011. 01946.x.

Brehmer, M., B. Lee, B. Bach, N. H. Riche, and T. Munzner (2017). "Timelines Revisited: A Design Space and Considerations for Expressive Storytelling". In: *IEEE Transactions on Visualization and Computer Graphics* 23.9, pp. 2151–2164. DOI: 10.1109/TVCG.2016.2614803.

Brehmer, M., B. Lee, J. Stasko, and C. Tominski (2021). "Interacting with Visualization on Mobile Devices". In: *Mobile Data Visualization*. Edited by Lee, B., Dachselt, R., Isenberg, P., and Choe, E. K. CRC Press, pp. 67–110. DOI: 10.1201/9781003090823-3.

Brehmer, M. and T. Munzner (2013). "A Multi-Level Typology of Abstract Visualization Tasks". In: *IEEE Transactions on Visualization and Computer Graphics* 19.12, pp. 2376–2385. DOI: 10.1109/TVCG.2013.124.

Brinton, W. C. (1914). *Graphic Methods for Presenting Facts*. New York, NY, USA: The Engineering Magazine Company.

Brinton, W. C. (1939). *Graphic Presentation*. New York, NY, USA: Brinton Associates.

Brockwell, P. J. and R. A. Davis (1991). *Time Series: Theory and Methods*. 2nd edition. 2009 reprint. Springer. DOI: 10.1007/978-1-4419-0320-4.

Brodbeck, D. and L. Girardin (2003). "Interactive Poster: Trend Analysis in Large Timeseries of High-Throughput Screening Data Using a Distortion-Oriented Lens with Semantic Zooming". In: *Poster Compendium of IEEE Symposium on Information Visualization (InfoVis)*. IEEE Computer Society, pp. 74–75. URL: https://download.macrofocus.com/publications/infovis2003-poster.pdf.

Bu, C., Q. Zhang, Q. Wang, J. Zhang, M. Sedlmair, O. Deussen, and Y. Wang (2021). "SineStream: Improving the Readability of Streamgraphs by Minimizing Sine Illusion Effects". In: *IEEE Transactions on Visualization and Computer Graphics* 27.2, pp. 1634–1643. DOI: 10.1109/TVCG.2020.3030404.

Buono, P., A. Aris, C. Plaisant, A. Khella, and B. Shneiderman (2005). "Interactive Pattern Search in Time Series". In: *Proceedings of the Conference on Visualization and Data Analysis (VDA)*. SPIE, pp. 175–186. DOI: 10.1117/12.587537.

Buono, P., C. Plaisant, A. Simeone, A. Aris, B. Shneiderman, G. Shmueli, and W. Jank (2007). "Similarity-Based Forecasting with Simultaneous Previews: A River Plot Interface for Time Series Forecasting". In: *Proceedings of the International Conference Information Visualisation (IV)*. IEEE Computer Society, pp. 191–196. DOI: 10.1109/IV.2007.101.

Burch, M., F. Beck, and S. Diehl (2008). "Timeline Trees: Visualizing Sequences of Transactions in Information Hierarchies". In: *Proceedings of the Conference on Advanced Visual Interfaces (AVI)*. ACM Press, pp. 75–82. DOI: 10.1145/1385569.1385584.

Byron, L. and M. Wattenberg (2008). "Stacked Graphs – Geometry & Aesthetics". In: *IEEE Transactions on Visualization and Computer Graphics* 14.6, pp. 1245–1252. DOI: 10.1109/TVCG.2008.166.

Card, S., J. Mackinlay, and B. Shneiderman (1999). *Readings in Information Visualization: Using Vision to Think.* Morgan Kaufmann Publishers.

Card, S. K., B. Suh, B. A. Pendleton, J. Heer, and J. W. Bodnar (2006). "Time Tree: Exploring Time Changing Hierarchies". In: *Proceedings of the IEEE Symposium on Visual Analytics Science and Technology (VAST).* IEEE Computer Society, pp. 3–10. DOI: 10.1109/VAST.2006.261450.

Carlis, J. V. and J. A. Konstan (1998). "Interactive Visualization of Serial Periodic Data". In: *Proceedings of the ACM Symposium on User Interface Software and Technology (UIST).* ACM Press, pp. 29–38. DOI: 10.1145/288392.288399.

Carvalho, A., A. S. Augusto de, C. Ribeiro, and E. Costa (2008). "A Temporal Focus + Context Visualization Model for Handling Valid-Time Spatial Information". In: *Information Visualization* 7.3-4, pp. 265–274. DOI: 10.1057/palgrave.ivs.9500188.

Ceneda, D., N. Andrienko, G. Andrienko, T. Gschwandtner, S. Miksch, N. Piccolotto, T. Schreck, M. Streit, J. Suschnigg, and C. Tominski (2020). "Guide Me in Analysis: A Framework for Guidance Designers". In: *Computer Graphics Forum* 39.6, pp. 269–288. DOI: 10.1111/cgf.14017.

Ceneda, D., T. Gschwandtner, T. May, S. Miksch, H.-J. Schulz, M. Streit, and C. Tominski (2017). "Characterizing Guidance in Visual Analytics". In: *IEEE Transactions on Visualization and Computer Graphics* 23.1, pp. 111–120. DOI: 10.1109/TVCG.2016.2598468.

Ceneda, D., T. Gschwandtner, and S. Miksch (2019). "A Review of Guidance Approaches in Visual Data Analysis: A Multifocal Perspective". In: *Computer Graphics Forum* 38.3, pp. 861–879. DOI: 10.1111/cgf.13730.

Ceneda, D., T. Gschwandtner, S. Miksch, and C. Tominski (2018). "Guided Visual Exploration of Cyclical Patterns in Time-series". In: *Proceedings of the IEEE Symposium on Visualization in Data Science (VDS).* IEEE Computer Society.

Chang, R., M. Ghoniem, R. Kosara, W. Ribarsky, J. Yang, E. Suma, C. Ziemkiewicz, D. Kern, and A. Sudjianto (2007). "WireVis: Visualization of Categorical, Time-Varying Data From Financial Transactions". In: *Proceedings of the IEEE Symposium on Visual Analytics Science and Technology (VAST).* IEEE Computer Society, pp. 155–162. DOI: 10.1109/VAST.2007.4389009.

Chapple, S. and R. Garofalo (1977). *Rock 'N' Roll is Here to Pay: The History and Politics of the Music Industry.* Chicago, IL, USA: Burnham Inc Pub.

Chen, C. (2006). "CiteSpace II: Detecting and Visualizing Emerging Trends and Transient Patterns in Scientific Literature". In: *Journal of the American Society for Information Science and Technology* 57.3, pp. 359–377. DOI: 10.1002/asi.20317.

Chen, H. (2004). "Compound Brushing Explained". In: *Information Visualization* 3.2, pp. 96–108. DOI: 10.1057/palgrave.ivs.9500068.

Chen, S., N. V. Andrienko, G. L. Andrienko, J. Li, and X. Yuan (2021). "Co-Bridges: Pair-wise Visual Connection and Comparison for Multi-item Data Streams". In:

IEEE Transactions on Visualization and Computer Graphics 27.2, pp. 1612–1622. DOI: 10.1109/TVCG.2020.3030411.

Cheng, X., D. Cook, and H. Hofmann (2016). "Enabling Interactivity on Displays of Multivariate Time Series and Longitudinal Data". In: *Journal of Computational and Graphical Statistics* 25.4, pp. 1057–1076. DOI: 10.1080/10618600.2015.1105749.

Chi, E. H. (2000). "A Taxonomy of Visualization Techniques Using the Data State Reference Model". In: *Proceedings of the IEEE Symposium Information Visualization (InfoVis)*. IEEE Computer Society, pp. 69–76. DOI: 10.1109/INFVIS.2000.885092.

Chittaro, L. and C. Combi (2003). "Visualizing Queries on Databases of Temporal Histories: New Metaphors and their Evaluation". In: *Data & Knowledge Engineering* 44.2, pp. 239–264. DOI: 10.1016/s0169-023x(02)00137-4.

Chittaro, L., C. Combi, and G. Trapasso (2003). "Data Mining on Temporal Data: A Visual Approach and its Clinical Application to Hemodialysis". In: *Journal of Visual Languages and Computing* 14.6, pp. 591–620. DOI: 10.1016/j.jvlc.2003.06.003.

Cho, M., B. Kim, H.-J. Bae, and J. Seo (2014). "Stroscope: Multi-Scale Visualization of Irregularly Measured Time-Series Data". In: *IEEE Transactions on Visualization and Computer Graphics* 20.5, pp. 808–821. DOI: 10.1109/TVCG.2013.2297933.

Chuah, M. C. (1998). "Dynamic Aggregation with Circular Visual Designs". In: *Proceedings of the IEEE Symposium Information Visualization (InfoVis)*. IEEE Computer Society, pp. 35–43. DOI: 10.1109/INFVIS.1998.729557.

Chuah, M. C. and S. G. Eick (1998). "Information Rich Glyphs for Software Management Data". In: *IEEE Computer Graphics and Applications* 18.4, pp. 24–29. DOI: 10.1109/38.689658.

Claessen, J. H. T. and J. J. van Wijk (2011). "Flexible Linked Axes for Multivariate Data Visualization". In: *IEEE Transactions on Visualization and Computer Graphics* 17.12, pp. 2310–2316. DOI: 10.1109/TVCG.2011.201.

Clancey, W. J. (1985). "Heuristic Classification". In: *Artificial Intelligence* 27.3, pp. 289–350. DOI: 10.1016/0004-3702(85)90016-5.

Cleveland, W. S. (1993). *Visualizing Data*. Hobart Press.

Cleveland, W. S., M. E. McGill, and R. McGill (1988). "The Shape Parameter of a Two-Variable Graph". In: *Journal of the American Statistical Association* 83.402, pp. 289–300. DOI: 10.1080/01621459.1988.10478598.

Cockburn, A., A. Karlson, and B. B. Bederson (2009). "A Review of Overview+Detail, Zooming, and Focus+Context Interfaces". In: *ACM Computing Surveys* 41.1, 2:1–2:31. DOI: 10.1145/1456650.1456652.

Cohen, P., B. Heeringa, and N. Adams (2002). "Unsupervised Segmentation of Categorical Time Series into Episodes". In: *Proceedings of the International Conference on Data Mining (ICDM)*. IEEE Computer Society, pp. 99–106. DOI: 10.1109/ICDM.2002.1183891.

Collins, C., N. Andrienko, T. Schreck, J. Yang, J. Choo, U. Engelke, A. Jena, and T. Dwyer (2018). "Guidance in the Human-Machine Analytics Process". In: *Visual Informatics* 2.3. DOI: 10.1016/j.visinf.2018.09.003.

Combi, C., E. Keravnou-Papailiou, and Y. Shahar (2010). *Temporal Information Systems in Medicine*. Springer. DOI: 10.1007/978-1-4419-6543-1.

Combi, C. and G. Pozzi (2001). "HMAP – A Temporal Data Model Managing Intervals with Different Granularities and Indeterminacy from Natural Language Sentences". In: *The VLDB Journal* 9.4, pp. 294–311. DOI: 10.1007/s007780100033.

Constantine, L. L. (2003). "Canonical Abstract Prototypes for Abstract Visual and Interaction Design". In: *Interactive Systems: Design, Specification, and Verification*. Edited by Jorge, J., Nunes, N. J., and e Cunha, J. F. Vol. 2844. Lecture Notes in Computer Science. Springer, pp. 1–15. DOI: 10.1007/978-3-540-39929-2_1.

Cooper, A., R. Reimann, and D. Cronin (2007). *About Face 3: The Essentials of Interaction Design*. Wiley Publishing, Inc.

Courage, C. and K. Baxter (2005). *Understanding Your Users*. Morgan Kaufmann. DOI: 10.1016/B978-1-55860-935-8.X5029-5.

Cousins, S. B. and M. G. Kahn (1991). "The Visual Display of Temporal Information". In: *Artificial Intelligence in Medicine* 3.6, pp. 341–357. DOI: 10.1016/0933-3657(91)90005-v.

Cuenca, E., A. Sallaberry, F. Y. Wang, and P. Poncelet (2018). "MultiStream: A Multiresolution Streamgraph Approach to Explore Hierarchical Time Series". In: *IEEE Transactions on Visualization and Computer Graphics* 24.12, pp. 3160–3173. DOI: 10.1109/TVCG.2018.2796591.

Cui, W., S. Liu, L. Tan, C. Shi, Y. Song, Z. Gao, H. Qu, and X. Tong (2011). "TextFlow: Towards Better Understanding of Evolving Topics in Text". In: *IEEE Transactions on Visualization and Computer Graphics* 17.12, pp. 2412–2421. DOI: 10.1109/TVCG.2011.239.

Cui, W., Y. Wu, S. Liu, F. Wei, M. X. Zhou, and H. Qu (2010). "Context-Preserving, Dynamic Word Cloud Visualization". In: *IEEE Computer Graphics and Applications* 30.6, pp. 42–53. DOI: 10.1109/MCG.2010.102.

Daassi, C., L. Nigay, and M.-C. Fauvet (2005). "A Taxonomy of Temporal Data Visualization Techniques". In: *Interaction Information Intelligence* 5.2, pp. 41–63. URL: https://www.irit.fr/journal-i3/volume05/numero02/revue_i3_05_02_02.pdf.

Dachselt, R., M. Frisch, and M. Weiland (2008). "FacetZoom: A Continuous Multi-Scale Widget for Navigating Hierarchical Metadata". In: *Proceedings of the SIGCHI Conference on Human Factors in Computing Systems (CHI)*. ACM Press, pp. 1353–1356. DOI: 10.1145/1357054.1357265.

Dachselt, R. and M. Weiland (2006). "TimeZoom: A Flexible Detail and Context Timeline". In: *CHI Extended Abstracts on Human Factors in Computing Systems*. ACM Press, pp. 682–687. DOI: 10.1145/1125451.1125590.

Dahnert, M., A. Rind, W. Aigner, and J. Kehrer (2019). *Looking Beyond the Horizon: Evaluation of Four Compact Visualization Techniques for Time Series in a Spatial*

Context. Tech. rep. arXiv:1906.07377 [cs.HC]. DOI: 10.48550/arXiv.1906.07377.

Davis, S. B. (2012). "History on the Line: Time as Dimension". In: *Design Issues* 28.4, pp. 4–17. DOI: 10.1162/DESI_a_00171.

Davis, S. B. (2017). "Early Visualizations of Historical Time". In: *Information Design*. Routledge. DOI: 10.4324/9781315585680-8.

Devaux, S., F. Bouali, and G. Venturini (2014). "DataTube4log: A Visual Tool for Mining Multi-threaded Software Logs". In: *Proceedings of the International Conference Information Visualisation (IV)*. IEEE Computer Society, pp. 189–195. DOI: 10.1109/IV.2014.73.

Dick, M. (2020). *The Infographic: A History of Data Graphics in News and Communications*. MIT Press.

Doleisch, H. and H. Hauser (2002). "Smooth Brushing for Focus+Context Visualization of Simulation Data in 3D". In: *Proceedings of the International Conference in Central Europe on Computer Graphics, Visualization and Computer Vision (WSCG)*. University of West Bohemia, pp. 147–154. URL: http://wscg.zcu.cz/wscg2002/Papers_2002/E71.pdf.

Dos Santos, S. and K. Brodlie (2004). "Gaining Understanding of Multivariate and Multidimensional Data through Visualization". In: *Computers & Graphics* 28, pp. 311–325. DOI: 10.1016/j.cag.2004.03.013.

Dragicevic, P. and S. Huot (2002). "SpiraClock: A Continuous and Non-Intrusive Display for Upcoming Events". In: *CHI Extended Abstracts on Human Factors in Computing Systems*. ACM Press, pp. 604–605. DOI: 10.1145/506443.506505.

Draper, G. M., Y. Livnat, and R. F. Riesenfeld (2009). "A Survey of Radial Methods for Information Visualization". In: *IEEE Transactions on Visualization and Computer Graphics* 15.5, pp. 759–776. DOI: 10.1109/TVCG.2009.23.

Dübel, S., M. Röhlig, H. Schumann, and M. Trapp (2014). "2D and 3D Presentation of Spatial Data: A Systematic Review". In: *Proceedings of the International Workshop on 3DVis (3DVis@IEEE VIS)*. IEEE Computer Society, pp. 11–18. DOI: 10.1109/3DVis.2014.7160094.

Duchowski, A. T. (2018). "Gaze-based Interaction: A 30 Year Retrospective". In: *Computers & Graphics* 73, pp. 59–69. DOI: 10.1016/j.cag.2018.04.002.

Dudek, I. and J.-Y. Blaise (2013). "Visualising Time with Multiple Granularities: A Generic Framework". In: *Proceedings of the Annual Conference of Computer Applications and Quantitative Methods in Archaeology*. Amsterdam University Press, pp. 470–481. DOI: 10.1515/9789048519590-050.

Dyreson, C. E., W. S. Evans, H. Lin, and R. T. Snodgrass (2000). "Efficiently Supporting Temporal Granularities". In: *IEEE Transactions on Knowledge and Data Engineering* 12.4, pp. 568–587. DOI: 10.1109/69.868908.

Eccles, R., T. Kapler, R. Harper, and W. Wright (2008). "Stories in GeoTime". In: *Information Visualization* 7.1, pp. 3–17. DOI: 10.1057/palgrave.ivs.9500173.

Eichner, C., H. Schumann, and C. Tominski (2020). "Making Parameter Dependencies of Time-Series Segmentation Visually Understandable". In: *Computer Graphics Forum* 39.1, pp. 607–622. DOI: 10.1111/cgf.13894.

Elmqvist, N. and P. Tsigas (2007). "A Taxonomy of 3D Occlusion Management Techniques". In: *Proceedings of the IEEE Conference on Virtual Reality (VR)*. IEEE Computer Society, pp. 51–58. DOI: 10.1109/vr.2007.352463.

Enge, K., A. Rind, M. Iber, R. Höldrich, and W. Aigner (2021). "It's about Time: Adopting Theoretical Constructs from Visualization for Sonification". In: *Proceedings of the International Audio Mostly Conference (AMI)*. ACM Press, pp. 64–71. DOI: 10.1145/3478384.3478415.

Enge, K., A. Rind, M. Iber, R. Höldrich, and W. Aigner (2022). "Towards Multimodal Exploratory Data Analysis: SoniScope as a Prototypical Implementation". In: *Proceedings of the Eurographics / IEEE Conference on Visualization (EuroVis) - Short Papers*. Eurographics Association, pp. 67–71. DOI: 10.2312/evs.20221095.

Ens, B., S. Goodwin, A. Prouzeau, F. Anderson, F. Y. Wang, S. Gratzl, Z. Lucarelli, B. Moyle, J. Smiley, and T. Dwyer (2021). "Uplift: A Tangible and Immersive Tabletop System for Casual Collaborative Visual Analytics". In: *IEEE Transactions on Visualization and Computer Graphics* 27.2, pp. 1193–1203. DOI: 10.1109/TVCG.2020.3030334.

Erbacher, R. F., K. L. Walker, and D. A. Frincke (2002). "Intrusion and Misuse Detection in Large-Scale Systems". In: *IEEE Computer Graphics and Applications* 22.1, pp. 38–48. DOI: 10.1109/38.974517.

Fails, J. A., A. Karlson, L. Shahamat, and B. Shneiderman (2006). "A Visual Interface for Multivariate Temporal Data: Finding Patterns of Events across Multiple Histories". In: *Proceedings of the IEEE Symposium on Visual Analytics Science and Technology (VAST)*. IEEE Computer Society, pp. 167–174. DOI: 10.1109/VAST.2006.261421.

Fanea, E., M. S. T. Carpendale, and T. Isenberg (2005). "An Interactive 3D Integration of Parallel Coordinates and Star Glyphs". In: *Proceedings of the IEEE Symposium Information Visualization (InfoVis)*. IEEE Computer Society, pp. 149–156. DOI: 10.1109/INFVIS.2005.1532141.

Fang, Y., H. Xu, and J. Jiang (2020). "A Survey of Time Series Data Visualization Research". In: *IOP Conference Series: Materials Science and Engineering* 782. DOI: 10.1088/1757-899x/782/2/022013.

Farquhar, A. B. and H. Farquhar (1891). *Economic and Industrial Solutions*. New York, NY: G. B. Putnam's Sons.

Fayyad, U., G. G. Grinstein, and A. Wierse, eds. (2001). *Information Visualization in Data Mining and Knowledge Discovery*. Morgan Kaufmann.

Fayyad, U., G. Piatetsky-Shapiro, and P. Smyth (1996). "From Data Mining to Knowledge Discovery in Databases". In: *AI Magazine* 17.3, pp. 37–54. DOI: 10.1609/aimag.v17i3.1230.

Federico, P., S. Hoffmann, A. Rind, W. Aigner, and S. Miksch (2014). "Qualizon Graphs: Space-Efficient Time-Series Visualization with Qualitative Abstractions". In: *Proceedings of the Conference on Advanced Visual Interfaces (AVI)*. ACM Press, pp. 273–280. DOI: 10.1145/2598153.2598172.

Federico, P., M. Wagner, A. Rind, A. Amor-Amoros, S. Miksch, and W. Aigner (2017). "The Role of Explicit Knowledge: A Conceptual Model of Knowledge-

Assisted Visual Analytics". In: *Proceedings of the IEEE Conference on Visual Analytics Science and Technology (VAST)*. IEEE Computer Society, pp. 92–103. DOI: 10.1109/VAST.2017.8585498.

Ferguson, S. (1991). "The 1753 Carte chronographique of Jacques Barbeu-Dubourg". In: *Princeton University Library Chronicle* 52.2, pp. 190–230. DOI: 10.2307/26404421.

Filho, J. A. W., W. Stuerzlinger, and L. P. Nedel (2020). "Evaluating an Immersive Space-Time Cube Geovisualization for Intuitive Trajectory Data Exploration". In: *IEEE Transactions on Visualization and Computer Graphics* 26.1, pp. 514–524. DOI: 10.1109/TVCG.2019.2934415.

Filipov, V., V. Schetinger, K. Raminger, N. Soursos, S. Zapke, and S. Miksch (2021). "Gone Full Circle: A Radial Approach to Visualize Event-Based Networks in Digital Humanities". In: *Visual Informatics* 5.1, pp. 45–60. DOI: 10.1016/j.visinf.2021.01.001.

Fischer, F., J. Fuchs, and F. Mansmann (2012). "ClockMap: Enhancing Circular Treemaps with Temporal Glyphs for Time-Series Data". In: *Proceedings of the Eurographics / IEEE Conference on Visualization (EuroVis) - Short Papers*. Eurographics Association, pp. 97–101. DOI: 10.2312/PE/EuroVisShort/EuroVisShort2012/097-101.

Foley, J. D. (2000). "Getting There: The Ten Top Problems Left". In: *IEEE Computer Graphics and Applications* 20.1, pp. 66–68. DOI: 10.1109/38.814569.

Forlines, C. and K. Wittenburg (2010). "Wakame: Sense Making of Multi-Dimensional Spatial-Temporal Data". In: *Proceedings of the Conference on Advanced Visual Interfaces (AVI)*. ACM Press, pp. 33–40. DOI: 10.1145/1842993.1843000.

Frank, A. U. (1998). "Different Types of "Times" in GIS". In: *Spatial and Temporal Reasoning in Geographic Information Systems*. Edited by Egenhofer, M. J. and Golledge, R. G. Oxford University Press, pp. 40–62.

Friendly, M. (2008). "A Brief History of Data Visualization". In: *Handbook of Data Visualization*. Edited by Chen, C.-h., Härdle, W., and Unwin, A. Springer, pp. 15–56.

Fry, B. (2000). "Organic Information Design". MA thesis. Massachusetts Institute of Technology. URL: https://benfry.com/organic/thesis-0522d.pdf.

Fuchs, G. and H. Schumann (2004a). "Intelligent Icon Positioning for Interactive Map-Based Information Systems". In: *Proceedings of the International Conference of the Information Resources Management Association (IRMA)*. Idea Group Inc., pp. 261–264. URL: https://www.irma-international.org/proceeding-paper/intelligent-icon-positioning-interactive-map/32349/.

Fuchs, G. and H. Schumann (2004b). "Visualizing Abstract Data on Maps". In: *Proceedings of the International Conference Information Visualisation (IV)*. IEEE Computer Society, pp. 139–144. DOI: 10.1109/IV.2004.1320136.

Fuchs, J., F. Fischer, F. Mansmann, E. Bertini, and P. Isenberg (2013). "Evaluation of Alternative Glyph Designs for Time Series Data in a Small Multiple Setting". In: *Proceedings of the SIGCHI Conference on Human Factors in Computing Systems (CHI)*. ACM Press, pp. 3237–3246. DOI: 10.1145/2470654.2466443.

Funkhouser, H. G. (1936). "A Note on a Tenth Century Graph". In: *Osiris* 1.1, pp. 260–262. DOI: 10.1086/368425.

Furia, C. A., D. Mandrioli, A. Morzenti, and M. Rossi (2010). "Modeling Time in Computing: A Taxonomy and a Comparative Survey". In: *ACM Computing Surveys* 42.2, 6:1–6:59. DOI: 10.1145/1667062.1667063.

Furia, C. A., D. Mandrioli, A. Morzenti, and M. Rossi (2012). *Modeling Time in Computing*. Springer. DOI: 10.1007/978-3-642-32332-4.

Gajos, K. Z., M. Czerwinski, D. S. Tan, and D. S. Weld (2006). "Exploring the Design Space for Adaptive Graphical User Interfaces". In: *Proceedings of the Conference on Advanced Visual Interfaces (AVI)*. ACM Press, pp. 201–208. DOI: 10.1145/1133265.1133306.

Gall, H., M. Jazayeri, and C. Riva (1999). "Visualizing Software Release Histories: The Use of Color and Third Dimension". In: *Proceedings of the International Conference on Software Maintenance (ICSM)*. IEEE Computer Society, pp. 99–108. DOI: 10.1109/icsm.1999.792584.

Gan, G., C. Ma, and J. Wu (2007). *Data Clustering: Theory, Algorithms, and Applications*. ASA-SIAM Series on Statistics and Applied Probability. Society for Industrial and Applied Mathematics. DOI: 10.1137/1.9780898718348.

Gannett, H. (1898). *Statistical Atlas of the United States, Based upon Results of the Eleventh Census (1890)*. United States Census Office.

Gantt, H. L. (1910). *Work, Wages, and Profits*. The Engineering Magazine. URL: https://archive.org/details/cu31924001636418.

Gapminder Foundation (2021). *Gapminder Tools*. URL: https://www.gapminder.org/tools.

Gatalsky, P., N. Andrienko, and G. Andrienko (2004). "Interactive Analysis of Event Data Using Space-Time Cube". In: *Proceedings of the International Conference Information Visualisation (IV)*. IEEE Computer Society, pp. 145–152. DOI: 10.1109/IV.2004.1320137.

Gautier, J., P.-A. Davoine, and C. Cunty (2016). "Helical Time Representation to Visualize Return-periods of Spatio-temporal Events". In: *Proceedings of the 19th AGILE International Conference on Geographic Information Science*. Helsinki, Finland. URL: https://agile-online.org/images/conferences/2016/documents/shortpapers/122_Paper_in_PDF.pdf.

Gershon, N. and W. Page (2001). "What Storytelling Can Do for Information Visualization". In: *Communications of the ACM* 44.8, pp. 31–37. DOI: 10.1145/381641.381653.

Gleicher, M. (2018). "Considerations for Visualizing Comparison". In: *IEEE Transactions on Visualization and Computer Graphics* 24.1, pp. 413–423. DOI: 10.1109/TVCG.2017.2744199.

Gleicher, M., D. Albers, R. Walker, I. Jusufi, C. D. Hansen, and J. C. Roberts (2011). "Visual Comparison for Information Visualization". In: *Information Visualization* 10.4, pp. 289–309. DOI: 10.1177/1473871611416549.

Goralwalla, I. A., M. T. Özsu, and D. Szafron (1998). "An Object-Oriented Framework for Temporal Data Models". In: *Temporal Databases: Research and Prac-*

tice. Edited by Etzion, O., Jajodia, S., and Sripada, S. Springer, pp. 1–35. DOI: `10.1007/bfb0053696`.

Gotz, D. and H. Stavropoulos (2014). "DecisionFlow: Visual Analytics for High-Dimensional Temporal Event Sequence Data". In: *IEEE Transactions on Visualization and Computer Graphics* 20.12, pp. 1783–1792. DOI: `10.1109/TVCG.2014.2346682`.

Graells, E. and A. Jaimes (2012). "Lin-spiration: Using a Mixture of Spiral and Linear Visualization Layouts to Explore Time Series". In: *Proceedings of the International Conference on Intelligent User Interfaces (IUI)*. ACM Press, pp. 237–240. DOI: `10.1145/2166966.2167006`.

Gruendl, H., P. Riehmann, Y. Pausch, and B. Fröhlich (2016). "Time-Series Plots Integrated in Parallel-Coordinates Displays". In: *Computer Graphics Forum* 35.3, pp. 321–330. DOI: `10.1111/cgf.12908`.

Gschwandtner, T., W. Aigner, K. Kaiser, S. Miksch, and A. Seyfang (2011). "CareCruiser: Exploring and Visualizing Plans, Events, and Effects Interactively". In: *Proceedings of the IEEE Pacific Visualization Symposium (PacificVis)*. IEEE Computer Society, pp. 43–50. DOI: `10.1109/pacificvis.2011.5742371`.

Gschwandtner, T., W. Aigner, S. Miksch, J. Gärtner, S. Kriglstein, M. Pohl, and N. Suchy (2014). "TimeCleanser: A Visual Analytics Approach for Data Cleansing of Time-oriented Data". In: *Proceedings of the International Conference on Knowledge Technologies and Data-driven Business (i-KNOW)*. ACM Press, pp. 1–8. DOI: `10.1145/2637748.2638423`.

Gschwandtner, T., M. Bögl, P. Federico, and S. Miksch (2016). "Visual Encodings of Temporal Uncertainty: A Comparative User Study". In: *IEEE Transactions on Visualization and Computer Graphics* 22.1, pp. 539–548. DOI: `10.1109/TVCG.2015.2467752`.

Gschwandtner, T., J. Gärtner, W. Aigner, and S. Miksch (2012). "A Taxonomy of Dirty Time-Oriented Data". In: *Multidisciplinary Research and Practice for Information Systems*. Edited by Quirchmayr, G., Basl, J., You, I., Xu, L., and Weippl, E. Springer, pp. 58–72. DOI: `10.1007/978-3-642-32498-7_5`.

Guo, D., J. Chen, A. M. MacEachren, and K. Liao (2006). "A Visualization System for Space-Time and Multivariate Patterns (VIS-STAMP)". In: *IEEE Transactions on Visualization and Computer Graphics* 12.6, pp. 1461–1474. DOI: `10.1109/TVCG.2006.84`.

Haber, R. B. and D. A. McNabb (1990). "Visualization Idioms: A Conceptual Model for Scientific Visualization Systems". In: *Visualization in Scientific Computing*. IEEE Computer Society, pp. 74–93.

Hackos, J. T. and J. C. Redish (1998). *User and Task Analysis for Interface Design*. John Wiley & Sons, Inc.

Hadlak, S., C. Tominski, H.-J. Schulz, and H. Schumann (2010). "Visualization of Attributed Hierarchical Structures in a Spatio-Temporal Context". In: *International Journal of Geographical Information Science* 24.10, pp. 1497–1513. DOI: `10.1080/13658816.2010.510840`.

Hägerstrand, T. (1970). "What About People in Regional Science?" In: *Papers of the Regional Science Association* 24, pp. 6–21. DOI: https://doi.org/10.1007/BF01936872.

Hajnicz, E. (1996). *Time Structures: Formal Description and Algorithmic Representation.* Vol. 1047. Lecture Notes in Computer Science. Springer. DOI: 10.1007/3-540-60941-5.

Hall, K. W., C. Perin, P. G. Kusalik, C. Gutwin, and S. Carpendale (2014). "Formalizing Emphasis in Information Visualization". In: *Computer Graphics Forum* 35.3, pp. 717–737. DOI: 10.1111/cgf.12936.

Han, J., M. Kamber, and J. Pei (2012). *Data Mining: Concepts and Techniques.* 3rd edition. Morgan Kaufmann. DOI: 10.1016/C2009-0-61819-5.

Hao, M., U. Dayal, D. Keim, and T. Schreck (2007). "Multi-Resolution Techniques for Visual Exploration of Large Time-Series Data". In: *Proceedings of the Joint Eurographics / IEEE VGTC Symposium on Visualization (VisSym)*. Eurographics Association, pp. 27–34. DOI: 10.2312/VisSym/EuroVis07/027-034.

Hao, M. C., U. Dayal, D. A. Keim, and T. Schreck (2005). "Importance-Driven Visualization Layouts for Large Time Series Data". In: *Proceedings of the IEEE Symposium Information Visualization (InfoVis)*. IEEE Computer Society, pp. 203–210. DOI: 10.1109/INFOVIS.2005.20.

Haroz, S., R. Kosara, and S. L. Franconeri (2016). "The Connected Scatterplot for Presenting Paired Time Series". In: *IEEE Transactions on Visualization and Computer Graphics* 22.9, pp. 2174–2186. DOI: 10.1109/TVCG.2015.2502587.

Harris, R. L. (1999). *Information Graphics: A Comprehensive Illustrated Reference.* Oxford University Press. URL: https://global.oup.com/academic/product/information-graphics-9780195135329.

Harrower, M. A. and C. A. Brewer (2003). "ColorBrewer.org: An Online Tool for Selecting Color Schemes for Maps". In: *The Cartographic Journal* 40.1, pp. 27–37. DOI: 10.4324/9781351191234-18.

Hart, G. (1996). "The Five W's: An Old Tool for the New Task of Task Analysis". In: *Technical Communication* 43.2, pp. 139–145.

Hashimoto, Y. and R. Matsushita (2012). "Heat Map Scope Technique for Stacked Time-series Data Visualization". In: *Proceedings of the International Conference Information Visualisation (IV)*, pp. 270–273. DOI: 10.1109/IV.2012.53.

Hassan, K. A., N. Rönnberg, C. Forsell, M. Cooper, and J. Johansson (2019). "A Study on 2D and 3D Parallel Coordinates for Pattern Identification in Temporal Multivariate Data". In: *Proceedings of the International Conference Information Visualisation (IV)*. IEEE Computer Society, pp. 145–150. DOI: 10.1109/IV.2019.00033.

Hauser, H., F. Ledermann, and H. Doleisch (2002). "Angular Brushing of Extended Parallel Coordinates". In: *Proceedings of the IEEE Symposium Information Visualization (InfoVis)*. IEEE Computer Society, pp. 127–130. DOI: 10.1109/INFVIS.2002.1173157.

Havre, S., E. Hetzler, and L. Nowell (2000). "ThemeRiver: Visualizing Theme Changes Over Time". In: *Proceedings of the IEEE Symposium Information Vi-

sualization (InfoVis). IEEE Computer Society, pp. 115–124. DOI: 10.1109/INFVIS.2000.885098.

Havre, S., E. Hetzler, P. Whitney, and L. Nowell (2002). "ThemeRiver: Visualizing Thematic Changes in Large Document Collections". In: *IEEE Transactions on Visualization and Computer Graphics* 8.1, pp. 9–20. DOI: 10.1109/2945.981848.

Healey, C. G. and J. T. Enns (2012). "Attention and Visual Memory in Visualization and Computer Graphics". In: *IEEE Transactions on Visualization and Computer Graphics* 18.7, pp. 1170–1188. DOI: 10.1109/TVCG.2011.127.

Heer, J. and M. Agrawala (2006). "Multi-Scale Banking to 45 Degrees". In: *IEEE Transactions on Visualization and Computer Graphics* 12.5, pp. 701–708. DOI: 10.1109/TVCG.2006.163.

Heer, J., N. Kong, and M. Agrawala (2009). "Sizing the Horizon: The Effects of Chart Size and Layering on the Graphical Perception of Time Series Visualizations". In: *Proceedings of the SIGCHI Conference on Human Factors in Computing Systems (CHI)*. ACM Press, pp. 1303–1312. DOI: 10.1145/1518701.1518897.

Heer, J. and G. Robertson (2007). "Animated Transitions in Statistical Data Graphics". In: *IEEE Transactions on Visualization and Computer Graphics* 13.6, pp. 1240–1247. DOI: 10.1109/tvcg.2007.70539.

Henriksen, K., J. Sporring, and K. Hornbæk (2004). "Virtual Trackballs Revisited". In: *IEEE Transactions on Visualization and Computer Graphics* 10.2, pp. 206–216. DOI: 10.1109/tvcg.2004.1260772.

Hinum, K., S. Miksch, W. Aigner, S. Ohmann, C. Popow, M. Pohl, and M. Rester (2005). "Gravi++: Interactive Information Visualization to Explore Highly Structured Temporal Data". In: *Journal of Universal Computer Science* 11.11, pp. 1792–1805. DOI: 10.3217/jucs-011-11-1792.

Hochheiser, H. and B. Shneiderman (2004). "Dynamic Query Tools for Time Series Data Sets: Timebox Widgets for Interactive Exploration". In: *Information Visualization* 3.1, pp. 1–18. DOI: 10.1057/palgrave.ivs.9500061.

Hoffswell, J., W. Li, and Z. Liu (2020). "Techniques for Flexible Responsive Visualization Design". In: *Proceedings of the SIGCHI Conference on Human Factors in Computing Systems (CHI)*. ACM Press, pp. 1–13. DOI: 10.1145/3313831.3376777.

Höhn, M., M. Wunderlich, K. Ballweg, J. Kohlhammer, and T. von Landesberger (2022). "Interactive Input and Visualization for Planning with Temporal Uncertainty". In: *Proceedings of the International Joint Conference on Computer Vision, Imaging and Computer Graphics Theory and Applications (VISIGRAPP)*. SCITEPRESS - Science and Technology Publications, pp. 27–37. DOI: 10.5220/0010761900003124.

Holme, P. and J. Saramäki (2012). "Temporal Networks". In: *Physics Reports* 519.3, pp. 97–125. DOI: 10.1016/j.physrep.2012.03.001.

Holz, C. and S. Feiner (2009). "Relaxed Selection Techniques for Querying Time-Series Graphs". In: *Proceedings of the ACM Symposium on User Interface Software and Technology (UIST)*. ACM Press, pp. 213–222. DOI: 10.1145/1622176.1622217.

Hong, S., Y.-S. Kim, J.-C. Yoon, and C. R. Aragon (2014). "Traffigram: Distortion for Clarification via Isochronal Cartography". In: *Proceedings of the SIGCHI Conference on Human Factors in Computing Systems (CHI)*. ACM Press, pp. 907–916. DOI: 10.1145/2556288.2557224.

Hong, S., R. Kocielnik, M.-J. Yoo, S. Battersby, J. Kim, and C. R. Aragon (2017). "Designing Interactive Distance Cartograms to Support Urban Travelers". In: *Proceedings of the IEEE Pacific Visualization Symposium (PacificVis)*. IEEE Computer Society, pp. 81–90. DOI: 10.1109/PACIFICVIS.2017.8031582.

Hong, S., M.-J. Yoo, B. Chinh, A. Han, S. Battersby, and J. Kim (2018). "To Distort or Not to Distort: Distance Cartograms in the Wild". In: *Proceedings of the SIGCHI Conference on Human Factors in Computing Systems (CHI)*. ACM Press. DOI: 10.1145/3173574.3174202.

Horak, T., S. K. Badam, N. Elmqvist, and R. Dachselt (2018). "When David Meets Goliath: Combining Smartwatches with a Large Vertical Display for Visual Data Exploration". In: *Proceedings of the SIGCHI Conference on Human Factors in Computing Systems (CHI)*. ACM Press. DOI: 10.1145/3173574.3173593.

Horn, W., C. Popow, and L. Unterasinger (2001). "Support for Fast Comprehension of ICU Data: Visualization using Metaphor Graphics". In: *Methods of Information in Medicine* 40.5, pp. 421–424. DOI: 10.1055/s-0038-1634202.

Huang, G., S. Govoni, J. Choi, D. M. Hartley, and J. M. Wilson (2008). "Geovisualizing Data With Ring Maps". In: *ArcUser* 11.1, pp. 54–55. URL: https://www.esri.com/news/arcuser/0408/files/ringmaps.pdf.

Huynh, D. (2021). *OpenRefine*. URL: https://openrefine.org/ (visited on 02/26/2021).

Imrich, P., K. Mueller, D. Imre, D. Zelenyuk, and W. Zhu (2003). "Interactive Poster: 3D ThemeRiver". In: *Poster Compendium of IEEE Symposium on Information Visualization (InfoVis)*. IEEE Computer Society. URL: https://www3.cs.stonybrook.edu/~mueller/papers/3DThemeriver.pdf.

Inselberg, A. and B. Dimsdale (1990). "Parallel Coordinates: A Tool for Visualizing Multi-Dimensional Geometry". In: *Proceedings of the IEEE Visualization Conference (Vis)*. IEEE Computer Society, pp. 361–378. DOI: 10.1109/VISUAL.1990.146402.

Jabbari, A., R. Blanch, and S. Dupuy-Chessa (2018). "Composite Visual Mapping for Time Series Visualization". In: *Proceedings of the IEEE Pacific Visualization Symposium (PacificVis)*. IEEE, pp. 116–124. DOI: 10.1109/PacificVis.2018.00023.

Jackson, J. E. (2003). *A User's Guide to Principal Components*. John Wiley & Sons.

Jain, A. K., M. N. Murty, and P. J. Flynn (1999). "Data Clustering: A Review". In: *ACM Computing Surveys* 31.3, pp. 264–323. DOI: 10.1145/331499.331504.

Jakobsen, M. R., Y. S. Haile, S. Knudsen, and K. Hornbæk (2013). "Information Visualization and Proxemics: Design Opportunities and Empirical Findings". In: *IEEE Transactions on Visualization and Computer Graphics* 19.12, pp. 2386–2395. DOI: 10.1109/TVCG.2013.166.

Jankun-Kelly, T. J., K.-L. Ma, and M. Gertz (2007). "A Model and Framework for Visualization Exploration". In: *IEEE Transactions on Visualization and Computer Graphics* 13.2, pp. 357–369. DOI: `10.1109/tvcg.2007.28`.

Javed, W., B. McDonnel, and N. Elmqvist (2010). "Graphical Perception of Multiple Time Series". In: *IEEE Transactions on Visualization and Computer Graphics* 16.6, pp. 927–34. DOI: `10.1109/tvcg.2010.162`.

Jensen, C. S., C. E. Dyreson, M. H. Böhlen, J. Clifford, R. Elmasri, S. K. Gadia, F. Grandi, P. J. Hayes, S. Jajodia, W. Käfer, N. Kline, N. A. Lorentzos, Y. G. Mitsopoulos, A. Montanari, D. A. Nonen, E. Peressi, B. Pernici, J. F. Roddick, N. L. Sarda, M. R. Scalas, A. Segev, R. T. Snodgrass, M. D. Soo, A. U. Tansel, P. Tiberio, and G. Wiederhold (1998). "The Consensus Glossary of Temporal Database Concepts – February 1998 Version". In: *Temporal Databases: Research and Practice*. Edited by Etzion, O., Jajodia, S., and Sripada, S. Springer, pp. 367–405. DOI: `10.1007/bfb0053710`.

Jeong, D. H., C. Ziemkiewicz, B. Fisher, W. Ribarsky, and R. Chang (2009). "iPCA: An Interactive System for PCA-based Visual Analytics". In: *Computer Graphics Forum* 28.3, pp. 767–774. DOI: `10.1111/j.1467-8659.2009.01475.x`.

Jiang, S., S. Fang, and S. Grannis (2016). "Spiral Theme Plot". In: *Proceedings of the Eurographics / IEEE Conference on Visualization (EuroVis) - Short Papers*. Eurographics Association. DOI: `10.2312/eurovisshort.20161170`.

Jo, J., J. Huh, J. Park, B. Kim, and J. Seo (2014). "LiveGantt: Interactively Visualizing a Large Manufacturing Schedule". In: *IEEE Transactions on Visualization and Computer Graphics* 20.12, pp. 2329–2338. DOI: `10.1109/TVCG.2014.2346454`.

Jolliffe, I. T. (2002). *Principal Component Analysis*. 2nd edition. Springer. DOI: `10.1007/b98835`.

Jones, C. P. (2020). *How Paintings Depict Time*. URL: `https://medium.com/thinksheet/how-paintings-depict-time-33850ff344f4`.

Juarez, J. M., J. M. Ochotorena, M. Campos, and C. Combi (2013). "Multiple Temporal Axes for Visualising the Behaviour of Elders Living Alone". In: *Proceedings of the IEEE International Conference on Healthcare Informatics*, pp. 387–395. DOI: `10.1109/ICHI.2013.54`.

Kandel, S., A. Paepcke, J. M. Hellerstein, and J. Heer (2011). "Wrangler: Interactive Visual Specification of Data Transformation Scripts". In: *Proceedings of the SIGCHI Conference on Human Factors in Computing Systems (CHI)*. ACM Press, pp. 3363–3372. DOI: `10.1145/1978942.1979444`.

Kapler, T., R. Eccles, R. Harper, and W. Wright (2008). "Configurable Spaces: Temporal Analysis in Diagrammatic Contexts". In: *Proceedings of the IEEE Symposium on Visual Analytics Science and Technology (VAST)*. IEEE Computer Society, pp. 43–50. DOI: `10.1109/VAST.2008.4677355`.

Kapler, T. and W. Wright (2005). "GeoTime Information Visualization". In: *Information Visualization* 4.2, pp. 136–146. DOI: `10.1057/palgrave.ivs.9500097`.

Keefe, D. F. and T. Isenberg (2013). "Reimagining the Scientific Visualization Interaction Paradigm". In: *Computer* 46.5, pp. 51–57. DOI: `10.1109/MC.2013.178`.

Keim, D., J. Kohlhammer, G. Ellis, and F. Mansmann, eds. (2010). *Mastering the Information Age – Solving Problems with Visual Analytics*. Eurographics Association. URL: https://diglib.eg.org/handle/10.2312/14803.

Keim, D. A., H.-P. Kriegel, and M. Ankerst (1995). "Recursive Pattern: A Technique for Visualizing Very Large Amounts of Data". In: *Proceedings of the IEEE Visualization Conference (Vis)*. IEEE Computer Society, pp. 279–286. DOI: 10.1109/VISUAL.1995.485140.

Keim, D. A., F. Mansmann, J. Schneidewind, and H. Ziegler (2006a). "Challenges in Visual Data Analysis". In: *Proceedings of the International Conference Information Visualisation (IV)*. IEEE Computer Society, pp. 9–16. DOI: 10.1109/IV.2006.31.

Keim, D. A., T. Nietzschmann, N. Schelwies, J. Schneidewind, T. Schreck, and H. Ziegler (2006b). "A Spectral Visualization System for Analyzing Financial Time Series Data". In: *Proceedings of the Joint Eurographics / IEEE VGTC Symposium on Visualization (VisSym)*. Eurographics Association, pp. 195–202. DOI: 10.2312/VisSym/EuroVis06/195-202.

Keim, D. A. and J. Schneidewind (2005). "Scalable Visual Data Exploration of Large Data Sets via MultiResolution". In: *Journal of Universal Computer Science* 11.11, pp. 1766–1779. DOI: 10.3217/jucs-011-11-1766.

Keim, D. A., J. Schneidewind, and M. Sips (2004). "CircleView: A New Approach for Visualizing Time-Related Multidimensional Data Sets". In: *Proceedings of the Conference on Advanced Visual Interfaces (AVI)*. ACM Press, pp. 179–182. DOI: 10.1145/989863.989891.

Kennedy, A. B. W. and M. H. P. R. Sankey (1898). "The Thermal Efficiency of Steam Engines. Report of the Committee appointed on the 31st March, 1896, to Consider and Report to the Council upon the Subject of the Definition of a Standard or Standards of Thermal Efficiency for Steam-Engines: with an Introductory Note." In: *Minutes of the Proceedings of the Institution of Civil Engineers*. Vol. 134. 1898, pp. 278–312. DOI: 10.1680/imotp.1898.19100.

Kerren, A., A. Ebert, and J. Meyer, eds. (2007). *Human-Centered Visualization Environments*. Vol. 4417. Lecture Notes in Computer Science. Springer. DOI: 10.1007/978-3-540-71949-6.

Kim, N. W., B. Bach, H. Im, S. Schriber, M. H. Gross, and H. Pfister (2018). "Visualizing Nonlinear Narratives with Story Curves". In: *IEEE Transactions on Visualization and Computer Graphics* 24.1, pp. 595–604. DOI: 10.1109/TVCG.2017.2744118.

Kim, N. W., S. K. Card, and J. Heer (2010). "Tracing Genealogical Data with TimeNets". In: *Proceedings of the Conference on Advanced Visual Interfaces (AVI)*. ACM Press, pp. 241–248. DOI: 10.1145/1842993.1843035.

Kim, W., B.-J. Choi, E.-K. Hong, S.-K. Kim, and D. Lee (2003). "A Taxonomy of Dirty Data". In: *Data Mining and Knowledge Discovery* 7.1, pp. 81–99. DOI: 10.1023/a:1021564703268.

Kirk, A. (2019). *Data Visualisation: A Handbook for Data Driven Design*. 2nd edition. SAGE.

Knight, D., B. Knight, M. Pearson, and M. Quintana (2018). *Microsoft Power BI Quick Start Guide: Build Dashboards and Visualizations to Make Your Data Come to Life*. 1st edition. Packt Publishing. URL: https://www.packtpub.com/product/microsoft-power-bi-quick-start-guide/9781789138221.

Kolence, K. W. and P. J. Kiviat (1973). "Software Unit Profiles & Kiviat Figures". In: *SIGMETRICS Performance Evaluation Review* 2.3, pp. 2–12. DOI: 10.1145/1041613.1041614.

Kondo, B. and C. Collins (2014). "DimpVis: Exploring Time-varying Information Visualizations by Direct Manipulation". In: *IEEE Transactions on Visualization and Computer Graphics* 20.12, pp. 2003–2012. DOI: 10.1109/TVCG.2014.2346250.

Kosara, R., F. Bendix, and H. Hauser (2004). "TimeHistograms for Large, Time-Dependent Data". In: *Proceedings of the Joint Eurographics / IEEE VGTC Symposium on Visualization (VisSym)*. Eurographics Association, pp. 45–54. DOI: 10.2312/VisSym/VisSym04/045-054.

Kosara, R. and S. Miksch (2001). "Metaphors of Movement - A Visualization and User Interface for Time-Oriented, Skeletal Plans". In: *Artificial Intelligence in Medicine* 22.2, pp. 111–131. DOI: 10.1016/s0933-3657(00)00103-2.

Kosara, R. and S. Miksch (2002). "Visualization Methods for Data Analysis and Planning". In: *International Journal of Medical Informatics* 68.1–3, pp. 141–153. DOI: 10.1016/S1386-5056(02)00072-2.

Kostitsyna, I., M. Nöllenburg, V. Polishchuk, A. Schulz, and D. Strash (2015). "On Minimizing Crossings in Storyline Visualizations". In: *Proceedings of the International Symposium on Graph Drawing and Network Visualization(GD)*. Springer, pp. 192–198. DOI: 10.1007/978-3-319-27261-0_16.

Kraak, M.-J. (2003). "The Space-Time Cube Revisited from a Geovisualization Perspective". In: *Proceedings of the 21st International Cartographic Conference (ICC)*. The International Cartographic Association (ICA), pp. 1988–1996. URL: https://icaci.org/files/documents/ICC_proceedings/ICC2003/Papers/255.pdf.

Kraak, M.-J. and F. Ormeling (2020). *Cartography: Visualization of Geospatial Data*. 4th edition. CRC Press. DOI: 10.1201/9780429464195.

Krasner, G. E. and S. T. Pope (1988). "A Cookbook for Using the Model-View-Controller User Interface Paradigm in Smalltalk-80". In: *Journal of Object-Oriented Programming* 1.3, pp. 26–49.

Kraus, M., K. Klein, J. Fuchs, D. A. Keim, F. Schreiber, M. Sedlmair, and T.-M. Rhyne (2021). "The Value of Immersive Visualization". In: *IEEE Computer Graphics and Applications* 41.4, pp. 125–132. DOI: 10.1109/MCG.2021.3075258.

Kriglstein, S., M. Pohl, and M. Smuc (2014). "Pep Up Your Time Machine: Recommendations for the Design of Information Visualizations of Time-Dependent Data". In: *Handbook of Human Centric Visualization*. Edited by Huang, W. Springer, pp. 203–225. DOI: 10.1007/978-1-4614-7485-2_8.

Kristensson, P. O., N. Dahlback, D. Anundi, M. Bjornstad, H. Gillberg, J. Haraldsson, I. Martensson, M. Nordvall, and J. Stahl (2009). "An Evaluation of Space

Time Cube Representation of Spatiotemporal Patterns". In: *IEEE Transactions on Visualization and Computer Graphics* 15.4, pp. 696–702. DOI: 10.1109/TVCG. 2008.194.

Krstajic, M., E. Bertini, and D. Keim (2011). "CloudLines: Compact Display of Event Episodes in Multiple Time-Series". In: *IEEE Transactions on Visualization and Computer Graphics* 17.12, pp. 2432–2439. DOI: 10.1109/TVCG.2011.179.

Krzywinski, M., J. Schein, I. Birol, J. Connors, R. Gascoyne, D. Horsman, S. J. Jones, and M. A. Marra (2009). "Circos: An Information Aesthetic for Comparative Genomics". In: *Genome Research* 19.9, pp. 1639–1645. DOI: 10.1101/gr. 092759.109.

Kulpa, Z. (1997). "Diagrammatic Representation of Interval Space in Proving Theorems about Interval Relations". In: *Reliable Computing* 3.3, pp. 209–217. DOI: 10.1023/A:1009919304728.

Kulpa, Z. (2006). "A Diagrammatic Approach to Investigate Interval Relations". In: *Journal of Visual Languages and Computing* 17.5, pp. 466–502. DOI: 10.1016/ j.jvlc.2005.10.004.

Kwan, M.-P. (2009). "Space-Time Paths". In: *Manual of Geographic Information Systems*. Edited by Madden, M. American Society for Photogrammetry and Remote Sensing. Chap. 25, pp. 427–442. URL: https://my.asprs.org/ ASPRSMember/Store/StoreLayouts/Item_Detail.aspx?iProductCode= 4650.

L'Yi, S., J. Jo, and J. Seo (2021). "Comparative Layouts Revisited: Design Space, Guidelines, and Future Directions". In: *IEEE Transactions on Visualization and Computer Graphics* 27.2, pp. 1525–1535. DOI: 10.1109/TVCG.2020.3030419.

La maison du cinema and Cinematheque Francaise (2000). *Étienne-Jules Marey: Movement in Light*. URL: https://web.archive.org/web/20060209080334/ http://www.expo-marey.com/ANGLAIS/home.html.

Lam, H. (2008). "A Framework of Interaction Costs in Information Visualization". In: *IEEE Transactions on Visualization and Computer Graphics* 14.6, pp. 1149–1156. DOI: 10.1109/TVCG.2008.109.

Lam, H., E. Bertini, P. Isenberg, C. Plaisant, and S. Carpendale (2012). "Empirical Studies in Information Visualization: Seven Scenarios". In: *IEEE Transactions on Visualization and Computer Graphics* 18.9, pp. 1520–1536. DOI: 10.1109/ TVCG.2011.279.

Lammarsch, T., W. Aigner, A. Bertone, J. Gärtner, E. Mayr, S. Miksch, and M. Smuc (2009). "Hierarchical Temporal Patterns and Interactive Aggregated Views for Pixel-based Visualizations". In: *Proceedings of the International Conference Information Visualisation (IV)*. IEEE Computer Society, pp. 44–49. DOI: 10. 1109/iv.2009.52.

Langner, R., L. Besançon, C. Collins, T. Dwyer, P. Isenberg, T. Isenberg, B. Lee, C. Perin, and C. Tominski (2021). "An Introduction to Mobile Data Visualization". In: *Mobile Data Visualization*. Edited by Lee, B., Dachselt, R., Isenberg, P., and Choe, E. K. CRC Press, pp. 1–32. DOI: 10.1201/9781003090823-1.

Laxman, S. and P. S. Sastry (2006). "A Survey of Temporal Data Mining". In: *Sādhanā* 31, pp. 173–198. DOI: 10.1007/bf02719780.

Lazar, J., J. H. Feng, and H. Hochheiser (2017). *Research Methods in Human-Computer Interaction*. 2nd edition. John Wiley & Sons, Ltd.

Lee, B., P. Isenberg, N. H. Riche, and S. Carpendale (2012). "Beyond Mouse and Keyboard: Expanding Design Considerations for Information Visualization Interactions". In: *IEEE Transactions on Visualization and Computer Graphics* 18.12, pp. 2689–2698. DOI: 10.1109/TVCG.2012.204.

Lee, B., N. H. Riche, A. K. Karlson, and S. Carpendale (2010). "SparkClouds: Visualizing Trends in Tag Clouds". In: *IEEE Transactions on Visualization and Computer Graphics* 16.6, pp. 1182–1189. DOI: 10.1109/TVCG.2010.194.

Lee, B., A. Srinivasan, P. Isenberg, and J. T. Stasko (2021). "Post-WIMP Interaction for Information Visualization". In: *Foundations and Trends in Human-Computer Interaction* 14.1, pp. 1–95. DOI: 10.1561/1100000081.

Lee, J. Y., R. Elmasri, and J. Won (1998). "An Integrated Temporal Data Model Incorporating Time Series Concept". In: *Data and Knowledge Engineering* 24.3, pp. 257–276. DOI: 10.1016/S0169-023X(97)00034-7.

Lenntorp, B. (1976). "Paths in Space-Time Environments: A Time Geographic Study of Movement Possibilities of Individuals". In: *Lund Studies in Geography*. Series B: Human Geography 44. Royal University of Lund.

Lenz, H. (2005). *Universalgeschichte der Zeit*. Wiesbaden, Germany: Marixverlag, p. 575.

Lévy, P. P., B. L. Grand, F. Poulet, M. Soto, L. Daragó, L. Toubiana, and J.-F. Vibert, eds. (2007). *Pixelization Paradigm*. Vol. 4370. Lecture Notes in Computer Science. Springer. DOI: 10.1007/978-3-540-71027-1.

Lin, J., E. Keogh, S. Lonardi, and P. Patel (2002). "Finding Motifs in Time Series". In: *Proceedings of the SIGKDD Workshop on Temporal Data Mining*, pp. 53–68. URL: https://cs.gmu.edu/~jessica/Lin_motif.pdf.

Lin, J., E. J. Keogh, and S. Lonardi (2005). "Visualizing and Discovering Non-Trivial Patterns in Large Time Series Databases". In: *Information Visualization* 4.2, pp. 61–82. DOI: 10.1057/palgrave.ivs.9500089.

Lin, J., E. J. Keogh, S. Lonardi, and B. Y.-c. Chiu (2003). "A Symbolic Representation of Time Series, with Implications for Streaming Algorithms". In: *Proceedings of the Workshop on Research Issues in Data Mining and Knowledge Discovery (DMKD)*. ACM Press, pp. 2–11. DOI: 10.1145/882082.882086.

Lin, J., E. J. Keogh, L. Wei, and S. Lonardi (2007). "Experiencing SAX: A Novel Symbolic Representation of Time Series". In: *Data Mining and Knowledge Discovery* 15.2, pp. 107–144. DOI: 10.1007/s10618-007-0064-z.

Liu, L. and M. T. Özsu, eds. (2018). *Encyclopedia of Database Systems*. 2nd edition. Springer. DOI: 10.1007/978-1-4614-8265-9.

Liu, S., Y. Wu, E. Wei, M. Liu, and Y. Liu (2013). "StoryFlow: Tracking the Evolution of Stories". In: *IEEE Transactions on Visualization and Computer Graphics* 19.12, pp. 2436–2445. DOI: 10.1109/TVCG.2013.196.

Liu, Z. and J. Heer (2014). "The Effects of Interactive Latency on Exploratory Visual Analysis". In: *IEEE Transactions on Visualization and Computer Graphics* 20.12, pp. 2122–2131. DOI: 10.1109/TVCG.2014.2346452.

Loth, A. (2019). *Visual Analytics with Tableau*. Wiley. DOI: 10.1002/9781119561996.

Luboschik, M., C. Maus, H.-J. Schulz, H. Schumann, and A. Uhrmacher (2012). "Heterogeneity-Based Guidance for Exploring Multiscale Data in Systems Biology". In: *Proceedings of the IEEE Symposium on Biological Data Visualization (BioVis)*. IEEE Computer Society, pp. 33–40. DOI: `10.1109/BioVis.2012.6378590`.

Luboschik, M., H. Schumann, and H. Cords (2008). "Particle-Based Labeling: Fast Point-feature Labeling Without Obscuring Other Visual Features". In: *IEEE Transactions on Visualization and Computer Graphics* 14.6, pp. 1237–1244. DOI: `10.1109/tvcg.2008.152`.

Luo, D., J. Yang, M. Krstajic, W. Ribarsky, and D. Keim (2012). "EventRiver: Visually Exploring Text Collections With Temporal References". In: *IEEE Transactions on Visualization and Computer Graphics* 18.1, pp. 93–105. DOI: `10.1109/TVCG.2010.225`.

Luz, S. and M. Masoodian (2004). "A Mobile System for Non-linear Access to Time-based Data". In: *Proceedings of the Conference on Advanced Visual Interfaces (AVI)*. ACM Press, pp. 454–457. DOI: `10.1145/989863.989950`.

Luz, S. and M. Masoodian (2007). "Visualisation of Parallel Data Streams with Temporal Mosaics". In: *Proceedings of the International Conference Information Visualisation (IV)*. IEEE Computer Society, pp. 197–202. DOI: `10.1109/IV.2007.127`.

Luz, S. and M. Masoodian (2010). "Improving Focus and Context Awareness in Interactive Visualization of Time Lines". In: *Proceedings of the British Computer Society Conference on Human-Computer Interaction (BCS-HCI)*. BCS Learning & Development, pp. 72–80. DOI: `10.14236/ewic/hci2010.11`.

MacEachren, A. M. (1995). *How Maps Work: Representation, Visualization, and Design*. Guilford Press.

Mackinlay, J. (1986). "Automating the Design of Graphical Presentations of Relational Information". In: *ACM Transactions on Graphics* 5.2, pp. 110–141. DOI: `10.1145/22949.22950`.

Mackinlay, J. D., G. G. Robertson, and S. K. Card (1991). "The Perspective Wall: Detail and Context Smoothly Integrated". In: *Proceedings of the SIGCHI Conference on Human Factors in Computing Systems (CHI)*. ACM Press, pp. 173–179. DOI: `10.1145/108844.108870`.

Mairena, A., C. Gutwin, and A. Cockburn (2022). "Which Emphasis Technique to Use? Perception of Emphasis Techniques with Varying Distractors, Backgrounds, and Visualization Types". In: *Information Visualization* 21.2, pp. 95–129. DOI: `10.1177/14738716211045354`.

Mannino, M. and A. Abouzied (2018). "Expressive Time Series Querying with Hand-Drawn Scale-Free Sketches". In: *Proceedings of the SIGCHI Conference on Human Factors in Computing Systems (CHI)*. ACM Press. DOI: `10.1145/3173574.3173962`.

Marey, É.-J. (1875). "La Méthode Graphique dans les Sciences Expérimentales (Suite)". In: *Physiologie Expérimentale: travaux du Laboratoire de M. Marey*. Vol. 1. Paris, France: Masson, G., pp. 255–278.

Marey, É.-J. (1894). *Le mouvement*. Paris, France: Masson, G.

Markovic, D. and M. Gelautz (2006). "Comics-Like Motion Depiction from Stereo". In: *Proceedings of the International Conference in Central Europe on Computer Graphics, Visualization and Computer Vision (WSCG)*. University of West Bohemia, pp. 155–160.

Marriott, K., F. Schreiber, T. Dwyer, K. Klein, N. H. Riche, T. Itoh, W. Stuerzlinger, and B. H. Thomas, eds. (2018). *Immersive Analytics*. Vol. 11190. Lecture Notes in Computer Science. Springer. DOI: 10.1007/978-3-030-01388-2.

Mathiesen, S. S. and H.-J. Schulz (2021). "Aesthetics and Ordering in Stacked Area Charts". In: *Proceedings of the International Conference on Theory and Application of Diagrams*. Springer, pp. 3–19. DOI: 10.1007/978-3-030-86062-2_1.

Matković, K., H. Hauser, R. Sainitzer, and E. Gröller (2002). "Process Visualization with Levels of Detail". In: *Proceedings of the IEEE Symposium Information Visualization (InfoVis)*. IEEE Computer Society, pp. 67–70. DOI: 10.1109/INFVIS.2002.1173149.

Matthews, G. and M. Roze (1997). "Worm Plots". In: *IEEE Computer Graphics and Applications* 17.6, pp. 17–20. DOI: 10.1109/38.626960.

McCloud, S. (1994). *Understanding Comics*. New York, NY, USA: HarperPerennial.

McCormick, B. H., T. A. DeFanti, and M. D. Brown (1987). "Visualization in Scientific Computing". In: *ACM SIGGRAPH Computer Graphics* 21.6, pp. 3–3. DOI: 10.1145/41997.41998.

McLachlan, P., T. Munzner, E. Koutsofios, and S. North (2008). "LiveRAC: Interactive Visual Exploration of System Management Time-Series Data". In: *Proceedings of the SIGCHI Conference on Human Factors in Computing Systems (CHI)*. ACM Press, pp. 1483–1492. DOI: 10.1145/1357054.1357286.

McNabb, L. and R. S. Laramee (2019). "Multivariate Maps – A Glyph-Placement Algorithm to Support Multivariate Geospatial Visualization". In: *Information* 10.10. DOI: 10.3390/info10100302.

Mennis, J. L., D. J. Peuquet, and L. Qian (2000). "A Conceptual Framework for Incorporating Cognitive Principles into Geographical Database Representation". In: *International Journal of Geographical Information Science* 14.6, pp. 501–520. DOI: 10.1080/136588100415710.

Miksch, S., W. Horn, C. Popow, and F. Paky (1996). "Utilizing Temporal Data Abstraction for Data Validation and Therapy Planning for Artificially Ventilated Newborn Infants". In: *Artificial Intelligence in Medicine* 8.6, pp. 543–576. DOI: 10.1016/s0933-3657(96)00355-7.

Miksch, S., A. Seyfang, W. Horn, and C. Popow (1999). "Abstracting Steady Qualitative Descriptions over Time from Noisy, High-Frequency Data". In: *Proceedings of the Joint European Conference on Artificial Intelligence in Medicine and Medical Decision Making (AIMDM)*. Springer, pp. 281–290. DOI: 10.1007/3-540-48720-4_31.

Miksch, S. and W. Aigner (2014). "A Matter of Time: Applying a Data-Users-Tasks Design Triangle to Visual Analytics of Time-Oriented Data". In: *Computers & Graphics* 38, pp. 286–290. DOI: 10.1016/j.cag.2013.11.002.

Miller, A. I. (2001). *Einstein, Picasso: Space, Time, and Beauty That Causes Havoc*. Basic Books.

Mintz, D., T. Fitz-Simons, and M. Wayland (1997). "Tracking Air Quality Trends with SAS/GRAPH". In: *Proceedings of the 22nd Annual SAS User Group International Conference (SUGI)*. SAS, pp. 807–812. URL: https://support.sas.com/resources/papers/proceedings/proceedings/sugi22/INFOVIS/PAPER173.PDF.

Mitsa, T. (2010). *Temporal Data Mining*. Chapman & Hall/CRC. DOI: 10.1201/9781420089776.

Mittelstädt, S., D. Jäckle, F. Stoffel, and D. A. Keim (2015). "ColorCAT: Guided Design of Colormaps for Combined Analysis Tasks". In: *Proceedings of the Eurographics / IEEE Conference on Visualization (EuroVis) - Short Papers*. Eurographics Association, pp. 115–119. DOI: 10.2312/eurovisshort.20151135.

Mittelstädt, S., A. Stoffel, and D. A. Keim (2014). "Methods for Compensating Contrast Effects in Information Visualization". In: *Computer Graphics Forum* 33.3, pp. 231–240. DOI: 10.1111/cgf.12379.

Monroe, M., R. Lan, J. M. del Olmo, B. Shneiderman, C. Plaisant, and J. Millstein (2013a). "The Challenges of Specifying Intervals and Absences in Temporal Queries: A Graphical Language Approach". In: *Proceedings of the SIGCHI Conference on Human Factors in Computing Systems (CHI)*. ACM Press, pp. 2349–2358. DOI: 10.1145/2470654.2481325.

Monroe, M., R. Lan, H. Lee, C. Plaisant, and B. Shneiderman (2013b). "Temporal Event Sequence Simplification". In: *IEEE Transactions on Visualization and Computer Graphics* 19.12, pp. 2227–2236. DOI: 10.1109/TVCG.2013.200.

Moritz, D. and D. Fisher (2018). *Visualizing a Million Time Series with the Density Line Chart*. Tech. rep. arXiv:1808.06019 [cs.HC]. CoRR. DOI: 10.48550/arXiv.1808.06019.

Moritz, D., C. Wang, G. L. Nelson, H. Lin, A. M. Smith, B. Howe, and J. Heer (2019). "Formalizing Visualization Design Knowledge as Constraints: Actionable and Extensible Models in Draco". In: *IEEE Transactions on Visualization and Computer Graphics* 25.1, pp. 438–448. DOI: 10.1109/TVCG.2018.2865240.

Mühlbacher, T., H. Piringer, S. Gratzl, M. Sedlmair, and M. Streit (2014). "Opening the Black Box: Strategies for Increased User Involvement in Existing Algorithm Implementations". In: *IEEE Transactions on Visualization and Computer Graphics* 20.12, pp. 1643–1652. DOI: 10.1109/TVCG.2014.2346578.

Müller, H. and J.-C. Freytag (2003). *Problems, Methods, and Challenges in Comprehensive Data Cleansing*. Tech. rep. HUB-IB-164. Humboldt University Berlin.

Müller, W. and H. Schumann (2003). "Visualization Methods for Time-Dependent Data - An Overview". In: *Proceedings of Winter Simulation Conference (WSC)*. IEEE Computer Society, pp. 737–745. DOI: 10.1109/WSC.2003.1261490.

Muniandy, K. (2001). "Visualizing Time-Related Events for Intrusion Detection". In: *Proceedings of the IEEE Symposium Information Visualization (InfoVis)*. Poster presentation. IEEE Computer Society.

Munzner, T. (2009). "A Nested Process Model for Visualization Design and Validation". In: *IEEE Transactions on Visualization and Computer Graphics* 15.6, pp. 921–928. DOI: 10.1109/TVCG.2009.111.

Munzner, T. (2014). *Visualization Analysis and Design*. A K Peters/CRC Press. DOI: 10.1201/b17511.

Nardini, P., M. Chen, F. Samsel, R. Bujack, M. Böttinger, and G. Scheuermann (2021). "The Making of Continuous Colormaps". In: *IEEE Transactions on Visualization and Computer Graphics* 27.6, pp. 3048–3063. DOI: 10.1109/TVCG.2019.2961674.

Newman, L. H. (1965). *Man and Insects*. London, UK: Aldus Books.

Nguyen, P. H., K. Xu, R. Walker, and B. W. Wong (2016). "TimeSets: Timeline Visualization with Set Relations". In: *Information Visualization* 15.3, pp. 253–269. DOI: 10.1177/1473871615605347.

Niederer, C., H. Stitz, R. Hourieh, F. Grassinger, W. Aigner, and M. Streit (2018). "TACO: Visualizing Changes in Tables Over Time". In: *IEEE Transactions on Visualization and Computer Graphics* 24.1, pp. 677–686. DOI: 10.1109/TVCG.2017.2745298.

Nocke, T., H. Schumann, and U. Böhm (2004). "Methods for the Visualization of Clustered Climate Data". In: *Computational Statistics* 19.1, pp. 75–94. DOI: 10.1007/bf02915277.

Noirhomme-Fraiture, M. (2002). "Visualization of Large Data Sets: The Zoom Star Solution". In: *Journal of Symbolic Data Analysis*. URL: https://www.researchgate.net/publication/228615915.

Nonnemann, L., M. Hogräfer, M. Röhlig, H. Schumann, B. Urban, and H.-J. Schulz (2022). "A Data-Driven Platform for the Coordination of Independent Visual Analytics Tools". In: *Computers & Graphics* 106, pp. 152–160. DOI: 10.1016/j.cag.2022.05.023.

Nonnemann, L., M. Hogräfer, H. Schumann, B. Urban, and H.-J. Schulz (2021). "Customizable Coordination of Independent Visual Analytics Tools". In: *Proceedings of the EuroVis Workshop on Visual Analytics (EuroVA)*. Eurographics Association. DOI: 10.2312/eurova.20211094.

Norman, D. A. (1993). *Things That Make Us Smart: Defending Human Attributes in the Age of the Machine*. Addison-Wesley Longman Publishing.

Norman, D. A. (2013). *The Design of Everyday Things*. Revised and expanded edition. Basic Books.

Paternò, F., C. Mancini, and S. Meniconi (1997). "ConcurTaskTrees: A Diagrammatic Notation for Specifying Task Models". In: *Proceedings of IFIP TC13 International Conference on Human-Computer Interaction (INTERACT)*. Springer, pp. 362–369. DOI: 10.1007/978-0-387-35175-9_58.

Perin, C., F. Vernier, and J.-D. Fekete (2013). "Interactive Horizon Graphs: Improving the Compact Visualization of Multiple Time Series". In: *Proceedings of the SIGCHI Conference on Human Factors in Computing Systems (CHI)*. ACM Press, pp. 3217–3226. DOI: 10.1145/2470654.2466441.

Perin, C., T. Wun, R. Pusch, and S. Carpendale (2018). "Assessing the Graphical Perception of Time and Speed on 2D+Time Trajectories". In: *IEEE Transactions on Visualization and Computer Graphics* 24.1, pp. 698–708. DOI: 10.1109/TVCG.2017.2743918.

Petzold, I. (2003). "Beschriftung von Bildschirmkarten in Echtzeit". PhD thesis. Rheinische Friedrich-Wilhelms-Universität Bonn. URL: https://hdl.handle.net/20.500.11811/1870.

Peuquet, D. J. (1994). "It's about Time: A Conceptual Framework for the Representation of Temporal Dynamics in Geographic Information Systems". In: *Annals of the Association of American Geographers* 84.3, pp. 441–461. DOI: 10.1111/j.1467-8306.1994.tb01869.x.

Peuquet, D. J. (2002). *Representations of Space and Time*. The Guilford Press.

Piccolotto, N., M. Bögl, and S. Miksch (2023). "Visual Parameter Space Exploration in Time and Space". In: *Computer Graphics Forum* 42.6. DOI: 10.1111/cgf.14785.

Piringer, H., C. Tominski, P. Muigg, and W. Berger (2009). "A Multi-Threading Architecture to Support Interactive Visual Exploration". In: *IEEE Transactions on Visualization and Computer Graphics* 15.6, pp. 1113–1120. DOI: 10.1109/TVCG.2009.110.

Pirolli, P. and S. Card (2005). "The Sensemaking Process and Leverage Points for Analyst Technology as Identified Through Cognitive Task Analysis". In: *Proceedings of the International Conference on Intelligence Analysis*. Vol. 5, pp. 2–4.

Plaisant, C. (2004). "The Challenge of Information Visualization Evaluation". In: *Proceedings of the Conference on Advanced Visual Interfaces (AVI)*. ACM Press, pp. 106–119. DOI: 10.1145/989863.989880.

Plaisant, C., B. Milash, A. Rose, S. Widoff, and B. Shneiderman (1996). "LifeLines: Visualizing Personal Histories". In: *Proceedings of the SIGCHI Conference on Human Factors in Computing Systems (CHI)*. ACM Press, pp. 221–227. DOI: 10.1145/238386.238493.

Plaisant, C., R. Mushlin, A. Snyder, J. Li, D. Heller, and B. Shneiderman (1998). "LifeLines: Using Visualization to Enhance Navigation and Analysis of Patient Records". In: *Proceedings of the American Medical Informatics Association Annual Fall Symposium*. American Medical Informatic Association (AMIA), pp. 76–80. URL: https://www.ncbi.nlm.nih.gov/pmc/articles/PMC2232192/.

Playfair, W. (1805). *An Inquiry into the Permanent Causes of the Decline and Fall of Powerful and Wealthy Nations*. London, UK: Greenland and Norris. URL: https://archive.org/details/inquiryintoperma00play.

Playfair, W. (1821). *A Letter on our Agricultural Distresses, their Causes and Remedies*. London, UK: William Sams.

Playfair, W. and J. Corry (1786). *The Commercial and Political Atlas: Representing, by Means of Stained Copper-Plate Charts, the Progress of the Commerce, Revenues, Expenditure and Debts of England during the Whole of the Eighteenth Century*. London, UK: printed for J. Debrett; G. G. et al.

Powsner, S. M. and E. R. Tufte (1994). "Graphical Summary of Patient Status". In: *The Lancet* 344.8919, pp. 386–389. DOI: 10.1016/S0140-6736(94)91406-0.

Priestley, J. (1765). *A Chart of Biography*. London, UK: Johnson, J.

Pulo, K. (2007). "Navani: Navigating Large-Scale Visualisations with Animated Transitions". In: *Proceedings of the International Conference Information Visualisation (IV)*. IEEE Computer Society, pp. 271–276. DOI: 10.1109/iv.2007.82.

Qiang, Y., S. H. Chavoshi, S. Logghe, P. D. Maeyer, and N. V. de Weghe (2014). "Multi-scale Analysis of Linear Data in a Two-dimensional Space". In: *Information Visualization* 13.3, pp. 248–265. DOI: 10.1177/1473871613477853.

Qiang, Y., M. Delafontaine, M. Versichele, P. de Maeyer, and N. van de Weghe (2012). "Interactive Analysis of Time Intervals in a Two-Dimensional Space". In: *Information Visualization* 11.4, pp. 255–272. DOI: 10.1177/1473871612436775.

Rahm, E. and H. H. Do (2000). "Data Cleaning: Problems and Current Approaches". In: *IEEE Data Engineering Bulletin* 23.4, pp. 3–13. URL: http://sites.computer.org/debull/A00DEC-CD.pdf.

Reijner, H. (2008). "The Development of the Horizon Graph". In: *Electronic Proceedings of the VisWeek Workshop From Theory to Practice: Design, Vision and Visualization*. URL: https://citeseerx.ist.psu.edu/viewdoc/summary?doi=10.1.1.363.5396.

Reinders, F., F. H. Post, and H. J. W. Spoelder (2001). "Visualization of Time-Dependent Data with Feature Tracking and Event Detection". In: *The Visual Computer* 17.1, pp. 55–71. DOI: 10.1007/pl00013399.

Rendgen, S. (2019). *History of Information Graphics*. Edited by Wiedemann, J. Taschen.

Reski, N., A. Alissandrakis, and A. Kerren (2020). "Exploration of Time-Oriented Data in Immersive Virtual Reality Using a 3D Radar Chart Approach". In: *Proceedings of the Nordic Conference on Human-Computer Interaction (NordiCHI)*. ACM Press, 33:1–33:11. DOI: 10.1145/3419249.3420171.

Riehmann, P., J. Reibert, J. Opolka, and B. Fröhlich (2018). "Touch the Time: Touch-Centered Interaction Paradigms for Time-Oriented Data". In: *Proceedings of the Eurographics/IEEE Conference on Visualization (EuroVis) - Short Papers*. Eurographics Association, pp. 113–117. DOI: 10.2312/eurovisshort.20181088.

Rind, A., W. Aigner, S. Miksch, S. Wiltner, M. Pohl, F. Drexler, B. Neubauer, and N. Suchy (2011a). "Visually Exploring Multivariate Trends in Patient Cohorts using Animated Scatter Plots". In: *Proceedings of the International Conference on Human-Computer Interaction (HCI-I)*. Springer. DOI: 10.1007/978-3-642-21716-6_15.

Rind, A., W. Aigner, S. Miksch, S. Wiltner, M. Pohl, T. Turic, and F. Drexler (2011b). "Visual Exploration of Time-Oriented Patient Data for Chronic Diseases: Design Study and Evaluation". In: *Information Quality in e-Health*. Springer, pp. 301–320. DOI: 10.1007/978-3-642-25364-5_22.

Rind, A., P. Federico, T. Gschwandtner, W. Aigner, J. Doppler, and M. Wagner (2017). "Visual Analytics of Electronic Health Records with a Focus on Time". In: *New Perspectives in Medical Records: Meeting the Needs of Patients and Practitioners*. Springer, pp. 65–77. DOI: 10.1007/978-3-319-28661-7_5.

Rind, A., T. Lammarsch, W. Aigner, B. Alsallakh, and S. Miksch (2013a). "TimeBench: A Data Model and Software Library for Visual Analytics of Time-

Oriented Data". In: *IEEE Transactions on Visualization and Computer Graphics* 19.12, pp. 2247–2256. DOI: 10.1109/TVCG.2013.206.

Rind, A., T. D. Wang, W. Aigner, S. Miksch, K. Wongsuphasawat, C. Plaisant, and B. Shneiderman (2013b). "Interactive Information Visualization to Explore and Query Electronic Health Records". In: *Foundations and Trends in Human-Computer Interaction* 5.3, pp. 207–298. DOI: 10.1561/1100000039.

Rit, J.-F. (1986). "Propagating Temporal Constraints for Scheduling". In: *Proceedings of the National Conference on Artificial Intelligence (AAAI)*. Morgan Kaufmann, pp. 383–388. URL: https://www.aaai.org/Papers/AAAI/1986/AAAI86-064.pdf.

Robertson, G., R. Fernandez, D. Fisher, B. Lee, and J. Stasko (2008). "Effectiveness of Animation in Trend Visualization". In: *IEEE Transactions on Visualization and Computer Graphics* 14.6, pp. 1325–1332. DOI: 10.1109/TVCG.2008.125.

Robertson, P. K. (1991). "A Methodology for Choosing Data Representations". In: *IEEE Computer Graphics and Applications* 11.3, pp. 56–67. DOI: 10.1109/38.79454.

Röhlig, M., M. Luboschik, and H. Schumann (2017). "Visibility Widgets for Unveiling Occluded Data in 3D Terrain Visualization". In: *Journal of Visual Languages & Computing* 42, pp. 86–98. DOI: 10.1016/j.jvlc.2017.08.008.

Rohrdantz, C., M. C. Hao, U. Dayal, L.-E. Haug, and D. A. Keim (2012). "Feature-Based Visual Sentiment Analysis of Text Document Streams". In: *ACM Transactions on Intelligent Systems and Technology* 3.2, 26:1–26:25. DOI: 10.1145/2089094.2089102.

Rosenberg, D. and A. Grafton (2010). *Cartographies of Time: A History of the Timeline*. Princeton Architectural Press.

Rosvall, M. and C. T. Bergstrom (2010). "Mapping Change in Large Networks". In: *PLOS ONE* 5.1. DOI: 10.1371/journal.pone.0008694.

Sadri, R., C. Zaniolo, A. Zarkesh, and J. Adibi (2004). "Expressing and Optimizing Sequence Queries in Database Systems". In: *ACM Transactions on Database Systems* 29.2, pp. 282–318. DOI: 10.1145/1005566.1005568.

Saito, T., H. N. Miyamura, M. Yamamoto, H. Saito, Y. Hoshiya, and T. Kaseda (2005). "Two-Tone Pseudo Coloring: Compact Visualization for One-Dimensional Data". In: *Proceedings of the IEEE Symposium Information Visualization (InfoVis)*. IEEE Computer Society, pp. 173–180. DOI: 10.1109/INFVIS.2005.1532144.

Schulz, H.-J., M. Angelini, G. Santucci, and H. Schumann (2016). "An Enhanced Visualization Process Model for Incremental Visualization". In: *IEEE Transactions on Visualization and Computer Graphics* 22.7, pp. 1830–1842. DOI: 10.1109/TVCG.2015.2462356.

Schulz, H.-J., T. Nocke, M. Heitzler, and H. Schumann (2013a). "A Design Space of Visualization Tasks". In: *IEEE Transactions on Visualization and Computer Graphics* 19.12, pp. 2366–2375. DOI: 10.1109/TVCG.2013.120.

Schulz, H.-J., T. Nocke, M. Heitzler, and H. Schumann (2017). "A Systematic View on Data Descriptors for the Visual Analysis of Tabular Data". In: *Information Visualization* 16.3, pp. 232–256. DOI: 10.1177/1473871616667767.

Schulz, H.-J., M. Streit, T. May, and C. Tominski (2013b). *Towards a Characterization of Guidance in Visualization*. Poster at IEEE Conference on Information Visualization (InfoVis). Atlanta, USA.

Schulze-Wollgast, P., C. Tominski, and H. Schumann (2005). "Enhancing Visual Exploration by Appropriate Color Coding". In: *Proceedings of the International Conference in Central Europe on Computer Graphics, Visualization and Computer Vision (WSCG)*. University of West Bohemia, pp. 203–210.

Schwab, M., S. Hao, O. Vitek, J. Tompkin, J. Huang, and M. A. Borkin (2019a). "Evaluating Pan and Zoom Timelines and Sliders". In: *Proceedings of the SIGCHI Conference on Human Factors in Computing Systems (CHI)*. ACM Press, pp. 1–12. DOI: 10.1145/3290605.3300786.

Schwab, M., J. Tompkin, J. Huang, and M. A. Borkin (2019b). "EasyPZ.js: Interaction Binding for Pan and Zoom Visualizations". In: *IEEE Visualization Conference, IEEE VIS 2019 - Short Papers*. IEEE Computer Society, pp. 31–35. DOI: 10.1109/VISUAL.2019.8933747.

Sedig, K., P. Parsons, and A. Babanski (2012). "Towards a Characterization of Interactivity in Visual Analytics". In: *Journal of Multimedia Processing and Technologies, Special issue on Theory and Application of Visual Analytics* 3 (1), pp. 12–28. URL: https://www.dline.info/jmpt/fulltext/v3n1/2.pdf.

Sedlmair, M., C. Heinzl, S. Bruckner, H. Piringer, and T. Möller (2014). "Visual Parameter Space Analysis: A Conceptual Framework". In: *IEEE Transactions on Visualization and Computer Graphics* 20.12, pp. 2161–2170. DOI: 10.1109/TVCG.2014.2346321.

Senay, H. and E. Ignatius (1994). "A Knowledge-Based System for Visualization Design". In: *IEEE Computer Graphics and Applications* 14.6, pp. 36–47. DOI: 10.1109/38.329093.

Seyfert, M. and I. Viola (2017). "Dynamic Word Clouds". In: *Proceedings of the Spring Conference on Computer Graphics (SCCG)*. ACM Press, 7:1–7:8. DOI: 10.1145/3154353.3154358.

Shaer, O. and E. Hornecker (2009). "Tangible User Interfaces: Past, Present and Future Directions". In: *Foundations and Trends in Human-Computer Interaction* 3.1-2, pp. 1–137. DOI: 10.1561/1100000026.

Shahar, Y., D. Goren-Bar, D. Boaz, and G. Tahan (2006). "Distributed, Intelligent, Interactive Visualization and Exploration of Time-Oriented Clinical Data and their Abstractions". In: *Artificial Intelligence in Medicine* 38.2, pp. 115–135. DOI: 10.1016/j.artmed.2005.03.001.

Shahar, Y., S. Miksch, and P. Johnson (1998). "The Asgaard Project: A Task-Specific Framework for the Application and Critiquing of Time-Oriented Clinical Guidelines". In: *Artificial Intelligence in Medicine* 14.1-2, pp. 29–51. DOI: 10.1016/s0933-3657(98)00015-3.

Shanbhag, P., P. Rheingans, and M. desJardins (2005). "Temporal Visualization of Planning Polygons for Effcient Partitioning of Geo-Spatial Data". In: *Proceedings of the IEEE Symposium Information Visualization (InfoVis)*. IEEE Computer Society, pp. 211–218. DOI: 10.1109/INFVIS.2005.1532149.

Shao, L., A. Mahajan, T. Schreck, and D. J. Lehmann (2017). "Interactive Regression Lens for Exploring Scatter Plots". In: *Computer Graphics Forum* 36.3, pp. 157–166. DOI: 10.1111/cgf.13176.

Sheidin, J., J. Lanir, P. Bak, and T. Kuflik (2017). "Time-Ray Maps: Visualization of Spatial and Temporal Evolution of News Stories". In: *Proceedings of the Eurographics / IEEE Conference on Visualization (EuroVis) - Short Papers*. Eurographics Association, pp. 85–89. DOI: 10.2312/eurovisshort.20171138.

Shi, C., W. Cui, S. Liu, P. Xu, W. Chen, and H. Qu (2012). "RankExplorer: Visualization of Ranking Changes in Large Time Series Data". In: *IEEE Transactions on Visualization and Computer Graphics* 18.12, pp. 2669–2678. DOI: 10.1109/TVCG.2012.253.

Shimabukuro, M. H., E. F. Flores, M. C. F. de Oliveira, and H. Levkowitz (2004). "Coordinated Views to Assist Exploration of Spatio-Temporal Data: A Case Study". In: *Proceedings of the International Conference on Coordinated and Multiple Views in Exploratory Visualization (CMV)*. IEEE Computer Society, pp. 107–117. DOI: 10.1109/CMV.2004.1319531.

Shneiderman, B. (1983). "Direct Manipulation: A Step Beyond Programming Languages". In: *IEEE Computer* 16.8, pp. 57–69. DOI: 10.1109/mc.1983.1654471.

Shneiderman, B. (1994). "Dynamic Queries for Visual Information Seeking". In: *IEEE Software* 11.6, pp. 70–77. DOI: 10.1109/52.329404.

Shneiderman, B. (1996). "The Eyes Have It: A Task by Data Type Taxonomy for Information Visualizations". In: *Proceedings of the IEEE Symposium on Visual Languages*. IEEE Computer Society, pp. 336–343. DOI: 10.1109/VL.1996.545307.

Shneiderman, B. (2022). *Human-Centered AI*. Oxford University Press.

Shneiderman, B. and C. Plaisant (2004). *Designing the User Interface: Strategies for Effective Human-Computer Interaction*. 4th edition. Pearson Addison Wesley.

Silva, S., J. Madeira, and B. S. Santos (2007). "There is More to Color Scales than Meets the Eye: A Review on the Use of Color in Visualization". In: *Proceedings of the International Conference Information Visualisation (IV)*. IEEE Computer Society, pp. 943–950. DOI: 10.1109/iv.2007.113.

Silva, S., B. S. Santos, and J. Madeira (2011). "Using Color in Visualization: A Survey". In: *Computers & Graphics* 35.2, pp. 320–333. DOI: 10.1016/j.cag.2010.11.015.

Silva, S. F. and T. Catarci (2000). "Visualization of Linear Time-Oriented Data: A Survey". In: *Proceedings of the International Conference on Web Information Systems Engineering (WISE)*. IEEE Computer Society, pp. 310–319. DOI: 10.1109/WISE.2000.882407.

Simons, D. J. and R. A. Rensink (2005). "Change Blindness: Past, Present, and Future". In: *Trends in Cognitive Sciences* 9.1, pp. 16–20. DOI: 10.1016/j.tics.2004.11.006.

Sips, M., P. Köthur, A. Unger, H.-C. Hege, and D. Dransch (2012). "A Visual Analytics Approach to Multiscale Exploration of Environmental Time Series". In: *IEEE Transactions on Visualization and Computer Graphics* 18.12, pp. 2899–2907. DOI: 10.1109/TVCG.2012.191.

Speeth, S. D. (1961). "Seismometer Sounds". In: *The Journal of the Acoustical Society of America* 33.7, pp. 909–916. DOI: 10.1121/1.1908843.

Spence, R. (2007). *Information Visualization: Design for Interaction*. 2nd edition. Prentice-Hall.

Spindler, M., C. Tominski, H. Schumann, and R. Dachselt (2010). "Tangible Views for Information Visualization". In: *Proceedings of the International Conference on Interactive Tabletops and Surfaces (ITS)*. ACM Press, pp. 157–166. DOI: 10.1145/1936652.1936684.

Srinivasan, A. and J. T. Stasko (2018). "Orko: Facilitating Multimodal Interaction for Visual Exploration and Analysis of Networks". In: *IEEE Transactions on Visualization and Computer Graphics* 24.1, pp. 511–521. DOI: 10.1109/TVCG.2017.2745219.

Stacey, M. and C. McGregor (2007). "Temporal Abstraction in Intelligent Clinical Data Analysis: A Survey". In: *Artificial Intelligence in Medicine* 39.1, pp. 1–24. DOI: 10.1016/j.artmed.2006.08.002.

Steele, J. and N. Iliinsky (2010). *Beautiful Visualization: Looking at Data through the Eyes of Experts*. O'Reilly Media, Inc.

Stein, K., R. Wegener, and C. Schlieder (2010). "Pixel-Oriented Visualization of Change in Social Networks". In: *Proceedings of the International Conference on Advances in Social Networks Analysis and Mining (ASONAM)*. IEEE Computer Society, pp. 233–240. DOI: 10.1109/asonam.2010.18.

Steiner, A. (1998). "A Generalisation Approach to Temporal Data Models and their Implementations". PhD thesis. Swiss Federal Institute of Technology.

Stitz, H., S. Gratzl, W. Aigner, and M. Streit (2016). "ThermalPlot: Visualizing Multi-Attribute Time-Series Data Using a Thermal Metaphor". In: *IEEE Transactions on Visualization and Computer Graphics* 22.12, pp. 2594–2607. DOI: 10.1109/TVCG.2015.2513389.

Stoiber, C., D. Ceneda, M. Wagner, V. Schetinger, T. Gschwandtner, M. Streit, S. Miksch, and W. Aigner (2022). "Perspectives of Visualization Onboarding and Guidance in VA". In: *Visual Informatics* 6 (1), pp. 68–83. DOI: 10.1016/j.visinf.2022.02.005.

Stoiber, C., A. Rind, F. Grassinger, R. Gutounig, E. Goldgruber, M. Sedlmair, Š. Emrich, and W. Aigner (2019). "netflower: Dynamic Network Visualization for Data Journalists". In: *Computer Graphics Forum* 38.3, pp. 699–711. DOI: 10.1111/cgf.13721.

Stolper, C. D., A. Perer, and D. Gotz (2014). "Progressive Visual Analytics: User-Driven Visual Exploration of In-Progress Analytics". In: *IEEE Transactions on Visualization and Computer Graphics* 20.12, pp. 1653–1662. DOI: 10.1109/TVCG.2014.2346574.

Sureau, F., F. Plantard, F. Bouali, and G. Venturini (2009). "Visual Mining of Web Logs with DataTube2". In: *Proceedings of the International Conference Web Information Systems Engineering (WISE)*. Springer, pp. 555–562. DOI: 10.1007/978-3-642-04409-0_53.

Tableau Software (2021). *Tableau Prep*. URL: https://www.tableau.com/products/prep (visited on 02/26/2021).

Talbot, J., J. Gerth, and P. Hanrahan (2012). "An Empirical Model of Slope Ratio Comparisons". In: *IEEE Transactions on Visualization and Computer Graphics* 18.12, pp. 2613–2620. DOI: 10.1109/TVCG.2012.196.

Tanahashi, Y., C.-H. Hsueh, and K.-L. Ma (2015). "An Efficient Framework for Generating Storyline Visualizations from Streaming Data". In: *IEEE Transactions on Visualization and Computer Graphics* 21.6, pp. 730–742. DOI: 10.1109/TVCG.2015.2392771.

Tanahashi, Y. and K.-L. Ma (2012). "Design Considerations for Optimizing Storyline Visualizations". In: *IEEE Transactions on Visualization and Computer Graphics* 18.12, pp. 2679–2688. DOI: 10.1109/TVCG.2012.212.

Tang, H., S. Wei, Z. Zhou, Z. C. Qian, and Y. V. Chen (2019a). "TreeRoses: Outlier-Centric Monitoring and Analysis of Periodic Time Series Data". In: *Journal of Visualization* 22.5, pp. 1005–1019. DOI: 10.1007/s12650-019-00586-1.

Tang, T., S. Rubab, J. Lai, W. Cui, L. Yu, and Y. Wu (2019b). "iStoryline: Effective Convergence to Hand-drawn Storylines". In: *IEEE Transactions on Visualization and Computer Graphics* 25.1, pp. 769–778. DOI: 10.1109/TVCG.2018.2864899.

Telea, A. C. (2014). *Data Visualization: Principles and Practice*. 2nd edition. A K Peters/CRC Press. DOI: 10.1201/b17217.

Thakur, S. and A. J. Hanson (2010). "A 3D Visualization of Multiple Time Series on Maps". In: *Proceedings of the International Conference Information Visualisation (IV)*. IEEE Computer Society, pp. 336–343. DOI: 10.1109/IV.2010.54.

Thakur, S. and T.-M. Rhyne (2009). "Data Vases: 2D and 3D Plots for Visualizing Multiple Time Series". In: *Proceedings of the International Symposium on Visual Computing (ISVC)*. Springer, pp. 929–938. DOI: 10.1007/978-3-642-10520-3_89.

Third Millennium Press (2001). *Zeittafel der Weltgeschichte. Den letzen 6000 Jahren auf der Spur*. Cologne, Germany: Könemann Verlagsgesellschaft mbH.

Thomas, J. J. and K. A. Cook (2005). *Illuminating the Path: The Research and Development Agenda for Visual Analytics*. IEEE Computer Society.

Thompson, J. R., Z. Liu, W. Li, and J. Stasko (2020). "Understanding the Design Space and Authoring Paradigms for Animated Data Graphics". In: *Computer Graphics Forum* 39.3, pp. 207–218. DOI: 10.1111/cgf.13974.

Thudt, A., D. Baur, and S. Carpendale (2013). "Visits: A Spatiotemporal Visualization of Location Histories". In: *Proceedings of the Eurographics / IEEE Conference on Visualization (EuroVis) - Short Papers*. Eurographics Association. DOI: 10.2312/PE.EuroVisShort.EuroVisShort2013.079-083.

Tominski, C. (2011). "Event-Based Concepts for User-Driven Visualization". In: *Information Visualization* 10.1, pp. 65–81. DOI: 10.1057/ivs.2009.32.

Tominski, C. (2015). *Interaction for Visualization*. Synthesis Lectures on Visualization 3. Morgan & Claypool. DOI: 10.2200/S00651ED1V01Y201506VIS003.

Tominski, C., J. Abello, and H. Schumann (2004). "Axes-Based Visualizations with Radial Layouts". In: *Proceedings of the ACM Symposium on Applied Computing (SAC)*. ACM Press, pp. 1242–1247. DOI: 10.1145/967900.968153.

Tominski, C., J. Abello, and H. Schumann (2005a). "Interactive Poster: 3D Axes-Based Visualizations for Time Series Data". In: *Poster Compendium of IEEE Symposium on Information Visualization (InfoVis)*. IEEE Computer Society, pp. 49–50. URL: https://www.researchgate.net/publication/228744875.

Tominski, C., G. Andrienko, N. Andrienko, S. Bleisch, S. I. Fabrikant, E. Mayr, S. Miksch, M. Pohl, and A. Skupin (2021). "Toward Flexible Visual Analytics Augmented through Smooth Display Transitions". In: *Visual Informatics* 5.3, pp. 28–38. DOI: 10.1016/j.visinf.2021.06.004.

Tominski, C., C. Forsell, and J. Johansson (2012a). "Interaction Support for Visual Comparison Inspired by Natural Behavior". In: *IEEE Transactions on Visualization and Computer Graphics* 18.12, pp. 2719–2728. DOI: 10.1109/TVCG.2012.237.

Tominski, C., G. Fuchs, and H. Schumann (2008). "Task-Driven Color Coding". In: *Proceedings of the International Conference Information Visualisation (IV)*. IEEE Computer Society, pp. 373–380. DOI: 10.1109/IV.2008.24.

Tominski, C., S. Gladisch, U. Kister, R. Dachselt, and H. Schumann (2017). "Interactive Lenses for Visualization: An Extended Survey". In: *Computer Graphics Forum* 36.6, pp. 173–200. DOI: 10.1111/cgf.12871.

Tominski, C. and H.-J. Schulz (2012). "The Great Wall of Space-Time". In: *Proceedings of the Workshop on Vision, Modeling & Visualization (VMV)*. Eurographics Association, pp. 199–206. DOI: 10.2312/PE/VMV/VMV12/199-206.

Tominski, C., P. Schulze-Wollgast, and H. Schumann (2003). "Visualisierung zeitlicher Verläufe auf geographischen Karten". In: *Kartographische Schriften, Band 7: Visualisierung und Erschließung von Geodaten*. Kirschbaum Verlag, pp. 47–57. URL: https://www.researchgate.net/publication/247160352.

Tominski, C., P. Schulze-Wollgast, and H. Schumann (2005b). "3D Information Visualization for Time Dependent Data on Maps". In: *Proceedings of the International Conference Information Visualisation (IV)*. IEEE Computer Society, pp. 175–181. DOI: 10.1109/IV.2005.3.

Tominski, C. and H. Schumann (2008). "Enhanced Interactive Spiral Display". In: *Proceedings of the Annual Conference of the Swedish Computer Graphics Association (SIGRAD)*. Linköping University Electronic Press, pp. 53–56. URL: https://www.ep.liu.se/ecp/034/013/ecp083413.pdf.

Tominski, C. and H. Schumann (2020). *Interactive Visual Data Analysis*. AK Peters Visualization Series. CRC Press. DOI: 10.1201/9781315152707.

Tominski, C., H. Schumann, G. Andrienko, and N. Andrienko (2012b). "Stacking-Based Visualization of Trajectory Attribute Data". In: *IEEE Transactions on Visualization and Computer Graphics* 18.12, pp. 2565–2574. DOI: 10.1109/TVCG.2012.265.

Tomitsch, M., W. Aigner, and T. Grechenig (2007). "A Concept to Support Seamless Spectator Participation in Sports Events Based on Wearable Motion Sensors". In: *Proceedings of the 2nd International Conference on Pervasive Computing and Applications (ICPCA)*. IEEE Computer Society, pp. 209–214. DOI: 10.1109/icpca.2007.4365441.

Tory, M. and T. Möller (2004). "Rethinking Visualization: A High-Level Taxonomy". In: *Proceedings of the IEEE Symposium Information Visualization (InfoVis)*. IEEE Computer Society, pp. 151–158. DOI: 10.1109/INFVIS.2004.59.

Trifacta (2021). *Trifacta Wrangler*. URL: https://www.trifacta.com/ (visited on 02/26/2021).

Tufte, E. R. (1983). *The Visual Display of Quantitative Information*. Graphics Press. URL: https://www.edwardtufte.com/tufte/books_vdqi.

Tufte, E. R. (1990). *Envisioning Information*. Graphics Press. URL: https://www.edwardtufte.com/tufte/books_ei.

Tufte, E. R. (1997). *Visual Explanations*. Graphics Press. URL: https://www.edwardtufte.com/tufte/books_visex.

Tufte, E. R. (2006). *Beautiful Evidence*. Graphics Press. URL: https://www.edwardtufte.com/tufte/books_be.

Turdukulov, U. D., M.-J. Kraak, and C. A. Blok (2007). "Designing a Visual Environment for Exploration of Time Series of Remote Sensing Data: In Search for Convective Clouds". In: *Computers & Graphics* 31.3, pp. 370–379. DOI: 10.1016/j.cag.2007.01.028.

Tversky, B., J. B. Morrison, and M. Betrancourt (2002). "Animation: Can It Facilitate?" In: *International Journal of Human-Computer Studies* 57.4, pp. 247–262. DOI: 10.1006/ijhc.2002.1017.

Udell, J. (2004). "Space, Time, and Data". In: *InfoWorld* 26, p. 32. URL: https://www.infoworld.com/article/2665760/space--time--and-data.html.

Unger, A. and H. Schumann (2009). "Visual Support for the Understanding of Simulation Processes". In: *Proceedings of the IEEE Pacific Visualization Symposium (PacificVis)*. IEEE Computer Society, pp. 57–64. DOI: 10.1109/PACIFICVIS.2009.4906838.

Vande Moere, A. (2004). "Time-Varying Data Visualization Using Information Flocking Boids". In: *Proceedings of the IEEE Symposium Information Visualization (InfoVis)*. IEEE Computer Society, pp. 97–104. DOI: 10.1109/INFVIS.2004.65.

Vande Moere, A. and A. Lau (2007). "In-Formation Flocking: An Approach to Data Visualization Using Multi-agent Formation Behavior". In: *Progress in Artificial Life*. Vol. 4828. Lecture Notes in Computer Science. Springer, pp. 292–304. DOI: 10.1007/978-3-540-76931-6_26.

Van Wijk, J. J. (2006). "Views on Visualization". In: *IEEE Transactions on Visualization and Computer Graphics* 12.4, pp. 421–433. DOI: 10.1109/TVCG.2006.80.

Van Wijk, J. J. and E. R. van Selow (1999). "Cluster and Calendar Based Visualization of Time Series Data". In: *Proceedings of the IEEE Symposium Information Visualization (InfoVis)*. IEEE Computer Society, pp. 4–9. DOI: 10.1109/INFVIS.1999.801851.

Viégas, F. B., D. Boyd, D. H. Nguyen, J. Potter, and J. Donath (2004a). "Digital Artifacts for Remembering and Storytelling: PostHistory and Social Network Fragments". In: *Proceedings of the Annual Hawaii International Conference on System Sciences (HICSS)*. IEEE Computer Society, pp. 109–118. DOI: 10.1109/hicss.2004.1265287.

Viégas, F. B., M. Wattenberg, and K. Dave (2004b). "Studying Cooperation and Conflict between Authors with history flow Visualizations". In: *Proceedings of the SIGCHI Conference on Human Factors in Computing Systems (CHI)*. ACM Press, pp. 575–582. DOI: 10.1145/985692.985765.

Voida, S., M. Tobiasz, J. Stromer, P. Isenberg, and S. Carpendale (2009). "Getting Practical with Interactive Tabletop Displays: Designing for Dense Data, Fat Fingers, Diverse Interactions, and Face-to-face Collaboration". In: *Proceedings of the International Conference on Interactive Tabletops and Surfaces (ITS)*. ACM Press, pp. 109–116. DOI: 10.1145/1731903.1731926.

Wagner, M., D. Slijepcevic, B. Horsak, A. Rind, M. Zeppelzauer, and W. Aigner (2019). "KAVAGait: Knowledge-Assisted Visual Analytics for Clinical Gait Analysis". In: *IEEE Transactions on Visualization and Computer Graphics* 25.3, pp. 1528–1542. DOI: 10.1109/TVCG.2017.2785271.

Wainer, H. (1997). *Visual Revelations: Graphical Tales of Fate and Deception from Napoleon Bonaparte to Ross Perot*. Copernicus.

Wainer, H. (2005). *Graphic Discovery: A Trout in the Milk and Other Visual Adventures*. Princeton University Press.

Waldner, M., A. Karimov, and M. E. Gröller (2017). "Exploring Visual Prominence of Multi-channel Highlighting in Visualizations". In: *Proceedings of the Spring Conference on Computer Graphics (SCCG)*. ACM Press, 8:1–8:10. DOI: 10.1145/3154353.3154369.

Walker, F. A. (1874). *Statistical Atlas of the United States Based on the Results of the Ninth Census 1870 with Contributions from Many Eminent Men of Science and Several Departments of the Government*. United States Census Office.

Wall, E., M. Agnihotri, L. Matzen, K. Divis, M. Haass, A. Endert, and J. Stasko (2019). "A Heuristic Approach to Value-Driven Evaluation of Visualizations". In: *IEEE Transactions on Visualization and Computer Graphics* 25.1, pp. 491–500. DOI: 10.1109/TVCG.2018.2865146.

Wang, F. Y., A. Sallaberry, K. Klein, M. Takatsuka, and M. Roche (2015). "Senti-Compass: Interactive Visualization for Exploring and Comparing the Sentiments of Time-Varying Twitter Data". In: *Proceedings of the IEEE Pacific Visualization Symposium (PacificVis)*, pp. 129–133. DOI: 10.1109/PACIFICVIS.2015.7156368.

Wang, T. D., C. Plaisant, B. Shneiderman, N. Spring, D. Roseman, G. Marchand, V. Mukherjee, and M. Smith (2009). "Temporal Summaries: Supporting Temporal Categorical Searching, Aggregation and Comparison". In: *IEEE Transactions on Visualization and Computer Graphics* 15, pp. 1049–1056. DOI: 10.1109/TVCG.2009.187.

Ward, M. O., G. Grinstein, and D. Keim (2015). *Interactive Data Visualization: Foundations, Techniques, and Applications*. 2nd edition. A K Peters/CRC Press. DOI: 10.1201/b18379.

Ware, C. (2000). *Information Visualization: Perception for Design*. Morgan Kaufmann.

Ware, C. (2008). *Visual Thinking for Design*. Morgan Kaufmann. DOI: 10.1016/B978-0-12-370896-0.X0001-7.

Warren Liao, T. (2005). "Clustering of Time Series Data – A Survey". In: *Pattern Recognition* 38.11, pp. 1857–1874. DOI: 10.1016/j.patcog.2005.01.025.

Watson, M. C. (2015). "Time Maps: A Tool for Visualizing Many Discrete Events Across Multiple Timescales". In: *Proceedings of the IEEE International Conference on Big Data (Big Data)*. IEEE Computer Society, pp. 793–800. DOI: 10.1109/BigData.2015.7363824.

Wattenberg, M. (2002). "Arc Diagrams: Visualizing Structure in Strings". In: *Proceedings of the IEEE Symposium Information Visualization (InfoVis)*. IEEE Computer Society, pp. 110–116. DOI: 10.1109/INFVIS.2002.1173155.

Weber, M., M. Alexa, and W. Müller (2001). "Visualizing Time-Series on Spirals". In: *Proceedings of the IEEE Symposium Information Visualization (InfoVis)*. IEEE Computer Society, pp. 7–14. DOI: 10.1109/INFVIS.2001.963273.

Wegner, P. (1997). "Why Interaction Is More Powerful Than Algorithms". In: *Communications of the ACM* 40.5, pp. 80–91. DOI: 10.1145/253769.253801.

Wehrend, S. and C. Lewis (1990). "A Problem-Oriented Classification of Visualization Techniques". In: *Proceedings of the IEEE Visualization Conference (Vis)*. IEEE Computer Society, pp. 139–143. DOI: 10.1109/VISUAL.1990.146375.

Weiss, D. J., A. Nelson, H. S. Gibson, W. Temperley, S. Peedell, A. Lieber, M. Hancher, E. Poyart, S. Belchior, N. Fullman, B. Mappin, U. Dalrymple, J. Rozier, T. C. D. Lucas, R. E. Howes, L. S. Tusting, S. Y. Kang, E. Cameron, D. Bisanzio, K. E. Battle, S. Bhatt, and P. W. Gething (2018a). "A Global Map of Travel Time to Cities to Assess Inequalities in Accessibility in 2015". In: *Nature* 553.7688, pp. 333–336. DOI: 10.1038/nature25181.

Weiss, D., H. Gibson, U. Dalrymple, J. Rozier, T. Lucas, R. Howes, L. Tusting, S. Kang, E. Cameron, K. Battle, S. Bhatt, and P. Gething (2018b). *Accessibility to Cities*. https://malariaatlas.org/research-project/accessibility-to-cities/. URL: https://malariaatlas.org/wp-content/uploads/2017/12/MAP_Accessibility_To_Cities_Press_Release.zip.

Wertheimer, M. (1938). "Laws of Organization in Perceptual Forms". In: *A Sourcebook of Gestalt Psychology*. Edited by Ellis, W. D. London, UK: Routledge and Kegan Paul, pp. 71–88.

Whitrow, G. J., J. T. Fraser, and M. P. Soulsby (2003). *What is Time? The Classic Account of the Nature of Time*. Oxford University Press.

Wills, G. (2012). *Visualizing Time - Designing Graphical Representations for Statistical Data*. Springer. DOI: 10.1007/978-0-387-77907-2.

Wills, G. and L. Wilkinson (2010). "AutoVis: Automatic Visualization". In: *Information Visualization* 9.1, pp. 47–69. DOI: 10.1057/ivs.2008.27.

Wolter, M., I. Assenmacher, B. Hentschel, M. Schirski, and T. Kuhlen (2009). "A Time Model for Time-Varying Visualization". In: *Computer Graphics Forum* 28.6, pp. 1561–1571. DOI: 10.1111/j.1467-8659.2008.01314.x.

Wong, P. C. and R. D. Bergeron (1997). "30 Years of Multidimensional Multivariate Visualization". In: *Scientific Visualization: Overviews, Methodologies, and Techniques*. Edited by Nielson, G. M., Hagen, H., and Müller, H. IEEE Computer Society, pp. 40–62.

Wongsuphasawat, K. and B. Shneiderman (2009). "Finding Comparable Temporal Categorical Records: A Similarity Measure with an Interactive Visualization". In: *Proceedings of the IEEE Symposium on Visual Analytics Science and Technology (VAST)*. IEEE Computer Society, pp. 27–34. DOI: 10.1109/VAST.2009.5332595.

Wongsuphasawat, K., D. Moritz, A. Anand, J. D. Mackinlay, B. Howe, and J. Heer (2016). "Voyager: Exploratory Analysis via Faceted Browsing of Visualization Recommendations". In: *IEEE Transactions on Visualization and Computer Graphics* 22.1, pp. 649–658. DOI: 10.1109/TVCG.2015.2467191.

Wongsuphasawat, K. and D. Gotz (2012). "Exploring Flow, Factors, and Outcomes of Temporal Event Sequences with the Outflow Visualization". In: *IEEE Transactions on Visualization and Computer Graphics* 18.12, pp. 2659–2668. DOI: 10.1109/TVCG.2012.225.

Wright, H. (2007). *Introduction to Scientific Visualization*. Springer. DOI: 10.1007/978-1-84628-755-8.

Wunderlich, C. A. (1870). *Das Verhalten der Eigenwärme in Krankheiten*. 2nd edition. Leipzig, Germany: Otto Wigand.

Wunderlich, M., K. Ballweg, G. Fuchs, and T. von Landesberger (2017). "Visualization of Delay Uncertainty and its Impact on Train Trip Planning: A Design Study". In: *Computer Graphics Forum* 36.3, pp. 317–328. DOI: 10.1111/cgf.13190.

Xing, Z., J. Pei, and E. Keogh (2010). "A Brief Survey on Sequence Classification". In: *SIGKDD Explorations Newsletter* 12.1, pp. 40–48. DOI: 10.1145/1882471.1882478.

Xiong, R. and J. Donath (1999). "PeopleGarden: Creating Data Portraits for Users". In: *Proceedings of the ACM Symposium on User Interface Software and Technology (UIST)*. ACM Press, pp. 37–44. DOI: 10.1145/320719.322581.

Xu, K., A. Ottley, C. Walchshofer, M. Streit, R. Chang, and J. Wenskovitch (2020a). "Survey on the Analysis of User Interactions and Visualization Provenance". In: *Computer Graphics Forum* 39.3, pp. 757–783. DOI: 10.1111/cgf.14035.

Xu, K., S. Salisu, P. H. Nguyen, R. Walker, B. L. W. Wong, A. Wagstaff, G. Phillips, and M. Biggs (2020b). "TimeSets: Temporal Sensemaking in Intelligence Analysis". In: *IEEE Computer Graphics and Applications* 40.3, pp. 83–93. DOI: 10.1109/MCG.2020.2981855.

Xu, P., H. Mei, L. Ren, and W. Chen (2017). "ViDX: Visual Diagnostics of Assembly Line Performance in Smart Factories". In: *IEEE Transactions on Visualization and Computer Graphics* 23.1, pp. 291–300. DOI: 10.1109/TVCG.2016.2598664.

Xu, R. and D. C. Wunsch II (2009). *Clustering*. John Wiley & Sons. DOI: 10.1002/9780470382776.

Yang, J., W. Wang, and P. S. Yu (2000). "Mining Asynchronous Periodic Patterns in Time Series Data". In: *Proceedings of the ACM SIGKDD International Conference on Knowledge Discovery and Data Mining (KDD)*. ACM Press, pp. 275–279. DOI: 10.1145/347090.347150.

Yi, J. S., Y. ah Kang, J. T. Stasko, and J. A. Jacko (2007). "Toward a Deeper Understanding of the Role of Interaction in Information Visualization". In: *IEEE*

Transactions on Visualization and Computer Graphics 13.6, pp. 1224–1231. DOI: 10.1109/TVCG.2007.70515.

Zhang, X., T. Dekel, T. Xue, A. Owens, Q. He, J. Wu, S. Mueller, and W. T. Freeman (2018). "MoSculp: Interactive Visualization of Shape and Time". In: *Proceedings of the ACM Symposium on User Interface Software and Technology (UIST)*. ACM Press, pp. 275–285. DOI: 10.1145/3242587.3242592.

Zhao, H., C. Plaisant, B. Shneiderman, and J. Lazar (2008a). "Data Sonification for Users with Visual Impairment: A Case Study with Georeferenced Data". In: *ACM Transactions on Computer-Human Interaction* 15.1, 4:1–4:28. DOI: 10.1145/1352782.1352786.

Zhao, J., N. Cao, Z. Wen, Y. Song, Y.-R. Lin, and C. Collins (2014). "#FluxFlow: Visual Analysis of Anomalous Information Spreading on Social Media". In: *IEEE Transactions on Visualization and Computer Graphics* 20 (12), pp. 1773–1782. DOI: 10.1109/TVCG.2014.2346922.

Zhao, J., F. Chevalier, and R. Balakrishnan (2011a). "KronoMiner: Using Multi-Foci Navigation for the Visual Exploration of Time-Series Data". In: *Proceedings of the SIGCHI Conference on Human Factors in Computing Systems (CHI)*. ACM Press, pp. 1737–1746. DOI: 10.1145/1978942.1979195.

Zhao, J., F. Chevalier, E. Pietriga, and R. Balakrishnan (2011b). "Exploratory Analysis of Time-Series with ChronoLenses". In: *IEEE Transactions on Visualization and Computer Graphics* 17.12, pp. 2422–2431. DOI: 10.1109/TVCG.2011.195.

Zhao, J., S. M. Drucker, D. Fisher, and D. Brinkman (2012). "TimeSlice: Interactive Faceted Browsing of Timeline Data". In: *Proceedings of the Conference on Advanced Visual Interfaces (AVI)*. ACM Press, pp. 433–436. DOI: 10.1145/2254556.2254639.

Zhao, J., P. Forer, and A. S. Harvey (2008b). "Activities, Ringmaps and Geovisualization of Large Human Movement Fields". In: *Information Visualization* 7.3, pp. 198–209. DOI: 10.1057/palgrave.ivs.9500184.

Zhou, L. and C. D. Hansen (2016). "A Survey of Colormaps in Visualization". In: *IEEE Transactions on Visualization and Computer Graphics* 22.8, pp. 2051–2069. DOI: 10.1109/TVCG.2015.2489649.

Zhu, Y. (2007). "Measuring Effective Data Visualization". In: *Proceedings of the International Symposium on Visual Computing (ISVC)*. Springer, pp. 652–661. DOI: 10.1007/978-3-540-76856-2_64.

Zhu, Y., J. Yu, and J. Wu (2016). "Chro-Ring: a Time-Oriented Visual Approach to Represent Writer's History". In: *The Visual Computer* 32, pp. 1133–1149. DOI: 10.1007/s00371-016-1213-4.

Ziegler, H., T. Nietzschmann, and D. A. Keim (2007). "Visual Exploration and Discovery of Atypical Behavior in Financial Time Series Data using Two-Dimensional Colormaps". In: *Proceedings of the International Conference Information Visualisation (IV)*, pp. 308–315. DOI: 10.1109/IV.2007.124.

Index

Printed in the United States
by Baker & Taylor Publisher Services